HANDBOOK
OF
TECHNICAL
WRITING

FIFTH EDITION

HANDBOOK
OF
TECHNICAL
WRITING

FIFTH EDITION

Charles T. Brusaw
NCR Corporation (retired)

Gerald J. Alred
University of Wisconsin—Milwaukee

Walter E. Oliu
U.S. Nuclear Regulatory Commission

ST. MARTIN'S PRESS ▪ New York

SPONSORING EDITOR: Barbara A. Heinssen
DIRECTOR OF DEVELOPMENT: Carla Kay Samodulski
DEVELOPMENT EDITOR: Alicia Minsky
MANAGER, PUBLISHING SERVICES: Emily Berleth
SENIOR EDITOR, PUBLISHING SERVICES: Douglas Bell
PROJECT MANAGEMENT: Omega Publishing Services, Inc.
PRODUCTION SUPERVISOR: Joe Ford
COVER DESIGN: Patricia McFadden
COVER ART: David Bishop

Library of Congress Catalog Card Number: 95-73170

Copyright © 1997 by St. Martin's Press, Inc.

Manufactured in the United States of America.

1   0   9
f   e   d   c   b

For information, write:
St. Martin's Press, Inc.
175 Fifth Avenue
New York, NY 10010

ISBN: 0-312-13289-1 (spiral-bound)
       0-312-16692-3 (paperback)
       0-312-16690-7 (hardcover)

CONTENTS

PREFACE

The *Handbook of Technical Writing*, Fifth Edition, is a comprehensive, thoroughly practical reference guide for students, as well as a desktop reference for professionals already working in industry or government. The fifth edition reflects the rapidly changing technologies that challenge instructors, students, and working professionals, yet retains a focus on the fundamental challenge—strategic and effective writing skills. In the classroom, this book can be used as a basic resource for courses in which instructors design their own course materials, or as a supplement to a standard textbook. St. Martin's Press offers an Instructor's Manual to help instructors make maximum use of the *Handbook* in their classes. In the workplace, the *Handbook* will serve professionals as an up-to-date reference for correct and effective writing strategies and forms.

The Unique Four-Way Access System Improved

The four-way access system in the *Handbook of Technical Writing* includes the following:

- The **alphabetically organized text** enables readers to go immediately to the topic at hand.
- The **Topical Key to the Alphabetical Entries** groups the entries into categories and helps readers pull together information on broad subjects covered by several different entries (it also helps correlate the *Handbook* with standard textbooks or writing guides—or with instructor-designed course packages).
- The **Checklist of the Writing Process** summarizes the essay, "Five Steps to Successful Writing," in list fashion, with page references to pertinent entries.

- The **comprehensive index** provides an exhaustive list of all the topics, indexed not only by the terms actually used in the entries, but also by various other terms that readers might think of instead.

The four-way access system is flexible and easy to use—and this edition also features the following reference aids:

- A **bolder, clearer visual design** provides even quicker and easier access.
- Handy new **alphabetical letter tabs** on the edge of each page help users locate sections.
- Numerous **cross references** appear throughout, with entry titles bold-faced in the text of other entries.
- **Entries that are helpful for readers who speak English as a second language** are highlighted with a symbol: ○.
- The **introductory essay, "Five Steps to Successful Writing,"** has been updated and expanded to reflect the challenges of writing in highly technological environments.
- **The index entries that are also *Handbook* entries are boldfaced** in yet one more effort to make the information more accessible.

A Comprehensive Resource for Today's Technical Communicators

The fifth edition of the *Handbook* is tailored to provide the greatest possible relevance for professionally oriented readers. Its comprehensive treatment of grammar and usage has been consolidated and reorganized so that the reader can search under a general topic (such as sentence construction) and find specific subtopics (such as subjects and predicates) as well. Discussions of style, format, and writing procedures (planning, research, outlining, methods of development) have been updated and expanded to reflect technological resources such as **on-line resources for research, hypertext,** and the **Internet,** as well as **computer-assisted design, format,** and **editing resources.**

In response to reader's requests, we have grouped certain related entries together to make searching for topics easier. All the different types of pronouns and conjunctions, for example, have been consolidated in the entries on pronouns and conjunctions. In addition to the parts of speech and other grammatical entries, all the different types of proposals—internal proposals, government proposals, and sales proposals—have been consolidated under the single entry on proposals, and Library Research and Reference Books have also been consolidated into a single entry. These consolidations make it even easier for the reader to find needed information.

Major revisions have been made to a number of entries, including the following:

- **design and layout**
- **word processing**
- **library research** and **reference books**
- **documenting sources**
- **outlining**
- **résumés** and **job searches**
- **proposals**
- **adjectives, adverbs, complements, conjunctions, nouns, objects, pronouns, verbs**
- **bibliographies**
- **copyright**

The New Electronic World of Writing

In preparing entries for the fifth edition, we have updated a great many, deleted a few that reviewers have told us are seldom used, and added others. But most importantly, this edition pays careful attention to new technological resources and how they facilitate effective communication in writing. The fifth edition includes new entries on the **Internet, hypertext,** and the use of **facsimile (fax)** transmission. It also includes discussions of **on-line resources for research and documentation, computer indexing, protocols for electronic correspondence,** electronic **résumés** and **design and layout in word processing and desktop publishing.**

Because our focus is on helping professionally oriented writers develop effective skills and strategies for communicating in a rapidly changing environment, we have revised our introductory essay on writing to discuss and clarify the challenges of writing in a global, increasingly electronic workplace. This essay, along with the revisions made to entries proper, a new design that increases access to topics, a more concise organization of grammar and usage topics, and a consolidation of other entries makes this fifth edition an accurate and up-to-date reflection of today's writing workplace.

Other Handbook *Versions for Other Writing Needs*

The *Handbook of Technical Writing* shares not only its format but a majority of the topics in the alphabetical entries with *The Business*

Writer's Handbook, also published in a fifth edition by St. Martin's Press. Within the alphabetical entries common to both books, however, the examples given, as well as portions of the text, are often different—illustrating the same writing principles but in language that reflects a technical context in one handbook and a business context in the other. In a course combining both technical and business writing, either book may be used, the choice depending upon the emphasis the instructor thinks most appropriate. Both handbooks are available in spiral-bound versions.

For those instructors who prefer a more concise treatment that groups related entries by subject tabs, St. Martin's has published *The Concise Handbook for Technical Writing* and *The Business Writer's Companion*. These books are approximately half the size of the parent handbooks and are spiral bound for ease of use.

Acknowledgments

We are deeply indebted to the many instructors, students, technical writers, and others who have helped us prepare the first four editions of the *Handbook of Technical Writing*. Specifically, we wish to acknowledge the contributions of the following people:

Sandra Balzo, First Wisconsin National Bank
Deborah Barrett, Houston Baptist University—Houston
Eleanor Berry, University of Wisconsin—Milwaukee
James H. Chaffee, Normandale Community College
Darlene Chang, Free-Lance Graphics Designer
Mary Coney, University of Washington
Edmund Dandridge, North Carolina State University
Paul Falon, University of Michigan
James Farrelly, University of Dayton
Pat Goldstein, University of Wisconsin—Milwaukee
Susan Hitchcock, University of Virginia
Patrick Kelley, New Mexico State University
Wayne Losano, Rensselaer Polytechnic Institute
Carol Mablekos, Temple University
Roger Masse, New Mexico State University
Judy McInish, Hornbake Library, University of Maryland
Susan K. McLaughlin, Business Communications Consultants
Mimi Mejac, Nuclear Regulatory Commission
Mary Mullins, First Wisconsin National Bank
Brian Murphy, NCR Corporation

Lee Newcomer, University of Wisconsin—Milwaukee
Wendy Osborne, Society of American Foresters
Keith Palmer, Harnischfeger Corporation
Thomas Pearsall, University of Minnesota
Diana Reep, University of Akron
Lois Rew, San Jose State University
Rayleona Sanders, Nuclear Regulatory Commission
Elizabeth A. Schulz, Olympic College
Charles Stratton, University of Idaho
Donald R. Swanson, Wright State University
John Taylor, University of Wisconsin—Milwaukee
Erik Thelen, University of Wisconsin—Milwaukee
Ann Thomas, Nuclear Regulatory Commission
Mary Thompson, Krannert Graduate School of Management,
    Purdue University
William Van Pelt, University of Wisconsin—Milwaukee
Thomas Warren, Oklahoma State University
Merrill Whitburn, Rensselaer Polytechnic Institute
Conrad R. Winterhalter, NCR Corporation
Arthur Young, Michigan Technological University

For helping us identify aspects of the text that needed to be improved, updated, added, or deleted in the fifth edition, we wish to express our appreciation to:

Beatrice Christiana Birchac, University of Houston–Downtown
Donna Lee Dowdney, De Anza College
David Mair, University of Oklahoma–Norman
Alan Rauch, Georgia Institute of Technology
Stuart Selber, Clarkson University
Kristin R. Woolever, Northeastern University

We would also like to thank the following organizations for permission to use examples of their technical writing: Biospherics, Inc., *Chemical Engineering,* First Wisconsin Bank, Harnischfeger Corporation, Johnson Service Company, and NCR Corporation.
We wish to acknowledge and express our appreciation to Stuart A. Selber, who wrote the entries on hypertext and the Internet, and to Susan K. McLaughlin, who wrote the entry on oral presentations.
For her patience and the endless hours she spent preparing the manuscript, we offer our heartfelt thanks to Barbara Brusaw.

Finally, we most gratefully acknowledge the impressive contributions of Barbara Heinssen, Carla Samodulski, Alicia Minsky, and Natalie Hart, of the staff at St. Martin's Press and Rich Wright of Omega Publishing Services.

C.T.B.
G.J.A.
W.E.O.

FIVE STEPS TO SUCCESSFUL WRITING

Many people in today's technological, fast-paced workplaces have difficulty writing because they approach the task haphazardly. They may sit down at a computer, for example, and hope to fill the screen without realizing that successful writing, like the technical skills in which they are proficient, requires a disciplined and organized approach.

Successful writing on the job is not the product of inspiration, nor is it merely the spoken word converted to print; it is the result of knowing how to structure information in words and in visual design so that the writer achieves an intended purpose. The best way to ensure that a writing task will be successful—whether it is an email message, a proposal, or a technical manual—is to divide the writing process into the following five steps:

Preparation
Research
Organization
Writing the Draft
Revision

At first, these five steps must be consciously—even self-consciously—followed. But so, at first, must the steps in using a new computer system, interviewing a candidate for a job, or chairing a committee meeting. With practice, the steps in each of these processes become nearly automatic. This is not to suggest that writing becomes easy; it does not because writing well requires that you exercise judgment during each step. But the easiest and most efficient way to do it is systematically. Dividing the writing process into steps is especially useful when you are collaborating with others to produce a document. Doing so allows collaborators to appropriately divide the work, keep track of a project, and avoid wasting time by duplicating effort. (See **collaborative writing.**) When you collaborate, you can use networked computers and

email to share data, exchange texts, suggest improvements, and generally keep everyone informed of your progress as you follow the steps in the writing process.

As you master these five steps, keep in mind that they are interrelated and overlap at points. For example, your reader's needs and your purpose developed in the first step affect decisions in each of the other steps. You may also find that you need to retrace steps. When you organize the information you gathered during the second step, as an example, you may discover a gap that will require further research.

Be aware also that the time required for each step varies with different writing tasks. For example, when writing a brief informal memo, you might follow the first three steps (preparation, research, and organization) by simply listing the points you want to cover in the order you want to cover them. In other words, you gather and organize information mentally as you consider your purpose in writing the memo. For a formal report, on the other hand, these three steps require well-organized research, careful note taking, and detailed outlining. And for a brief, informal email message, the first four steps merge as you type the information on the screen. In short, the five steps expand, contract, and at times must be reviewed or even repeated to fit the complexity or context of the writing task.

The Checklist of the Writing Process that follows this summary can also help you detect and solve problems by referring you to the specific entry that explains the cause of the problem and shows you how to solve it. In the following discussion, as elsewhere throughout this book, words and phrases printed in **boldface type** refer to specific alphabetical entries.

Step 1. Preparation

Writing, like most technical tasks, requires solid **preparation.** In fact, adequate preparation is as important as writing the draft. Preparation for writing consists of the following steps:

1. Establishing your purpose
2. Assessing your reader
3. Determining the scope of your coverage

You establish your **purpose** by simply determining what you want your readers to know or be able to do when they have finished reading what you have written. Be precise; it is all too common for a writer to state a purpose in terms so broad that it is almost useless. A purpose such as "To report on possible locations for a new engineering facility" is too general to be of real help. But "To compare the relative advantages of Paris, Hong Kong, and San Francisco as possible locations for a new engineering facility so that top management can choose the best location" gives you a purpose that can guide you throughout the writing process.

The next task is to assess your **reader.** Again, be precise. What are your reader's needs in relation to your subject? What does your reader already know about your subject? You need to know, for example, whether you must define basic terminology or whether such definitions will merely bore, or even impede your reader. Is your "reader" actually several readers with different interests and levels of technical knowledge? Are you communicating with an international audience and thus need to deal with the issues inherent in writing **international correspondence.** For the purpose stated in the previous paragraph, the reader was described as "top management." But *who* is included in this category? Will one of the people evaluating the report be the personnel manager? If so, that person is likely to have a special interest in the availability of qualified personnel in each city, the presence of technical training that would be available for employees, housing conditions, and perhaps even recreational facilities. The engineering director will also be concerned about technical support services needed by the facility. The marketing manager will give priority to the facility's proximity to the primary markets for its products and the transportation facilities that are available. The financial vice-president will want to know about land and building costs and each country's tax structure. The president may be interested in all these things and even more; for example, he or she might also be concerned about the convenience of personal travel between corporate headquarters and the new facility.

In addition to knowing the needs and interests of these readers, you should know as much as you can about their background knowledge. For example, have they visited all three cities? Have they already seen other studies of the three cities? Is this the first new facility, or have they chosen locations for new facilities before? If you have many readers, one way to accommodate their needs is to combine all your readers into one composite reader and write for *that* reader. Another way is to aim parts of a document to different readers: an **executive summary** to a top manager, an **appendix** with detailed data to technical specialists, and the body of the report to middle managers.

Determining your purpose and assessing your readers will help you decide what to include and not to include in your writing. Then you will have established the scope of your writing project. If you do not clearly define the **scope** before beginning your research (the next step), you will spend needless extra hours on research because you will not be sure what kind of information you need, or even how much. For example, given the purpose and reader established in the preceding two paragraphs, the scope of your report on facility locations would include such information as land and building costs, available labor force, cultural issues, transportation facilities, proximity to sources of supply, and so forth; however, it probably would not include the early history of the cities being considered or their climate and geological features (unless these were directly related to your particular business).

Step 2. Research

The purpose of technical writing often is to explain something—usually something that is complex. In order to explain a complex subject, you must understand it. The only way to be sure that you can deal adequately with a complex subject is to compile a complete set of notes during your **research** and then to create a working outline from the notes. Numerous sources of information are available to you:

- Your own knowledge (see **research**)
- Printed sources (reports, manuals, letters, etc.)
- The library (see **library research**)
- Personal interviews (see **interviewing for information** and **questionnaires**)

The computer and the **Internet** can facilitate your research in many ways.

- You can record notes on the computer.
- You can retrieve already existing documents to incorporate with new ones.
- You can reach the library and important data bases through Internet connections.
- You can contact experts directly to ask questions or set up interviews.

Consider all sources of information when you begin your research, keep alert to new technologies as they become available, and use those that fit your needs. Of course, the amount of research you will need to do depends on your project; for a brief memo or letter, your "research" may amount to nothing more than jotting down the pertinent ideas before you begin to organize them.

Step 3. *Organization*

Without **organization,** the material gathered during your research would be incomprehensible to your reader. To organize your information effectively, you must determine the best sequence in which your ideas should be presented—that is, you must choose a **method of development.**

An appropriate method of development is the writer's tool for keeping information under control and the reader's means of following the writer's presentation. Your subject may lend itself readily to a particular method of development. For example, if you were writing instructions for operating a fax machine, you would naturally present the steps of the process in the order of their occurrence. For this task, the obvious method of development would be sequential. If you were writing about the history of an organization, your account would go from the beginning to the present, using the chronological method of development. If your subject naturally lends itself to a certain method of development, use it—don't attempt to impose another method on it. This book includes all the methods of development that are likely to be used by technical people: chronological, sequential, spatial, increasing or decreasing order of importance, comparison, division and classification, analysis, general to specific, specific to general, and cause and effect. As the writer, you must choose the method that best suits your subject, your reader, and your purpose. You are then ready to prepare your outline.

Outlining makes large or complex subjects easier for you to organize by breaking them into manageable parts, and it ensures that your finished writing will move logically from idea to idea without omitting anything important. It also enables you to emphasize your key points by placing them in the positions of greatest importance. Finally, by forcing you to structure your thinking at an early stage, a good outline allows you to concentrate exclusively on writing when you begin the rough draft. Creating the outline with word processing is especially helpful and saves time since you can fill it in when you begin the draft. Even if the task is only a short letter or memo, successful writing needs the logic and structure that a method of development and an outline provide; although for such simple projects, the method of development and outline may be in your mind rather than on screen or on paper.

Make sure at this point that you begin to consider a **design and layout** that will be helpful to your reader and a **format** appropriate to your subject and purpose. If you intend to include **computer graphics** and other **illustrations** with your writing, a good time to think about them is when you have completed your outline—especially if they need to be prepared by someone else while you are writing and revising the draft. If your outline is reasonably detailed, you should be able to determine which ideas will benefit from graphic support.

Step 4. Writing the Draft

When you have established your purpose, reader's needs, and scope, and have completed your research and your outline, you will be well prepared to write your first draft. To do so, simply expand the notes from your outline into **paragraphs,** without worrying about **grammar,** refinements of language, or such aspects of writing as **punctuation.** Refinements will come with revision.

Write the rough draft quickly, concentrating entirely on converting your outline to sentences and paragraphs. To facilitate this step, write as though you were explaining your subject to someone sitting across the desk from you. Don't worry about a good opening. Just start. There is no need in the rough draft to be concerned about an **introduction** or exact **word choices** unless they come quickly and easily—concentrate instead on *ideas.* Don't attempt to polish or revise. Writing and revising are different activities. Keep writing quickly so as to achieve **unity** and proportion. If your project is an online document, as is common for various **instructions** and **technical manuals,** consider using **hypertext** tools to link various parts of the document together or to link your document to related online information.

Even with good preparation, writing the draft remains a chore for many writers. The entry on **writing the draft** describes tactics used by experienced writers to get started and keep moving. Try several to discover the ones that are the most helpful and appropriate for you.

However, the most effective way to start and keep going is to use a good outline as a map for your first draft. Above all, don't wait for inspiration—you need to treat writing a draft in technical writing as you would any on-the-job task.

The computer is as important a tool for writing as it is for research. Using word processing software, you can develop your outline and from it create drafts of your work, formatting and printing out a professional-looking finished product. You can incorporate computer graphics from other software programs or from sources on the Internet—being sure to obtain permission before using copyrighted information. (See **copyright.**)

Step 5. Revision

Chances are that the clearer a piece of writing seems to the reader, the more effort that the writer has put into **revision.** If you have followed the steps of the writing process to this point, you will have a very rough draft. Revision, the obvious final step, requires a different frame of mind than does writing the draft. Read and evaluate the draft from the reader's point of view. Be eager to find and correct faults, and be honest. Be hard on yourself for the reader's convenience; never be easy on yourself at the reader's expense.

Do not try to do all your revising at once. Read your rough draft several times, each time looking for and correcting a different set of problems or errors. Concentrate on larger issues first; save more specific and mechanical ones until later.

Check your draft for accuracy and completeness. Your draft should give readers exactly what they need, but it should not burden them with unnecessary information or sidetrack them into loosely related subjects.

If you have not yet written an **opening** or **introduction,** now is the time to do it. Your introduction should serve as a frame into which readers can fit the information that follows in the body of your draft; check your introduction to see that it does, in fact, provide readers with such assistance. Your opening should both announce the subject and give the reader essential background information about it. Finally, check your draft for a number of important features.

- *Unity, coherence, and transition.* If a paragraph has unity, all of the sentences in it will contribute to the development of that paragraph's central idea (expressed in the topic sentence), and all of the paragraphs will contribute to the development of the main topic. If the draft has **coherence,** the sentences and the paragraphs will flow smoothly from one to the next, and the relationship of each sentence or paragraph to the one before it will be clear. You can achieve coherence in many ways, but especially by using **transition** devices and by maintaining a consistent **point of view.** Also check your draft for the proper **emphasis** and **subordination** of your ideas. Now is also a good time to adjust the **pace:** if you find places where too many ideas are jammed together, space out the ideas and slow the pace; conversely, if you find a series of simple ideas expressed in short, choppy sentences, combine some of the sentences to quicken the pace.
- *Clarity.* Although your revisions so far will already have improved the clarity, check now for any terms that must be defined or explained (see

defining terms). In addition, check for **ambiguity.** Is your writing free of **affectation?** Is it free of **jargon** that your readers may not understand? Are there **abstract words** that could be replaced by concrete words? Check your entire draft for appropriate **word choice.**

- *Style.* By now you will have already done much to improve the style, but there is more you can do. Check for **conciseness:** can you eliminate useless words or phrases—what some writers call "deadwood"? It is almost always possible to do so, and your reader will benefit greatly. Get rid of **clichés** and other **trite language.** Is your writing active? Resist the temptation to use the passive **voice** when the active voice would be far stronger and more concise. Replace negative writing with **positive writing.** Check, too, for **parallel structure.** Examine your **sentence construction,** and look for ways to create **sentence variety.**

- *Awkwardness and departures from the appropriate tone.* Awkwardness and tone are hard to define because they are the result of a great many different factors, most of which you will have already dealt with in your revision. Read your draft aloud, or have someone read it while you listen. Listen for passages that sound forced or clumsy—something you would never say if you were talking to your reader because it would sound unnatural. Listen as well for words, phrases, or statements that are inappropriate to the relationship between you and your reader. For example, avoid phrases or statements that belittle or patronize someone you're writing to or about. Assume that your readers are serious about your subject and avoid treating it lightly or humorously. Correcting the tone is frequently a matter of replacing one word with another that has more appropriate **connotations.** Finally, check slowly and carefully for problems of **grammar, punctuation,** and mechanics (**spelling, abbreviations, capital letters, acronyms and initialisms,** and the like).

Word processing software helps speed the revision process. The block, copy, and delete features allow you to try alternate ways of organizing the information without the time-consuming tedium of cutting and pasting paper drafts. Networked computers also allow you to collaborate with others. You can exchange and comment on drafts during the drafting and revision stages of a writing project. The spell checker and **thesaurus** features are helpful as a final review of your draft for typos, spelling errors, and **word choice** decisions.

The following Checklist of the Writing Process is designed to help you follow the five steps to successful writing. It can help you diagnose any writing problem you may have, and it refers you to the entry that explains the causes of the problem and shows you how to correct it.

CHECKLIST OF THE WRITING PROCESS

This checklist arranges the key entries of the *Handbook* according to a recommended sequence of steps useful for any writing project. The exact titles of the entries are in **boldface type** for quick reference, followed by the page numbers. When you turn to the entries themselves, you will find boldface cross-references to other entries that may be helpful. (You may also wish to refer to the Topical Key to the Alphabetical Entries and to the Index at the back of the book.)

1. PREPARATION 448

Establishing your **purpose** 495
Identifying your **readers/audience** 507
Determining your **scope** of coverage 538

2. RESEARCH 515

Using the **Internet** 287
Interviewing for information 291
Doing **library research** 317
Taking notes **(note taking)** 379
Creating and using **questionnaires** 497

3. ORGANIZATION 411

Choosing the best **method of development** 359
Outlining 413
Creating and using **illustrations** 264
Selecting an appropriate **format** 231
Using **hypertext** 258

4. WRITING A DRAFT 626

Choosing a **point of view** 442
Constructing sentences **(sentence construction)** 540

5. REVISION   533

Topical Key to the Alphabetical Entries

This list arranges the alphabetical entries for types of technical writing, the writing process, and document design into subject categories. Within each category the entries are listed mainly, but not strictly, in alphabetical order. For a list of the key entries in a recommended sequence of steps for any writing project, see the Checklist of the Writing Process (p. xxi). For entries on grammar and usage, see pages xxv and xxvi.

Topical Key to the Alphabetical Entries

This list arranges the alphabetical entries for style, grammar, and punctuation into subject categories. Within each category the entries are listed mainly, but not strictly, in alphabetical order. Excluded from this list are the numerous entries concerned with the usage of particular words or phrases (*imply/infer, already/all ready,* and the like). For entries on types of technical writing, the writing process, and document design, see pages xxiii and xxiv.

LANGUAGE, STYLE, AND USAGE

PARAGRAPHS, SENTENCES, CLAUSES, AND PHRASES

A

a/an

A and *an* are indefinite articles; "indefinite" implies that the **noun** designated by the article is not a specific person, place, or thing, but is one of a group.

> **EXAMPLE** The computer operator ran *a* program. (Not a specific program, but an unnamed program.)

Use *a* before words beginning with a consonant or consonant sound (including *y* or *w* sounds).

> **EXAMPLES** The year's activities are summarized in *a* one-page report. (Although the *o* in *one* is a vowel, the first *sound* is *w,* a consonant sound. Hence the article *a* precedes the word.)
>
> *A* manual has been written on that subject.
>
> It was *a* historic event for the laboratory.
>
> The office manager felt that it was *a* difficult situation.
>
> He took *a* TWA flight.

Use *an* before words beginning with a vowel or vowel sound.

> **EXAMPLES** The report is *an* overview of the year's activities.
>
> He seems *an* unlikely candidate for the job.
>
> The interviewer arrived *an* hour early. (Although the *h* in *hour* is a consonant, the word begins on a vowel *sound;* hence it is preceded by *an.*)
>
> He bought *an* SLR camera.

Be careful not to use unnecessary indefinite articles in a sentence.

> **CHANGE** I will meet you in *a* half *an* hour.
> **TO** I will meet you in *a* half hour.
> **OR** I will meet you in half *an* hour.

a lot/alot

A *lot* is often incorrectly written as one word *(alot)*. Write the **phrase** as two words: *a lot.* The phrase *a lot* is very informal, however, and should not normally be used in technical writing.

> CHANGE   The peer review group had *a lot* of objections.
> TO   The peer review group had many objections.

abbreviations

Abbreviations may be shortened versions of words, or they may be formed by the first letters of words. (See also **acronyms and initialisms.**)

> EXAMPLES   company/co.
> boulevard/blvd.
> electromotive force/emf
> horsepower/hp
> National Institute of Standards and Technology/NIST
> cash on delivery/c.o.d.

Abbreviations, like **symbols,** can be important space savers in technical writing, as it is often necessary to provide the maximum amount of information in a limited amount of space. Use abbreviations, however, only if you are certain that your **reader** will understand them as readily as he or she would the terms for which they stand. Remember also that a **memorandum** or **report** addressed to a specific person may be read by others and that you must consider those readers as well. Take into account your reader's level of knowledge of your subject when deciding whether to use abbreviations. Do not use them if they might become an inconvenience to the reader. A good rule of thumb: When in doubt, spell it out.

In general, use abbreviations only in charts, **tables, graphs,** footnotes, **bibliographies,** and other places where space is at a premium. Normally you should not make up your own abbreviations, for they will probably confuse your reader. Except for commonly used abbreviations (U.S.A., a.m.), a term to be abbreviated should be spelled out the first time it is used, with the abbreviation enclosed in **parentheses** following the term. Thereafter, the abbreviation may be used alone.

> EXAMPLE   The National Aeronautics and Space Administration (NASA) has accomplished much in its short history. NASA will now develop a permanent orbiting space laboratory.

Do not add an additional period at the end of a sentence that ends with an abbreviation.

> **EXAMPLE**   The official name of the company is Data Base, Inc.

For abbreviations specific to your profession, use the style guide provided by your company or professional organization. (A listing of style guides appears at the end of **documenting sources.**)

Forming Abbreviations

Measurements. When you abbreviate terms that refer to measurement, be sure your reader is familiar with the abbreviated form. The following list contains some common abbreviations used with units of measurement. Notice that except for *in.* (inch), *tan.* (tangent), and others that form words, abbreviations of measurements do not require periods.

amp, ampere	Hz, hertz
atm, atmosphere	in., inch
bbl, barrel	kc, kilocycle
bhp, brake horsepower	kg, kilogram
Btu, British thermal unit	km, kilometer
bu, bushel	lb, pound
cal, calorie	lm, lumen
cd, candela	min, minute
cm, centimeter	oz, ounce
cos, cosine	pa, pascal
ctn, cotangent	ppm, parts per million
doz or dz, dozen	pt, pint
emf or EMF, electromotive force	qt, quart
F, Fahrenheit	rad, radian
fig., figure (illustration)	rev, revolution
ft, foot (or feet)	sec, second or secant
gal., gallon	tan., tangent
hp, horsepower	yd, yard
hr, hour	yr, year

Abbreviations of units of measure are identical in the singular and plural: one *cm* and three *cm* (not three *cms*).

Personal Names and Titles.   Personal names should generally not be abbreviated.

> **CHANGE**   Chas., Thos., Wm., Geo.
> **TO**   Charles, Thomas, William, George

An academic, civil, religious, or military title should be spelled out when it does not precede a name.

EXAMPLE The doctor asked for the patient's chart.

When they precede names, some titles are customarily abbreviated. (See also **Ms./Miss/Mrs.**)

EXAMPLES Dr. Smith, Mr. Mills, Mrs. Katz

An abbreviation of a title may follow the name; however, be certain that it does not duplicate a title before the name.

CHANGE Dr. William Smith, Ph.D.
TO Dr. William Smith
OR William Smith, Ph.D.

When addressing **correspondence** and including names in other documents, you should normally spell out titles.

EXAMPLES The Honorable Mary J. Holt
Professor Charles Matlin
Captain Juan Ramirez
the Reverend James MacIntosh, D.D.

The following is a list of common abbreviations for personal and professional titles:

Atty.	Attorney
B.A.	Bachelor of Arts
B.S.	Bachelor of Science
B.S.E.E.	Bachelor of Science in Electrical Engineering
D.D.	Doctor of Divinity
D.D.S.	Doctor of Dental Science (or Surgery)
Dr.	Doctor (used with any doctor's degree)
Drs.	Plural of Dr.
Ed.D.	Doctor of Education
Hon.	Honorable
Jr.	Junior (used when a father with the same name is living)
LL.B.	Bachelor of Law
LL.D.	Doctor of Law
M.A.	Master of Arts
M.B.A.	Master of Business Administration
M.D.	Doctor of Medicine
Messrs.	Plural of Mr.
Mr.	Mister (spelled out only in the most formal contexts)

Mrs.	Married woman
Ms.	Woman of unspecified marital status
M.S.	Master of Science
Ph.D.	Doctor of Philosophy
Rev.	Reverend
Sr.	Senior (used when a son with the same name is living)

Names of Organizations. Many companies include in their names such terms as *Brothers, Incorporated, Corporation,* and *Company.* If these terms appear as abbreviations in the official company names, use them in their abbreviated forms: *Bros., Inc., Corp.,* and *Co.* If not abbreviated in the official names, such terms should be spelled out in most writing other than addresses, footnotes, bibliographies, and **lists,** where abbreviations may be used. A similar guideline applies for use of an **ampersand** (&); this symbol should be used only if it appears in an official company name. Titles of divisions within organizations, such as Department (Dept.) and Division (Div.) should be abbreviated only when space is limited.

Dates and Time. The following are common abbreviations.

A.D.	*anno Domini* (beginning of calendar time—A.D. 1790)	Jan.	January
B.C.	before Christ (before the beginning of calendar time— 647 B.C.)	Feb.	February
		Mar.	March
		Apr.	April
		Aug.	August
B.C.E.	before the common era	Sept.	September
C.E.	common era	Oct.	October
a.m.	*ante meridiem,* "before noon"	Nov.	November
p.m.	*post meridiem,* "after noon"	Dec.	December

Months should always be spelled out when only the month and year are given. The standard and alternative forms of abbreviations appear in current **dictionaries,** either in regular alphabetical order (by the letters of the abbreviation) or in a separate **index.**

Common Scholarly Abbreviations. The following is a partial list of abbreviations that are commonly used in reference books and for documenting sources in research papers and reports. (Handbook entries appear in **boldface.**)

| anon. | anonymous |
| assn. | association |

bibliog.	**bibliography,** bibliographer, bibliographic, bibliographical
©	**copyright**
c., ca.	*circa,* "about" (used with approximate dates: c. 1756)
cf.	*confer,* "compare"
ch., chs.	chapter(s)
diss.	dissertation
ed., eds.	edited by, editor(s), edition(s)
e.g.	*exempli gratia,* "for example" (see **e.g./i.e.**)
enl.	enlarged (as in "rev. and enl. ed.")
esp.	especially
et al.	*et alii,* "and others"
etc.	*et cetera,* "and so forth" (see **etc.**)
f., ff.	and the following page(s) or line(s)
fig.	figure
fwd.	**foreword,** foreword by
GPO	Government Printing Office, Washington DC
i.e.	*id est,* "that is"
l., ll.	line, lines
MS, MSS	manuscript, manuscripts (see **MS/MSS**)
n., nn.	note, notes (used immediately after page number: 56n., 56n.3, 56nn.3-5)
NB	*Nota Bene,* "take notice, mark well" (always capitalized)
n.d.	no date (of publication)
n.p.	no place (of publication); no publisher
n. pag.	no pagination
p., pp.	page, pages
pref.	preface, preface by
proc.	proceedings
pseud.	pseudonym
pub (publ.)	published by, publisher, publication
rept.	reported by, **report**
rev.	revised by, revision; review, reviewed by (spell out "review" where "rev." might be ambiguous)
rpt.	reprinted by, reprint
sec., secs.	section, sections
soc.	society
supp.	supplement
trans.	translated by, translator, translation
UP	University Press (used in MLA style of **documenting sources:** Oxford UP)
viz.	*videlicet,* "namely"

| vol., vols. | volume, volumes |
| vs., v. | versus, "against" (v. preferred in titles of legal cases) |

above

Avoid using *above* to refer to a preceding passage, **illustration,** or **table** in your writing. Its reference is often vague, and if the passage referred to is far from the current one, the use of *above* is distracting because your **reader** must go back to the earlier passage to understand your reference. The same is true of *aforesaid, aforementioned, above mentioned, the former,* and *the latter.* In addition to distracting the reader, these words also contribute to a heavy, wooden **style.** If you must refer to something previously mentioned, either repeat the **noun** or **pronoun** or construct your **paragraph** so that your reference is obvious (see also **former/latter** and **aforesaid**).

CHANGE   Please fill out and submit the *above* by March 1.
TO   Please fill out and submit the vacation schedule on the previous page by March 1.

absolute words

Absolute words (such as *round, unique, exact,* and *perfect*) are not logically subject to **comparison** *(rounder, roundest);* nevertheless, these words are sometimes used comparatively.

CHANGE   Phase-locked loop circuits make the FM tuner performance *more exact* by decreasing tuner distortion.
TO   Phase-locked loop circuits make the FM tuner performance *more accurate* by decreasing tuner distortion.

Absolute words should be used comparatively only with the greatest caution in technical writing, where accuracy and precision are crucial.

absolutely

Absolutely means "definitely," "entirely," "completely," or "unquestionably." It is not an **intensifier** and should not be used to mean "very" or "much." When used as an intensifier, it can be deleted.

CHANGE   A new analysis is *absolutely* impossible.
TO   A new analysis is impossible.

abstracts

An *abstract* summarizes and highlights the major points of a longer report. Abstracts are written for many **formal reports, trade journal articles,** and most dissertations, as well as for many other long works. Their primary purpose is to enable **readers** to decide whether to read the work in full. (For a discussion of how summaries differ from abstracts, see **executive summaries.**)

Although abstracts are published with the longer works they condense, they are also often published independently in computer-retrievable periodical indexes. These periodicals, such as the *American Statistics Index, Chemical Abstracts,* and *Metal Abstracts,* are devoted exclusively to abstracting information in specific fields of study. They enable researchers to review a great deal of **literature** in a short time. (For a discussion of computer-retrievable indexes, see **library research,** and for a listing of typical periodicals that publish abstracts, see **documenting sources.**)

The abstracts for **reports** and articles are also frequently the source of terms (called *keywords*) used to **index** the original document by subject for computerized information-retrieval systems. Because they are the source of keywords, abstracts must accurately but concisely describe the original work so that researchers in the field will not miss valuable information. Accordingly, they should contain no information not discussed in the original. Abstracts do vary, however, in the amount of information they provide about the original work. Depending on their **scope,** abstracts are usually classified as either *descriptive* or *informative.*

Descriptive Abstracts

A descriptive abstract includes information about the **purpose, scope,** and methods used to arrive at the findings contained in the original document. It is almost an expanded **table of contents** in sentence form. A descriptive abstract need not be longer than several sentences if it adequately summarizes the information.

The following descriptive abstract comes from a 30-page report that describes how a select group of foreign countries provide engineering expertise to their control-room operators for round-the-clock shift work at nuclear power plants:

ABSTRACT

Purpose
and scope
This report describes the practices of selected foreign countries for providing engineering expertise on shift in nuclear power plants. The report discusses the extent to which engineering

Methods used

expertise is made available and the alternative models of providing such expertise. The implications of foreign practices for U.S. consideration are discussed, with particular reference to the shift technical adviser position and to a proposed shift engineer position. The relevant information for this study came from the open literature, interviews with utility staff and officials, and governmental and nuclear utility reports.

Informative Abstracts

The informative abstract is an expanded version of the descriptive abstract. In addition to information about the purpose, scope, and methods of the original document, the informative abstract includes the results, **conclusions,** and recommendations, if any. The informative abstract retains the **tone** and essential scope of the original work while omitting its details.

The following informative abstract expands the scope of the sample descriptive abstract by including the report's findings, conclusions, and recommendation:

ABSTRACT

Purpose and scope

This report describes the practices of selected foreign countries for providing engineering expertise on shift in nuclear power plants. The report discusses the extent to which engineering expertise is made available and the alternative models of providing such expertise. The implications of foreign practices for U.S. consideration are discussed, with particular reference to the shift technical adviser position and to a proposed shift engineer position. The relevant information for this study came from the open literature, interviews with utility staff and officials, and governmental and nuclear utility reports.

Methods used

Finding

General conclusions

Recommendation

The countries studied used two approaches to provide engineering expertise on shift: (1) employing a graduate engineer in a line management operations position and (2) creating a specific engineering position to provide expertise to the operations staff. The comparison of these two models did not indicate that one system inherently functions more effectively than does the other for safe operations. However, the alternative modes are likely to affect crew relationships and performance; labor supply, recruitment, and retention; and system implementation. Of the two systems, the nonsupervisory engineering position seems more advantageous within the context of current recruitment and career-path practices.

Which of the two types of abstracts should you write? Informative abstracts satisfy the widest possible readership, including those who must index the original document based on the abstract. Descriptive abstracts, on the other hand, are good for information surveys, conference proceedings, **progress reports** that combine information from more than one project, and other publications that compile a variety of information.

Length of Abstracts

A long abstract defeats the purpose of an abstract. For this reason, abstracts are usually no longer than 200 to 250 words. Descriptive abstracts may be considerably shorter, of course.

Scope

Include in your abstract the following kinds of information, bearing in mind that your readers know nothing, except what your title announces, about your document:

- the subject of the study
- the scope of the study
- the purpose of the study
- the methods used
- the results obtained (informative abstract only)
- the recommendations made, if any (informative abstract only)

Do not include the following kinds of information:

- the background of the study
- a detailed discussion or explanation of the methods used
- administrative details about how the study was undertaken, who funded it, who worked on it, and the like
- references to figures, **tables,** charts, **maps,** and bibliographic references published in the body of the document (they will make no sense if the abstract is published independently)
- any information that does not appear in the original

Writing Style

Write the abstract after completing the report, article, or other work. Otherwise, your abstract may not accurately reflect the original document. Use the major and minor **headings** of your outline to help distinguish primary from secondary ideas.

Begin with a topic sentence that announces at least the subject and scope of the original document. Then combine the other relevant material for conciseness and clarity, eliminating unnecessary words and ideas. Write in complete sentences, but instead of stringing a group of short sentences end to end, combine ideas by using **subordination** and **parallel structure.** As a rule, spell out all **acronyms and initialisms** and all but the most common **abbreviations** (C°, F°, mph). In your attempts to write concisely, do not slip into a **telegraphic style** in which you omit articles *(a, an, the)* and important transitional words *(however, therefore, but, in summary).* (For guidance about the placement of abstracts in **reports,** see **formal report.**)

abstract words/concrete words

The difference between abstract and concrete words is the difference between *durability* (abstract) and *stone* (concrete).

Abstract words refer to general ideas, qualities, conditions, acts, or relationships. Abstract words refer to something that is intangible, something that cannot be discerned by the five senses.

> EXAMPLES   work, courage, crime, kindness, idealism, love, hate, fantasy, sportsmanship

Abstract words must frequently be qualified by other words.

> CHANGE   What the members of the Research and Development Department need is *freedom.*
>
> TO   What the members of the Research and Development Department need is *freedom to explore the problem further.*

Concrete words refer to specific persons, places, objects, and acts that can be perceived by the senses.

> EXAMPLES   wrench, book, house, scissors, gold, water
> *Skiing* (concrete) is a strenuous *sport* (abstract).

Concrete words are easier to understand, for they create images in the mind of your **reader.** Still, you cannot express ideas without using some abstract words. Actually, the two kinds of words are usually used best together, in support of each other. For example, the abstract idea of *transportation* is made clearer with the use of specific concrete words, such as *subways, jets,* or *automobiles.*

In fact, a word that is concrete in one context can be less concrete in another. The same word may even be abstract in one context and concrete in another. The choice between abstract and concrete terms always depends on the context. Just how concrete a particular context might require you to be is shown in Figure 1, which goes from the most abstract, on the left, to the most concrete, on the right.

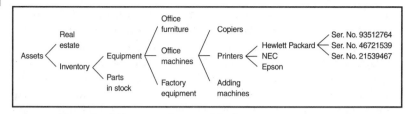

FIGURE 1   Example of Abstract-to-Concrete Words

The example represents seven levels of abstraction; which one would be appropriate depends on your purpose in writing and on the context in which you are using the word. For example, a company's annual report might logically use the most abstract term, *assets,* to refer to all the property and goods the firm owns; shareholders would probably not require a less abstract term. Interoffice **memorandums** between the company's accounting and legal departments would appropriately call the firm's holdings *real estate* and *inventory.* To the company's inventory control department, however, the word *inventory* is much too abstract to be useful, and a report on inventory might contain the more concrete terms *equipment* and *parts in stock.* But to the assistant inventory control manager in charge of equipment, that term is still too abstract; he or she would speak of several particular kinds of equipment: *office furniture, factory equipment,* and *office machines.* The breakdown of the types of office machines for which the inventory control assistant is responsible might include *copiers, adding machines,* and *printers.* But even this classification would not be concrete enough to enable the company's purchasing department to obtain service contracts for the normal maintenance of its printers. Because the department must deal with different printer manufacturers, *printers* would have to be listed by brand name: *Hewlett Packard, NEC,* and *Epson.* And the Epson technician who performs the maintenance must go one step further and identify each Epson printer by serial number. As the example in Figure 1 shows, then, a term that is sufficiently concrete in one context may be too abstract in another. (See also **word choice.**)

accept/except

Accept is a **verb** meaning "consent to," "agree to take," or "admit willingly."

> **EXAMPLE**   I *accept* the responsibility that goes with the appointment.

Except is normally a **preposition** meaning "other than" or "excluding."

EXAMPLE   We agreed on everything *except* the schedule.

acceptance letters

When you have received an offer of a job that you want to accept, begin by accepting, with pleasure, the job you have been offered. Identify the job you are accepting, and state the salary so that there is no confusion on these two important points.

The second **paragraph** might go into detail about moving dates and reporting for work. The details will vary depending on what

---

2647 Patterson Road
Beechwood, OH 45432
March 6, 19–

Mr. F. E. Cummins
Personnel Manager
Calcutex Industries, Inc.
3275 Commercial Park Drive
Bintonville, MI 49474

Dear Mr. Cummins:

State the
job and
salary
you are
accepting.

I am pleased to accept your offer of a position as assistant personnel manager at a salary of $X,XXX per month.

Specify the
date on
which you
will report
to work.

Since graduation is August 30, I plan to leave Dayton on Tuesday, September 2. I should be able to locate suitable living accommodations within a few days and be ready to report for work on the following Monday, September 8. Please let me know if this date is satisfactory to you.

State your
pleasure at
joining the
firm.

I look forward to a rewarding future with Calcutex.

Sincerely,

Craig Adderly

Craig Adderly

FIGURE 1   Acceptance Letter

occurred during your job interview. Complete the letter with a statement that you are looking forward to working for your new employer. Figure 1 is an example of an acceptance letter. (See also **correspondence** and **refusal letters.**)

accumulative/cumulative

Accumulative and *cumulative* are **synonyms** that mean "massed" or "added up over a period of time." *Accumulative* is rarely used except with reference to accumulated property or wealth.

> **EXAMPLES**   Last month's X-ray exposures were acceptable, but we must be careful so that *cumulative* doses do not exceed our guidelines.
>
> The corporation's *accumulative* holdings include 19 real-estate properties.

acknowledgment letters

It is sometimes necessary (and always considerate), as well as good public relations, to let someone know that you have received something sent to you. An *acknowledgment letter* serves such a function. It should usually be a short, polite note that mentions when the item arrived and that expresses thanks. See the **correspondence** entry for general advice on letter writing; study the example (Figure 1 on page 15) for an idea of how to phrase an acknowledgment letter.

acronyms and initialisms

An *acronym* is an **abbreviation** that is formed by combining the first letter or letters of several words. Acronyms are pronounced as words and are written without periods.

> **EXAMPLES**   radio *d*etecting *a*nd *r*anging/radar
> *a*lternating *g*radient *s*ynchrotron (AGS)
> *r*andom-*a*ccess *m*emory (RAM)
> *r*ecirculation *a*ctuation *s*ymbol (RAS)
> *r*ead-*o*nly *m*emory (ROM)

An *initialism* is an abbreviation that is formed by combining the initial letter of each word in a multiword term. Initialisms are pronounced as separate letters.

Energy Savings Systems
501 North Springfield
Phoenix, AZ 85302

▬▬▬▬▬▬▬▬▬▬▬▬▬▬▬

November 8, 19--

Ms. Wanda Evans, Consultant
936 East Avenue
Phoenix, AZ 85301

Dear Ms. Evans:

I received your report today; it appears to be complete and well done. Thank you for sending it so promptly.

When I finish studying the report, I will send you our cost estimate for the installation of the Mark II Energy-Saving System.

Sincerely,

Robert A. Martinez

Robert A. Martinez
Administrative Assistant

RAM/mo

Telephone (791) 823-6920   **Fax: (791) 823-6926**

FIGURE 1 Acknowledgment Letter

EXAMPLES   *e*nd *o*f *m*onth/e.o.m.
*S*ociety for *T*echnical *C*ommunication/STC
*c*ash *o*n *d*elivery/c.o.d.
*p*ost *m*eridiem/p.m.

In business, industry, and government, acronyms and initialisms are often used by people working together on particular projects or having the same specialties—as, for example, mechanical engineers or organic chemists. As long as such people are communicating with one another, the abbreviations are easily recognized and understood. If the same acronyms or initialisms were used in **correspondence** to someone outside the group, however, they might be incomprehensible to that **reader** and should be explained.

Acronyms and initialisms can be convenient—for the reader and the writer alike—if they are used appropriately. Technical people, however, often overuse them, either as an **affectation** or in a misguided attempt to make their writing concise.

When to Use Acronyms and Initialisms

The following are two sample guidelines to apply in deciding whether to use acronyms and initialisms:

1. If you must use a multiword term as much as once each **paragraph,** you should instead use its acronym or initialism. For example, a **phrase** such as "primary software overlay area" can become tiresome if repeated again and again in one piece of writing; it would be better, therefore, to use *PSOA.*
2. If something is better known by its acronym or initialism than by its formal term, you should use the abbreviated form. The initialism *a.m.,* for example, is much more common than the formal *ante meridiem.*

If these conditions do not exist, however, always spell out the full term.

How to Use Acronyms and Initialisms

The first time an acronym or initialism appears in a written work, write the complete term, followed by the abbreviated form in **parentheses.**

> **EXAMPLE**   The transaction processing monitor (TPM) controls all operations in the processor.

Thereafter, you may use the acronym or initialism alone. In a long document, however, you will help your reader greatly by repeating the full term in parentheses after the abbreviation at regular intervals so that he or she does not have to search back to the first time the acronym or initialism was used to find its meaning.

> **EXAMPLE**   Remember that the TPM (transaction processing monitor) controls all operations in the processor.

Write acronyms in **capital letters** without **periods.** The only exceptions are those acronyms that have become accepted as common **nouns,** which are written in lowercase letters.

NASA, HUD, laser, scuba

Initialisms may be written either uppercase or lowercase. Generally, do not use periods when they are uppercase, but use periods when they are lowercase. Two exceptions are geographic names and academic degrees.

EXAMPLES   EDP/e.d.p., EOM/e.o.m., OD/o.d.
U.S.A., R.O.K.
B.A., M.B.A.

Form the plural of an acronym or initialism by adding an s. Do not use an **apostrophe.**

EXAMPLES   MIRVs, CRTs

activate/actuate

Even linguists disagree on the distinction between these two words, although both mean "make active." *Actuate* is usually applied only to mechanical processes.

EXAMPLES   The relay *actuates* the trip hammer. (mechanical process)
The electrolyte *activates* the battery. (chemical process)
The governor *activated* the National Guard. (legal process)

active voice (see **voice**)

actually

Actually is an **adverb** meaning "really" or "in fact." Although it is often used for **emphasis** in speech, such use of the word should be avoided in writing. (See also **intensifiers.**)

CHANGE   Did he *actually* finish the report on time?
TO   Did he finish the report on time?

adapt/adept/adopt

Adapt is a **verb** meaning "adjust to a new situation." *Adept* is an **adjective** meaning "highly skilled." *Adopt* is a verb meaning "take or use as one's own."

EXAMPLE   The company will *adopt* a policy of finding engineers who are *adept* administrators and who can *adapt* to new situations.

ad hoc

Ad hoc is Latin for "for this" or "for this particular occasion." An ad hoc committee is one set up to consider a particular issue, as opposed to a permanent committee. This term has been fully assimilated into English and thus does not have to be italicized (or underlined). (See also **foreign words in English.**)

adjectives

An adjective makes the meaning of a **noun** or **pronoun** more specific by pointing out one of its qualities (descriptive adjective) or by imposing boundaries on it (limiting adjective).

> **EXAMPLES**   a *hot* iron (descriptive)
> He is *cold.* (descriptive)
> *ten* automobiles (limiting)
> *his* desk (limiting)

Limiting adjectives include these categories:

> *Articles* (a, an, the)
> *Demonstrative adjectives* (this, that, these, those)
> *Possessive adjectives* (my, his, her, your, our, their)
> *Numeral adjectives* (two, first)
> *Indefinite adjectives* (all, none, some, any)

As a **part of speech,** *articles* are considered to be adjectives because they modify the items they designate by either limiting them or making them more precise. There are two kinds of articles, *indefinite* and *definite.*
Indefinite: *a* and *an* (denotes an unspecified item)

> **EXAMPLE**   *A* program was run on our new computer. (Not a specific program, but an unspecified program. Therefore, the article is indefinite.)

Definite: *the* (denotes a particular item)

> **EXAMPLE**   *The* program was run on the computer. (Not just any program, but *the* specific program. Therefore, the article is definite.)

The choice between *a* and *an* depends on the sound rather than the letter following the article. Use *a* before words beginning with a consonant sound (*a* person, *a* happy person, *a* historical event). Use *an* before words beginning with a vowel sound (*an* uncle, *an* hour). With

abbreviations, use *a* before initial letters having a consonant sound: *a* TWA flight. Use *an* before initial letters having a vowel sound: *an* SLN report (note that the first sound is "*ess*").

Do not omit all articles from your writing. This is, unfortunately, an easy habit to develop. To include the articles costs nothing; to eliminate them makes the reading more difficult. (See also **telegraphic style.**) On the other hand, don't overdo it. An article can be superfluous.

> EXAMPLES   I'll meet you in *a* half *an* hour. (Choose one and eliminate the other.)
>
> Fill with *a* half *a* pint of fluid. (Choose one and eliminate the other.)

Do not capitalize articles when they appear in titles except as the first word of the title. (See also **capital letters.**)

> EXAMPLE   *Time* magazine reviewed *The Old World's New Order.*

The *demonstrative* adjectives "point to" the thing they modify, specifying its position in space or time. *This* and *these* specify a closer position; *that* and *those* specify a more remote position.

> EXAMPLES   *This* proposal is the one we accepted.
> *That* proposal would have been impracticable.
> *These* problems remain to be solved.
> *Those* problems are not insurmountable.

Demonstrative adjectives often cause problems when they modify the nouns *kind*, *type*, and *sort*. Demonstrative adjectives used with these nouns should agree with them in **number.**

> EXAMPLES   this kind/these kinds
> that type/those types
> this sort/these sorts

Confusion often develops when the **preposition** *of* is added ("this kind *of*," "these kinds *of*") and the **object** of the preposition is not made to conform in number to the demonstrative adjective and its **noun.**

> CHANGE   *This kind* of hydraulic *cranes* is best.
> TO   *This kind* of hydraulic *crane* is best.
>
> CHANGE   *These kinds* of hydraulic *crane* are best.
> TO   *These kinds* of hydraulic *cranes* are best.

Using demonstrative adjectives with words like *kind*, *type*, and *sort* can easily lead to vagueness. It is better to be more specific.

> CHANGE   It was *kind of* a bad year for the firm.
> TO   The firm's profits were down 10 percent from last year.

Because *possessive* adjectives *(my, your, our, his, her, its, their)* directly modify **nouns,** they function as adjectives, even though they are **pronoun** forms. Anything that modifies a noun functions as an adjective.

> **EXAMPLE**   The *proposal's* virtues outweighed *its* defects.

Numeral adjectives identify quantity, degree, or place in a sequence. They always modify count **nouns.** Numeral adjectives are divided into two subclasses: cardinal and ordinal.

A *cardinal adjective* expresses an exact quantity.

> **EXAMPLES**   *one* pencil, *two* computers, *three* airplanes

An *ordinal adjective* expresses degree or sequence.

> **EXAMPLES**   *first* quarter, *second* edition, *third* degree, *fourth* year

In most writing, an ordinal adjective should be spelled out if it is a single word *(tenth)* and written in figures if it is more than one word *(312th).* Ordinal numbers can also function as adverbs.

> **EXAMPLE**   John arrived *first.*

(See also **first/firstly.**)

Indefinite adjectives are so called because they do not designate anything specific about the nouns they modify.

> **EXAMPLES**   *some* circuit boards, *any* branches, *all* aircraft engines

Notice that the articles *a* and *an* are included among the indefinite adjectives. (See also **a/an.**)

> **EXAMPLE**   It was *an* hour before the needle of the gauge finally approached the danger area.

Comparison of Adjectives

Most adjectives add the **suffix** *-er* to show **comparison** with one other item and the suffix *-est* to show comparison with two or more other items.

> **EXAMPLES**   The first ingot is *bright.* (positive form)
> The second ingot is *brighter.* (comparative form)
> The third ingot is *brightest.* (superlative form)

However, many two-syllable adjectives and most three-syllable adjectives are preceded by the words *more* or *most* to form the comparative or the superlative.

> **EXAMPLES**   The new facility is *more* impressive than the old one.
> The new facility is the *most* impressive in the city.

A few adjectives have irregular forms of comparison (*much, more, most, little, less, least*).

   Absolute words (*round, unique*) are not logically subject to comparison (*rounder, roundest*); nevertheless, these words are sometimes used comparatively.

> CHANGE   Phase-locked loop circuits make FM tuner performance *more exact* by decreasing tuner distortion.
>
> TO   Phase-locked loop circuits make FM tuner performance *more accurate* by decreasing tuner distortion.

Placement of Adjectives

When limiting and descriptive adjectives appear together, the limiting adjectives precede the descriptive adjectives, with the articles usually in the first position.

> EXAMPLE   *The ten gray* cars were parked in a row. (article, limiting adjective, descriptive adjective)

Within a sentence, adjectives may appear before the nouns they modify (the attributive position) or after the nouns they modify (the predicative position).

> EXAMPLES   The *small* jobs are given priority. (attributive position)
> Tests are taken even when exposure is *brief*. (predicative position)

Use of Adjectives

Because of the need for precise qualification in technical writing, it is often necessary to use nouns as adjectives.

> EXAMPLE   The *test* conclusions led to a redesign of the system.

When adjectives modifying the same noun can be reversed and still make sense or when they can be separated by *and* or *or*, they should be separated by **commas.**

> EXAMPLE   The company is seeking a *young, energetic, creative* engineering team.

When an adjective modifies a phrase, no comma is required.

> EXAMPLE   He was wearing his *old cotton tennis hat* (*old* modifies the phrase *cotton tennis hat*).

Never separate a final adjective from its noun.

> CHANGE   He is a conscientious, honest, *reliable,* worker.
> TO   He is a conscientious, honest, *reliable* worker.

Technical people frequently string together a series of nouns to form a unit modifier, thereby creating jammed **modifiers.** Be aware of this problem when using nouns as adjectives.

CHANGE   The test control group meeting was held on Wednesday.
TO   The meeting of the test control group was held on Wednesday.

As a rule of thumb, it is better to avoid general *(nice, fine, good)* and trite (a *fond* farewell) adjectives; in fact, it is good practice to question the need for most adjectives in your writing. Often, your writing will not only read as well without an adjective, but it may be even better without it. If an adjective is needed, try to find one that is as precise as possible for your meaning.

adjustment letters

An *adjustment letter* is written in response to a **complaint letter** and tells the customer what your firm intends to do about his or her complaint. You should settle claims quickly and courteously, trying always to satisfy the customer at a reasonable cost to your company.

Although it is sent in response to a problem, an adjustment letter actually provides an excellent opportunity to build goodwill for the firm. An effective adjustment letter both repairs the damage that has been done and restores the customer's confidence in your firm.

Grant adjustments graciously, for a settlement made grudgingly will do more harm than good. **Tone** is critical. No matter how unpleasant or unreasonable the complaint letter, your response must remain both respectful and positive. Avoid emphasizing the unfortunate situation at hand; put your **emphasis** instead on what you are doing to correct it. Not only must you be gracious, but also you must admit your error in such a way that the buyer will not lose confidence in your firm.

Before granting an adjustment to a claim for which your company is at fault, you must investigate what happened and decide what you can do to satisfy the customer. Be certain that you know your company's policy regarding adjustments before you attempt to write an adjustment letter. Also, be careful about how you put certain words together; for example, "we have just received your letter of May 7 about our *defective product*" could be ruled in a court of law as an admission that the product is in fact defective. Treat every claim individually, and lean toward giving the customer the benefit of the doubt. The following guidelines might help you write adjustment letters:

ELECTRONIC Parts Inc.
One South Park Avenue
Columbia, AL 36319

January 17, 19--

Mr. Paul E. Denlinger
Denlinger Television Services, Inc.
4873 Wenton Way
Birmingham, AL 35214

Dear Mr. Denlinger:

We have received your letter regarding your order for nine TR-5771-3 tuners and have shipped the correct tuners by United Parcel. You should receive them shortly after you receive this letter. I have also instructed our Billing Department to charge you our preferred-customer rate—normally reserved for orders of more than $2,000. Please accept our apologies for not sending the proper tuners and for incorrectly billing you.

Evidently, when your package arrived at our loading docks, a dock worker failed to see your letter in the container. We set the box aside with several boxes of parts destined for our Rebuilt Parts Department; therefore, your note did not come to the attention of our parts manager.

We appreciate your business and hope we have resolved your problem to your satisfaction.

Sincerely,

Jonathan L. Pennington

Jonathan L. Pennington
Office Manager

JLP/jq

e-mail: epicol:juno.com
Telephone: (569) 237-8228   Fax: (569) 237-8232

FIGURE 1 Adjustment Letter Granting the Claim

1. Open with whatever you believe the **reader** will consider good news:
   Grant the adjustment for uncomplicated situations ("Enclosed is a replacement for the damaged part").
   Reveal that you intend to grant the adjustment by admitting that the customer was in the right ("Yes, you were incorrectly billed for the

General Television, Inc.
5521 West 23rd Street
New York, NY 10062
Customer Relations
(212) 574-3894

September 28, 19--

Mr. Fred J. Swesky
7811 Ranchero Drive
Tucson, AZ 85761

Dear Mr. Swesky:

We have received your letter regarding the replacement of your KL-71 television set and have investigated the situation.

You stated in your letter that you used the set on an uncovered patio. As our local service representative pointed out, this model is not designed to operate in extreme heat conditions. As the instruction manual accompanying your new set stated, such exposure can produce irreparable damage to this model. Since your set was used in such extreme heat conditions, therefore, we cannot honor the two-year replacement warranty.

However, we are enclosing a certificate entitling you to a trade-in allowance equal to your local GTI dealer's markup for the set. This certificate will enable you to purchase a new set from the dealer at the wholesale price when you return the original set to your local dealer.

Sincerely yours,

Susan Siegel

Susan Siegel
Assistant Director

SS/mr
Enclosure

FIGURE 2 Adjustment Letter Granting a Partial Adjustment

oil delivery"). Then later explain the specific details of the adjustment. This method is good for adjustments that require detailed explanations.

Apologize for the error ("Please accept our apologies for not acting sooner to correct your account"). This method is effective when the customer's inconvenience is as much an issue as money.

SWELCO Coffee Maker, Inc._____

9025 North Main Street
Butte, MT 59702
Telephone: (297) 542-9664
Fax: (297) 544-0696

August 26, 19—

Mr. Carlos Ortiz
638 McSwaney Drive
Butte, MT 59702

Dear Mr. Ortiz:

Enclosed is your SWELCO Coffee Maker, which you sent to us on August 17.

In various parts of the country, tap water may contain a high mineral content. If you fill your SWELCO Coffee Maker with water for breakfast coffee before going to bed, a mineral scale will build up on the inner wall of the water tube—as explained on page 2 of your SWELCO Instruction Booklet.

We have removed the mineral scale from the water tube of your coffee maker and thoroughly cleaned the entire unit. To ensure the best service from your coffee maker in the future, clean it once a month by operating it with four ounces of white vinegar and eight cups of water. To rinse out the vinegar taste, operate the unit twice with clear water.

With proper care your SWELCO Coffee Maker will serve you faithfully and well for many years to come.

Sincerely,

Helen Upham

Helen Upham
Customer Services

HU/mo
Enclosure

FIGURE 3   Educational Adjustment Letter

Use a combination of these techniques. Often, situations requiring an adjustment are unique and do not fit a single pattern.

2. Explain what caused the problem—if such an explanation will help restore your reader's confidence or goodwill.
3. Explain specifically how you intend to make the adjustment—if it is not obvious in your opening.

4. Express appreciation to the customer for calling your attention to the situation, explaining that this helps your firm keep the quality of its product or service high.

5. Point out any steps you may be taking to prevent a recurrence of whatever went wrong, giving the customer as much credit as the facts allow.

6. Close pleasantly—looking forward, not back. Avoid recalling the problem in your closing ("Again, we apologize . . ."). Figure 1 is a typical adjustment letter.

Sometimes you may decide to grant a partial adjustment, even though the claim is not really justified, as in Figure 2.

You may sometimes need to educate your reader about the use of your product or service. Customers sometimes submit claims that are not justified, even though they honestly believe them to be. The customer may actually be at fault—for not following maintenance instructions properly for example. Such a claim is granted only to build goodwill. When you write a letter of adjustment in such a situation, it is wise to give the explanation before granting the claim—otherwise, your reader may never get to the explanation. If your explanation establishes customer responsibility, be sure that it is by implication rather than by outright statement. Look at the example of such an educational adjustment letter (see Figure 3).

adverbs

An adverb modifies the action or condition expressed by a **verb.**

> **EXAMPLE**   The recording head hit the surface of the disc *hard.* (The adverb tells *how* the recording head hit the disc.)

An adverb may also modify an **adjective,** another adverb, or a **clause.**

> **EXAMPLES**   The graphics department used *extremely* bright colors. (modifying an adjective)
>
> The redesigned brake pad lasted *much* longer. (modifying another adverb)
>
> *Surprisingly,* the machine failed. (modifying a clause)

An adverb answers one of the following questions:

WHERE? (adverb of place)

> **EXAMPLE**   Move the throttle *forward* slightly.

WHEN? (adverb of time)

> **EXAMPLE**   Replace the thermostat *immediately.*

HOW? (adverb of manner)

> **EXAMPLE**   Add the agent *cautiously.*

HOW MUCH? (adverb of degree)

> **EXAMPLE**   The *nearly* completed report was lost in the move.

Types of Adverbs

An adverb may be a common, a conjunctive, an interrogative, or a numeric **modifier.**

Typical *common adverbs* are *almost, seldom, down, also, now, ever,* and *always.*

> **EXAMPLE**   I *rarely* work on the weekend.

A *conjunctive adverb* is an adverb because it modifies the **clause** that it introduces; it operates as a **conjunction** because it joins two independent clauses. The most common conjunctive adverbs are *however, nevertheless, moreover, therefore, further, then, consequently, besides, accordingly, also,* and *thus.*

> **EXAMPLE**   The engine performed well in the laboratory; *however,* it failed under road conditions.

Two independent clauses that are joined by a conjunctive adverb require a **semicolon** before and a **comma** after the conjunctive adverb because the conjunctive adverb is introducing the independent clause that follows it and is indicating its relationship with the independent clause preceding it. The conjunctive adverb connects ideas but does not grammatically connect the clause stating the ideas; the semicolon must do that. The conjunctive adverb both connects and modifies. As a **modifier,** it is part of one of the two clauses that it connects. The use of the semicolon makes it a part of the clause it modifies.

> **EXAMPLES**   The building was not finished on the scheduled date; *nevertheless,* Rogers moved in and began to conduct the business of the department from the unfinished offices.
>
> The new project is almost completed and is already over cost estimates; *however,* we must emphasize the importance of adequate drainage.

Interrogative adverbs ask questions. Common interrogative adverbs include *where, when, why,* and *how.*

EXAMPLES *How* many hours did you work last week?
Where are we going when the new project is finished?
Why did it take so long to complete the job?

Typical *numeric adverbs* are *once* and *twice.*

EXAMPLE I have worked overtime *twice* this week.

Comparison of Adverbs

Adverbs are normally compared by adding *-er* or *-est* to them or by inserting *more* or *most* in front of them. One-syllable adverbs use the comparative ending *-er* and the superlative ending *-est.*

EXAMPLES This copier is *faster* than the old one.
This copier is the *fastest* of the three tested.

Most adverbs with two or more **syllables** end in *-ly,* and most adverbs ending in *-ly* are compared by inserting the comparative *more* or *less* or the superlative *most* or *least* in front of them.

EXAMPLES He moved *more quickly* than the other company's salesman.
Most surprisingly, the engine failed during the test phase.
This filter can be changed *less quickly* than the previous filter.
Least surprisingly, the engine failed during the test phase.

There are a few irregular adverbs that require a change in form to indicate **comparison.**

EXAMPLES The training program functions *well.*

Our training program functions *better* than most others in the industry.

Many consider our training program the *best* in the industry.

Placement of Adverbs

An adverb is normally placed in front of the verb it modifies.
EXAMPLE The pilot *meticulously* performed the preflight check.

An adverb may, however, follow the verb (or the verb and its **object**) that it modifies.

EXAMPLES The gauge dipped *suddenly.*
They repaired the computer *quickly.*

An adverb also may be placed between a helping verb and a main verb.

EXAMPLE In this temperature range, the pressure will *quickly* drop.

To place **emphasis** on an adverb, put it before the subject of the sentence.

EXAMPLES   *Clearly,* he was ready for the promotion when it came.
Unfortunately, fuel rationing has been unavoidable.

Place such adverbs as *nearly, only, almost, just,* and *hardly* immediately before the words they limit.

CHANGE   The punch press with the auxiliary equipment *only* costs $47,000.

TO   The punch press with the auxiliary equipment costs *only* $47,000.

The first sentence is ambiguous because it might be understood to mean that only the punch press with auxiliary equipment costs $47,000. (See also **modifiers.**)

advice/advise

Advice is a **noun** that means "counsel" or "suggestion."

EXAMPLE   My *advice* is to sign the contract immediately.

Advise is a **verb** that means "give advice."

EXAMPLE   I *advise* you to sign the contract immediately.

affect/effect

Affect is a **verb** that means "influence."

EXAMPLE   The public utility commission's decisions *affect* all state utilities.

Effect can function either as a verb that means "bring about" or "cause" or as a **noun** that means "result." It is best, however, to avoid using *effect* as a verb. A less formal word, such as *made*, is usually preferable.

CHANGE   The new manager *effected* several changes that had a good *effect* on morale.

TO   The new manager *made* several changes that had a good *effect* on morale.

affectation

Affectation is the use of language that is more formal or showy than is necessary to communicate information to the **reader.** A writer who is unnecessarily ornate, pompous, or pretentious is usually attempting to impress the reader by showing off a repertoire of fancy or flashy words and phrases. Using unnecessarily formal words (such as *herewith*) and outdated phrases (such as *please find enclosed*) is another cause of affectation.

EXAMPLES   pursuant to (instead of *about* or *regarding*)
in view of the foregoing (instead of *therefore*)
in view of the fact that (instead of *because*)
it is interesting to note that (omit)
it may be said that (omit)

Affected writing forces the reader to work harder to understand the writer's meaning. Affected writing typically contains abstract, highly technical, pseudotechnical, pseudolegal, or foreign words and is often liberally sprinkled with **vogue words. Jargon** can become affectation if it is misused. **Euphemisms** can contribute to affected writing if their purpose is to hide the facts of a situation rather than treat them with dignity or restraint.

The easiest kind of affectation to be lured into is the use of **long variants:** words created by adding **prefixes** and **suffixes** to simpler words *(analyzation* for *analysis, telephonic communication* for *telephone call).* The practice of **elegant variation**—attempting to avoid repeating the same word in the same **paragraph** by substituting pretentious **synonyms**—is also a form of affectation. Another contributor to affectation is **gobbledygook,** which is wordy, roundabout writing that has many pseudolegal and psuedoscientific terms sprinkled throughout.

Attempts to make the trivial seem important can also cause affectation. This attempt is apparent in the first version of the following example, taken from a specification soliciting bids from local merchants for the operation of a television repair shop in an Air Force Post Exchange.

CHANGE   In addition to performing interior housekeeping services, the concessionaire shall perform custodial maintenance on the exterior of the facility and grounds. Where a concessionaire shares a facility with one or more other concessionaires, exterior custodial maintenance responsibilities will be assigned by Post Exchange management on a fair and equitable basis. In those instances where the concessionaire's activity is located in a Post Exchange complex wherein predominant tenancy is by Post Exchange-operated activities, then Post Exchange management shall be responsible for exterior custodial maintenance except for those described in 1, 2, 3, and 4 below. The necessary equipment and labor to perform exterior custodial maintenance, when such a responsibility has been assigned to the concessionaire, shall be furnished by the concessionaire. Exterior custodial maintenance shall include the following tasks:

1. Clean entrance door and exterior of storefront windows daily.
2. Sweep and clean the entrance and customer walks daily.
3. Empty and clean waste and smoking receptacles daily.
4. Check exterior lighting and report failures to the contracting officer's representative daily.

TO   The merchant will, with his or her own equipment, perform the following duties daily:

1. Maintain a clean and neat appearance inside the store.
2. Clean the entrance door and the outsides of the store windows.
3. Sweep the entrance and the sidewalk.
4. Empty and clean wastebaskets and ashtrays.
5. Check the exterior lighting, and report failures to the representative of the contracting officer.

The merchant will also maintain the grounds surrounding the store. Where two or more merchants share the same building, Post Exchange management will assign responsibility for the grounds. Where the merchant's store is in a building that is occupied predominantly by Post Exchange operations, Post Exchange management will be responsible for the grounds.

Affectation is a widespread problem in technical writing because many people apparently feel that affectation lends a degree of formality, and hence authority, to their writing. Nothing could be further from the truth. Affectation puts up a smoke screen that the reader must penetrate to get to the writer's meaning. It can alienate the customer or client, making him or her feel that the writer is difficult to communicate with. (See also **conciseness/wordiness, clichés** and **nominalizations.**)

affinity

Affinity refers to the attraction of two persons or things to each other.

EXAMPLE   The *affinity* between these two elements can be explained in terms of their valence electrons.

Affinity should never be used to mean "ability" or "aptitude."

CHANGE   She has an *affinity* for chemistry.
TO   She has an *aptitude* for chemistry.
OR   She has a *talent* for chemistry.

aforesaid

Aforesaid means "stated previously," but it is legal **jargon** and should be avoided in writing. (See also **above.**)

> **CHANGE** The *aforesaid* problems can be solved in six months.
> **TO** The three problems explained in the preceding paragraphs can be solved in six months.

agree to/agree with

When you *agree to* something, you are "giving consent."

> **EXAMPLE** I *agree to* a road test of the new model by August 1.

When you *agree with* something, you are "in accord" with it.

> **EXAMPLE** I *agree with* the recommendations of the advisory board.

agreement

Agreement, grammatically, means the correspondence in form between different elements of a sentence to indicate **person, number, gender,** and **case.** A **pronoun** must agree with its antecedent, and a **verb** must agree with its subject.

A subject and its verb must agree in number and in person.

> **EXAMPLES** The *design is* an acceptable one. (The first person singular subject, *design,* requires the first person singular verb, *is.*)
>
> The new *products are* going into production soon. (The third person plural subject, *products,* requires the third person plural form of the verb, *are.*)

A pronoun and its antecedent must agree in person, number, and gender.

> **EXAMPLES** The *employees* report that *they* become less efficient as the humidity rises. (The third person plural subject, *employees,* requires the third person plural pronoun, *they.*)
>
> *Mr. Joiner* said that *he* would serve as a negotiator. (The third person singular, masculine subject, *Mr. Joiner,* requires the third person singular, masculine form of the pronoun, *he.*)

(See also **agreement of pronouns and antecedents** and **agreement of subjects and verbs.**)

agreement of pronouns and antecedents

Every **pronoun** must have an antecedent, or a **noun** to which it refers. In the following example, the pronoun *it* has no noun to which it could logically refer. The solution is to use a noun instead of the pronoun or to provide a noun to which it could logically refer.

> **CHANGE** Electronics technicians must constantly struggle to keep up with the professional literature because *it* is a dynamic science.
>
> **TO** Electronics technicians must constantly struggle to keep up with the professional literature because *electronics* is a dynamic science.
>
> **OR** Electronics technicians must continue to study *electronics* because *it* is a dynamic science.

Using the relative **pronoun** *which* to refer to an idea instead of a specific noun can be confusing.

> **CHANGE** He acted independently on the advice of his consultant, *which* the others thought unjust. (Was it the fact that he acted independently or was it the advice that the others thought unjust?)
>
> **TO** The others thought it unjust for him to act independently.
>
> **OR** The others thought it unjust for him to act on the advice of his consultant.

Using the pronouns *it, they, these, those, that,* and *this* can also lead to vague or uncertain references.

> **CHANGE** Studs and thick treads make snow tires effective. *They* are implanted with an air gun. (Which are implanted with an air gun: studs or thick treads?)
>
> **TO** Studs and thick treads make snow tires effective. *The studs* are implanted with an air gun.
>
> **CHANGE** The inadequate quality-control procedure has resulted in an equipment failure. *This* is our most serious problem at present. (Which is our most serious problem at present: the procedure or the failure?)
>
> **TO** The inadequate quality-control procedure has resulted in an equipment failure. *Quality control* is our most serious problem at present.
>
> **OR** The inadequate quality-control procedure has resulted in an equipment failure. *The equipment failure* is our most serious problem at present.

Gender

A pronoun must agree in **gender** with its antecedent.

> EXAMPLE   *Mr. Swivet* in the Accounting Department acknowledges *his* share of the responsibility for the misunderstanding, just as *Ms. Barkley* in the Research Division must acknowledge *hers.*

Traditionally, a masculine, singular pronoun has been used to agree with such indefinite antecedents as *anyone* and *person.*

> EXAMPLE   *Each* may stay or go as *he* chooses.

It is now recognized that there is an implied sexual bias in such usage. When alternatives are available, use them. One solution is to use the plural. Another is to use both feminine and masculine pronouns, although this combination is clumsy when used too often.

> CHANGE   Every *employee* must sign *his* time card.
> TO   All *employees* must sign *their* time cards.
> OR   Every *employee* must sign *his or her* time card.

Do not, however, attempt to avoid expressing gender by resorting to a plural pronoun when the antecedent is singular.

> CHANGE   A *technician* can expect to advance on *their* merit.
> TO   *Technicians* can expect to advance on *their* merit.
> OR   A *technician* can expect to advance on *his or her* merit.

Also be careful to avoid gender-related stereotypes, as in "the nurse . . . *she,*" or "the doctor . . . *he.*"

Number

A pronoun must agree with its antecedent in **number.** Many problems of agreement are caused by expressions that are not clear in number.

> CHANGE   Although the typical *engine* runs well in moderate temperatures, *they* often stall in extreme cold.
> TO   Although the typical *engine* runs well in moderate temperatures, *it* often stalls in extreme cold.

Use singular pronouns with the antecedents *everybody* and *everyone* unless to do so would be illogical because the meaning is obviously plural. (See also **everybody/everyone.**)

> EXAMPLES   *Everyone* pulled *his* share of the load.
>
> *Everyone* laughed at my sales slogan, and I really couldn't blame *them.*

The demonstrative adjectives sometimes cause problems with agreement of number. *This* and *that* are used with singular nouns, and *these* and *those* are used with plural nouns. Demonstrative adjectives often cause problems when they modify the nouns *kind, type,* and *sort*. Demonstrative adjectives used with these nouns should agree with them in number.

EXAMPLES   this kind / these kinds
                      that type / those types
                      this sort / these sorts

Confusion often develops when the **preposition** *of* is added ("this kind *of*," "these kinds *of*") and the **object** of the preposition is not made to conform in number to the demonstrative **adjective** and its noun.

CHANGE   *This kind* of hydraulic *cranes* is best.
TO   *This kind* of hydraulic *crane* is best.

CHANGE   *These kinds* of hydraulic *crane* are best.
TO   *These kinds* of hydraulic *cranes* are best.

The error can be avoided by remembering to make the demonstrative adjective, the noun, and the object of the preposition—all three—agree in number. This **agreement** makes the sentence not only correct but also more precise.

Using demonstrative adjectives with words like *kind, type,* and *sort* can easily lead to vagueness. It is better to be more specific.

CHANGE   *These kinds* of hydraulic cranes are best.
TO   *Computer-controlled* hydraulic cranes are best.

A compound antecedent joined by *or* or *nor* is singular if both elements are singular, and plural if both elements are plural.

EXAMPLES   Neither the *engineer* nor the *draftsman* could do *his* job until *he* understood the new concept.

                      Neither the *executives* nor the *directors* were pleased at the performance of *their* company.

When one of the antecedents connected by *or* or *nor* is singular and the other plural, the pronoun agrees with the nearer antecedent.

EXAMPLES   Either the *supervisor* or the *operators* will have *their* licenses suspended.

                      Either the *operators* or the *supervisor* will have *his* license suspended.

A compound antecedent with its elements joined by *and* requires a plural pronoun.

> **EXAMPLE**   Martha and Joan took *their* layout drawings with *them*.

If both elements refer to the same person, however, use the singular pronoun.

> **EXAMPLE**   The respected *economist* and *author* departed from *her* prepared speech.

Collective nouns may be singular or plural, depending on meaning.

> **EXAMPLES**   The *committee* arrived at the recommended solutions only after *it* had deliberated for days.
>
> The *committee* quit for the day and went to *their* respective homes.

agreement of subjects and verbs

A **verb** must agree with its subject. Do not let intervening **phrases** and **clauses** mislead you.

> **CHANGE**   The *program*, despite the new instructional materials, still *take* several days to install.
>
> **TO**   The *program*, despite the new instructional materials, still *takes* several days to install. (The verb *takes* must agree in **number** with the subject, *program*, rather than *materials*.)
>
> **CHANGE**   The *use* of insecticides, fertilizers, and weed killers, although they offer unquestionable benefits, often *result* in unfortunate side effects.
>
> **TO**   The *use* of insecticides, fertilizers, and weed killers, although they offer unquestionable benefits, often *results* in unfortunate side effects. (The verb *results* must agree with the subject of the sentence, *use*, rather than with the subject of the preceding clause, *they*.)

Be careful to avoid making the verb agree with the **noun** immediately in front of it if that noun is not its subject. This problem is especially likely to occur when a plural noun falls between a singular subject and its verb.

> **EXAMPLES**   Only *one* of the emergency lights *was* functioning when the accident occurred. (The subject of the verb is *one*, not *lights*.)
>
> *Each* of the switches *controls* a separate circuit. (The subject of the verb is *each*, not *switches*.)
>
> *Each* of the managers *supervises* a very large region. (The subject of the verb is *each*, not *managers*.)

Be careful not to let modifying **phrases** obscure a simple subject.

> **EXAMPLE** The *advice* of two engineers, one lawyer, and three executives *was* obtained prior to making a commitment. (The subject of the verb is *advice,* not *executives.*)

Sentences with inverted word order can cause problems with agreement between subject and verb.

> **EXAMPLE** From this work *have come* several important *improvements.* (The subject of the verb is *improvements,* not *work.*)

Such words as *type, part, series,* and *portion* take singular verbs even when they precede a phrase containing a plural noun.

> **EXAMPLES** A *series* of meetings *was* held about the best way to market the new product.
>
> A large *portion* of most industrial annual reports *is* devoted to promoting the corporate image.

Subjects expressing measurement, weight, mass, or total often take singular verbs even when the subject word is plural in form. Such subjects are treated as a unit.

> **EXAMPLES** *Four years is* the normal duration of the training program.
> *Twenty dollars is* the wholesale price of each unit.

Indefinite **pronouns** such as *some, none, all, more,* and *most* may be singular or plural depending upon whether they are used with a mass noun (*oil* in the following examples) or with a count noun (*drivers* in the following examples). Mass nouns are singular and count nouns are plural.

> **EXAMPLES** *None* of the oil *is* to be used.
> *None* of the truck drivers *are* to go.
> *Most* of the oil *has* been used.
> *Most* of the drivers *know* why they are here.
> *Some* of the oil *has* leaked.
> *Some* of the drivers *have* gone.

One and *each* are normally singular.

> **EXAMPLES** *One* of the brake drums *is* still scored.
> *Each* of the original founders *is* scheduled to speak at the dedication ceremony.

A verb following the relative pronouns *who* and *that* agrees in number with the noun to which the pronoun refers (its antecedent).

EXAMPLES   Steel is one of those *industries* that *are* hardest hit by high energy costs. *(That* refers to *industries.)*

This is one of those engineering *problems* that *require* careful analysis. *(That* refers to *problems.)*

She is one of those *employees* who *are* rarely absent. *(Who* refers to *employees.)*

The word *number* sometimes causes confusion. When used to mean a specific number, it is singular.

EXAMPLE   *The number* of committee members *was* six.

When used to mean an approximate number, it is plural.

EXAMPLE   *A number* of people *were* waiting for the announcement.

Relative pronouns *(who, which, that)* may take either singular or plural verbs, depending upon whether the antecedent is singular or plural. (See also **who/whom.**)

EXAMPLES   He is a chemist *who takes* work home at night.
He is one of those chemists *who take* work home at night.

Some abstract nouns are singular in meaning though plural in form: *mathematics, news, physics,* and *economics.*

EXAMPLES   *News* of the merger *is* on page 4 of the *Chronicle.*
Textiles is an industry in need of import quotas.

Some words are always plural, such as *trousers* and *scissors.*

EXAMPLE   His *trousers were* torn by the machine.
BUT   *A pair* of trousers *is* on order.

Collective subjects take singular verbs when the group is thought of as a unit, plural verbs when the individuals are thought of separately.

A singular subject that is followed by a phrase or clause containing a plural noun still requires a singular verb.

CHANGE   *One* in twenty transistors we receive from our suppliers *are* faulty.
TO   *One* in twenty transistors we receive from our suppliers *is* faulty. (The subject is *one,* not *transistors.)*

The number of a subjective **complement** does not affect the number of the verb—the verb must always agree with the subject.

EXAMPLE   The *topic* of his report *was* rivers. (The subject of the sentence is *topic,* not *rivers.)*

A book with a plural **title** requires a single verb.

> EXAMPLE   Romig's *Binomial Tables* is a useful source.

Compound Subjects

A compound subject is one that is composed of two or more elements joined by a **conjunction** such as *and, or, nor, either . . . or,* or *neither . . . nor.* Usually, when the elements are connected by *and,* the subject is plural and requires a plural verb.

> EXAMPLE   *Chemistry and accounting are* both prerequisites for this position.

One exception occurs when the elements connected by *and* form a unit or refer to the same thing. In this case, the subject is regarded as singular and takes a singular verb.

> EXAMPLES   *Bacon and eggs is* a high-cholesterol meal.
> The *red, white, and blue flutters* from the top of the Capitol.
> Our greatest *technical challenge and business opportunity is* the Model MX Calculator.

A compound subject with a singular and a plural element joined by *or* or *nor* requires that the verb agree with the element nearest to it.

> EXAMPLES   Neither the office manager nor the *secretaries were* there.
> Neither the secretaries nor the office *manager was* there.
> Either they or *I am* going to write the report.
> Either I or *they are* going to write the report.

If *each* or *every* modifies the elements of a compound subject, use the singular verb.

> EXAMPLES   *Each* manager and supervisor *has* a production goal to meet.
> *Every* manager and supervisor *has* a production goal to meet.

all around/all-around/all-round

All-round and *all-around* both mean "comprehensive" or "versatile."

> EXAMPLE   The company started an *all-round* training program.

Do not confuse these words with the two-word **phrase** *all around,* as in "The fence was installed *all around* the building."

all right/all-right/alright

All right means "all correct," as in "The answers were *all right.*" In formal writing it should not be used to mean "good" or "acceptable." It is always written as two words, with no **hyphen;** *all-right* and *alright* are incorrect.

CHANGE   The decision that the committee reached was *all right.*
TO   The decision that the committee reached was acceptable.

all together/altogether

All together means "all acting together," or "all in one place."

EXAMPLE   The necessary instruments were *all together* on the tray.

Altogether means "entirely" or "completely."

EXAMPLE   The trip was *altogether* unnecessary.

allude/elude/refer

Allude means to make an indirect reference to something not specifically mentioned.

EXAMPLE   The report simply *alluded* to the problem, rather than stating it clearly.

Elude means to escape notice or detection.

EXAMPLE   The discrepancy in the account *eluded* the auditor.

Refer is used to indicate a direct reference to something.

EXAMPLE   He *referred* to the chart three times during his speech.

allusion

The use of allusion (implied or indirect reference) promotes economical writing because it is a shorthand way of referring to a body of material in a few words, or of helping to explain a new and unfamiliar process in terms of one that is familiar. Be sure, however, that your **reader** is familiar with the material to which you allude. In the following **paragraph,** the writer sums up the argument he has been developing with an allusion to the Bible. The biblical story is well known, and the allusion, with its implicit reference to "right standing up to might," concisely emphasizes the writer's point.

> **EXAMPLE** As it presently exists, the review process involves the consumer's attorney sitting alone, usually without adequate technical assistance, faced by two or three government attorneys, two or three attorneys from the XYZ Corporation, and large teams of experts who support the government and corporation's attorneys. The entire proceeding is reminiscent of David versus Goliath.

Allusions should be used with restraint. If overdone, they can lead to **affectation.** (See also **analogy** and **international correspondence.**)

allusion/illusion

An *allusion* is an indirect reference to something not specifically mentioned.

> **EXAMPLE** The report made an *allusion* to metal fatigue in support structures.

An *illusion* is a mistaken perception or a false image.

> **EXAMPLE** County officials are under the *illusion* that the landfill will last indefinitely.

almost/most

Do not use *most* as a colloquial substitute for *almost* in your writing.

> **CHANGE** New shipments arrive *most* every day.
> **TO** New shipments arrive *almost* every day.

If you can substitute *almost* for *most* in a sentence, *almost* is the word you need.

already/all ready

Already is an **adverb** expressing time.

> **EXAMPLE** We had *already* shipped the transistors when the stop order arrived.

All ready is a two-word **phrase** meaning "completely prepared."

> **EXAMPLE** He was *all ready* to start work on the project when it was suddenly canceled.

also

Also is an **adverb** that means "additionally."

> **EXAMPLE** Two 500,000-gallon tanks have recently been constructed on site. Several 10,000-gallon tanks are *also* available, if needed.

It should not be used as a connective in the sense of "and."

> **CHANGE** He brought the reports, letters, *also* the section supervisor's recommendations.
> **TO** He brought the reports, letters, *and* the section supervisor's recommendations.

Avoid opening sentences with *also*. It is a weak transitional word that suggests an afterthought rather than planned writing.

> **CHANGE** *Also* he brought statistical data to support his proposal.
> **TO** *In addition,* he brought statistical data to support his proposal.
> **OR** He *also* brought statistical data to support his proposal.

ambiguity

A word or passage is ambiguous when it is susceptible to two or more interpretations, yet provides the **reader** with no certain basis for choosing among the alternatives.

> **EXAMPLE** Mathematics is more valuable to an engineer than a computer. (Does this mean that an engineer is more in need of mathematics than a computer is? Or does it mean that mathematics is more valuable to an engineer than a computer is?)

Ambiguity can take many forms: ambiguous **pronoun reference,** misplaced **modifiers, dangling modifiers,** ambiguous coordination, ambiguous juxtaposition, incomplete **comparison,** incomplete **idiom,** ambiguous **word choice,** and so on.

> **CHANGE** Inadequate quality-control procedures have resulted in more equipment failures. *This* is our most serious problem at present. (ambiguous pronoun reference; does *this* refer to "quality-control procedures" or "equipment failures"?)
> **TO** Inadequate quality-control procedures have resulted in more equipment failures. *These failures* are our most serious problem at present.
> **OR** Inadequate quality-control procedures have resulted in more equipment failures. *Quality control* is our most serious problem at present.

CHANGE	Ms. Jones values rigid quality-control standards more than Mr. Rosenblum. (incomplete comparison)
TO	Ms. Jones values rigid quality-control standards more than Mr. Rosenblum *does.*
CHANGE	His hobby was cooking. He was especially fond of cocker spaniels. (missing modifier)
TO	His hobby was cooking. He was *also* especially fond of cocker spaniels.
CHANGE	All navigators are *not* talented in mathematics. (misplaced modifier; the implication is that *no* navigator is talented in mathematics)
TO	*Not* all navigators are talented in mathematics.

Ambiguity is also often caused by thoughtless **word choice.**

| CHANGE | The general manager has denied reports that the plant's recent fuel allocation cut will be *restored.* (inappropriate word choice) |
| TO | The general manager has denied reports that the plant's recent fuel allocation cut will be *rescinded.* |

amount/number

Amount is used with things thought of in bulk (mass **nouns**).

| EXAMPLES | The *amount* of electricity available for industrial use is limited. The *amount* of oxygen was insufficient for combustion. |

Number is used with things that can be counted as individual items (count nouns).

| EXAMPLES | A large *number* of stockholders attended the meeting. |
| | The *number* of employees who are qualified for early retirement has increased in recent years. |

Avoid using *amount* when referring to countable items.

CHANGE	The *amount* of people in the room gradually increased.
TO	The *number* of people in the room gradually increased.
CHANGE	I was surprised at the *amount* of errors in the report.
TO	I was surprised at the *number* of errors in the report.
CHANGE	Because the *amount* of thefts has increased, the doors will be locked in the evening.
TO	Because the *number* of thefts has increased, the doors will be locked in the evening.

ampersands

The ampersand (&) is a **symbol** sometimes used to represent the word *and*, especially in the names of organizations.

> EXAMPLES Chicago & Northwestern Railway
> Watkins & Watkins, Inc.

The ampersand is appropriate for footnotes, **bibliographies, lists,** and references if the ampersand appears in the name being listed. When writing the name of an organization in sentences or in an address, however, spell out the word *and* unless the ampersand appears in the official name of the company. (See also **abbreviations.**)

analogy (see **figures of speech**)

and/or

And/or means that either both circumstances are possible or only one of two circumstances is possible; however, it is clumsy and awkward because it makes the **reader** stop to puzzle over your distinction.

> CHANGE Use A *and/or* B.
> TO Use A or B or both.

ante-/anti-

Ante- means "before" or "in front of."

> EXAMPLES *Ante*room, *ante*date, *ante*diluvian (before the Flood)

Anti- means "against" or "opposed to."

> EXAMPLES *anti*body, *anti*clerical, *anti*social

Anti- is hyphenated when joined to proper nouns or to words beginning with the letter *i.*

> EXAMPLES *anti*-American, *anti*-intellectual

When in doubt, consult your **dictionary.** (See also **prefixes.**)

antonyms

An *antonym* is a word that is nearly the opposite, in meaning, of another word.

EXAMPLES   good/bad, well/ill, fresh/stale

Many pairs of words that look as if they are *antonyms* are not. Be careful not to use these words incorrectly.

EXAMPLES   famous/infamous, flammable/inflammable, limit/delimit

apostrophes

The *apostrophe* (') is used to show possession, to mark the omission of letters, and sometimes to indicate the plural of Arabic **numbers,** letters, and **acronyms.** Do not confuse the apostrophe used to show the plural with the apostrophe used to show possession.

EXAMPLES   The entry required five *7's* in the appropriate columns. (The apostrophe is used here to indicate the plural, not possession.)

The *letter's* purpose was clearly evident in its opening paragraph. (The apostrophe here is used to show possession, not the plural.)

To Show Possession

An apostrophe is used with an *s* to form the possessive **case** of some **nouns.**

EXAMPLE   A recent scientific analysis of *New York City's* atmosphere concluded that a New Yorker on the street took into his or her lungs the equivalent in toxic materials of 38 cigarettes a day.

With coordinate nouns, the last noun takes the possessive form to show joint possession.

EXAMPLE   Michelson and *Morley's* famous experiment on the velocity of light was made in 1887.

To show individual possession with coordinate nouns, each noun should take the possessive form.

EXAMPLE   The difference between *Tom's* and *Mary's* test results is statistically insignificant.

Singular nouns ending in *s* may form the possessive either by an apostrophe alone or by *'s.* Whichever way you do it, however, be consistent.

EXAMPLES   a *waitress'* uniform, an *actress'* career
a *waitress's* uniform, an *actress's* career

Singular nouns of one syllable form the possessive by adding *'s*.

>   EXAMPLE   The *boss's* desk was cluttered.

Use only an apostrophe with plural nouns ending in *s*.

>   EXAMPLES   a *managers'* meeting, the *technicians'* handbook, the *waitresses'*
>   lounge

When adding *'s* would result in a noun ending in multiple consecutive *s* sounds, add only an apostrophe.

>   EXAMPLES   The *Joneses'* test result

Do not use the apostrophe with possessive **pronouns.**

>   EXAMPLES   yours, its, his, ours, whose, theirs

It's is a **contraction** of *it is; its* is the possessive form of the pronoun. Be careful not to confuse the two words. (See **its/it's.**)

>   EXAMPLE   *It's* important that the sales force meet *its* quota.

In names of places and institutions, the apostrophe is usually omitted.

>   EXAMPLES   Harpers Ferry, Writers Book Club

To Show Omission

An apostrophe is used to mark the omission of letters in a word or date.

>   EXAMPLES   can't, I'm, I'll
>   the class of '61

To Form Plurals

An apostrophe and an *s* may be added to show the plural of a word as a word. (The word itself is underlined, or italicized, to call attention to its use.)

>   EXAMPLE   There were five *and's* in his first sentence.

If the term is in all **capital letters** or ends with a capital letter, however, the apostrophe is not required to form the plural. (See also **acronyms and initialisms.**)

>   EXAMPLES   The university awarded seven *Ph.D.s* in engineering last year.
>   He had included 43 *ADDs* in his computer program.

Use an apostrophe to indicate the plural of numbers and letters only if confusion would result without one.

>   EXAMPLES   5s, 30s, two 100s, seven *I's*

appendix/appendixes/appendices

An appendix contains material at the end of a **formal report** or document that supplements or clarifies. (The plural form of the word may be either *appendixes* or *appendices.*)

Although not a mandatory part of a **report,** an appendix can be useful for explanations that are too long for notes but that could be helpful to the reader seeking further assistance or a clarification of points made in the report. Information placed in an appendix is too detailed or voluminous to appear in text without impeding the orderly presentation of ideas. This information typically includes passages from documents and laws that reinforce or illustrate the text, long lists of charts and **tables,** letters and other supporting documents, calculations (in full), computer printouts of raw data, and case histories. An appendix, however, should not be used for miscellaneous bits and pieces of information you were unable to work into the text.

Generally, each appendix contains only one type of information. The contents of each appendix should be identifiable without the reader having to refer to the body of the report. An introductory paragraph describing the contents of the appendix, therefore, is necessary for some appendixes, especially those containing computer printouts of data, long tables, or similar information.

When the report contains more than one appendix, arrange them in the order in which they are referred to in the text. Thus, a reference in the text to Appendix A should precede the first text reference to Appendix B.

Each appendix begins on a new page. (For guidance on where to locate appendixes in reports, see **formal reports.**) Identify each with a **title** and a **heading:**

Appendix A
Sample Questionnaire

Appendixes are ordinarily labeled *Appendix A, Appendix B,* and so on. If your report has only one appendix, label it Appendix, followed by the title. To call it Appendix A implies that an Appendix B will follow.

The titles and beginning page numbers of the appendixes are listed in the t**able of contents** of the report in which they appear.

application letters

The letter of application is essentially a sales letter. You are marketing your skills, abilities, and knowledge. Remember that you may be competing with many other applicants. The immediate **purpose** of an application letter is to get the attention of the person who screens and hires job applicants. Your ultimate goal is to obtain a job interview (see **interviewing for a job**).

The successful application letter does three things: catches the **reader's** favorable attention, convinces the reader that you are qualified for consideration, and requests an interview. It should be concisely written.

A letter of application should provide the following information:

1. Identify an employment area or state a specific job title.
2. Point out your source of information about the job.
3. Summarize your qualifications for the job, specifically education, work experience, and activities showing leadership skills.
4. Refer the reader to your **résumé.**
5. Ask for an interview, stating where you can be reached and when you will be available for an interview.

If you are applying for a specific job, include information pertinent to the position—details not included on the more general résumé.

Personnel directors review many letters each week. To save them time, you should state your job objective directly at the beginning of the letter.

> EXAMPLE   I am seeking a position in an engineering department where I can use my computer science training to solve engineering problems.

If you have been referred to a prospective employer by one of its employees, a placement counselor, a professor, or someone else, however, you might say so before stating your job objective.

> EXAMPLE   During the recent NOMAD convention in Washington, D.C., a member of your sales staff, Mr. Dale Jarrett, informed me of a possible opening for a manager in your Dealer Sales Division. My extensive background in the office machine industry, I believe, makes me highly qualified for the position.

In succeeding **paragraphs** expand upon the qualifications you mentioned in your **opening.** Add any appropriate details, highlighting the experience listed on your résumé that is especially pertinent to the job you are seeking. Close your letter with a request for an interview.

Prepare your letter with utmost care and **proofread** it very carefully.

See the accompanying three sample letters of application. The first one is written by a recent college graduate, the second by a college stu-

6819 Locustview Drive
Topeka, Kansas 66614
June 14, 19—

Loudons, Inc.
4619 Drove Lane
Kansas City, Kansas 63511

Dear Personnel Manager:

The *Kansas Dispatch* recently reported that Loudons is building a new data processing center just north of Kansas City. I would like to apply for a position as an entry-level programmer at the center.

I am a recent graduate of Fairview Community College in Topeka, with an Associate Degree in Computer Science. In addition to taking a broad range of courses, I have served as a computer consultant at the college's computer center, where I helped train novice computer users. Since I understand Loudons produces both in-house and customer documentation, my technical writing skills (as described in the enclosed resume) may be particularly useful.

I will be happy to meet with you at your convenience and provide any additional information you may need. You can reach me either at my home address or at (913) 233-1552 (telephone) or (913) 233-7643 (fax).

Sincerely,

David B. Edwards

David B. Edwards

Enclosure: Resume

FIGURE 1 Sample Application Letter

dent about to graduate, and the third by a writer with many years of work experience.

Figure 1 is in response to a local newspaper article about a company's plan to build a new plant. The writer is not applying for a spe-

273 East Sixth Street
Bloomington, IN 47401
May 29, 19—

Ms. Laura Goldman
Personnel Manager
Acton, Inc.
80 Roseville Road
St. Louis, MO 63130

Dear Ms. Goldman:

I am seeking a responsible position as a financial research assistant in which I may use my training to solve financial problems. I would be interested in exploring the possibility of obtaining such a position within your firm.

I expect to receive a Bachelor of Business Administration degree in finance from Indiana University in June. Since September 19-- I have been participating, through the university, in the Professional Training Program at Computer Systems International, in Indianapolis. In the program I was assigned, on a rotating basis, to several staff sections in apprentice positions. Most recently I have been a financial trainee in the Accounting Department and have gained a great deal of experience. Details of the academic courses I have taken are contained in the enclosed resume.

I look forward to meeting you soon in an interview. I can be contacted at my office (812-866-7000, ext. 312) or at home (812-256-6320). My fax number is (812) 493-1918 and my e-mail address is caw.aol.edu.

Sincerely yours,

Carol Ann Walker

Carol Ann Walker

Enclosure: Resume

FIGURE 2   Sample Application Letter

cific job opening but describes the position he is looking for. In Figure 2 the writer does not specify where she learned of the opening because she does not know whether a position is actually available. Figure 3 opens with an indication of where the writer learned of the job vacancy. (See also **job search** and **reference letters.**)

522 Beethoven Drive
Roanoke, Virginia 24017
November 15, 19—

Ms. Cecilia Smathers
Vice President, Dealer Sales
Hamilton Office Machines, Inc.
6194 Main Street
Hampton, Virginia 23661

Dear Ms. Smathers:

During the recent NOMAD convention in Washington, a member of your sales staff, Mr. Dale Jarrett, informed me of a possible opening for a Manager in your Dealer Sales Division. I believe that my extensive background in the office machine industry qualifies me for the position.

I was with the Technology, Inc., Dealer Division from its formation in 19— to its phase-out last year. During this period, I was involved in all areas of dealer sales, both wtihin Technology, Inc., and through personal contact with a number of independent dealers. Between 19— and 19— I served as Assistant to the Dealer Sales Manager as a Special Representative. My education and work experience are detailed in the enclosed resume.

I would like to discuss my qualifications in an interview at your convenience. Please write to me or telephone me at 703-449-6743 any weekday.

Sincerely,

Gregory Mindukakis

Gregory Mindukakis

Enclosure: Resume

FIGURE 3   Sample Application Letter

appositives

An *appositive* is a **noun** or noun **phrase** that follows and amplifies another noun or noun phrase. It has the same grammatical function as the noun it complements.

EXAMPLES   Dennis Gabor, *a British scientist,* experimented with coherent light in the 1940s.

The British scientist *Dennis Gabor* experimented with coherent light in the 1940s.

George Thomas, *head of the Economic and Planning Branch of PRC,* summarized the president's speech in a confidential memo to the advertising staff.

For detailed information on the use of **commas** with appositives, see **restrictive and nonrestrictive elements.**

When in doubt about the **case** of an appositive, you can check it by substituting the appositive for the noun it modifies.

EXAMPLE   My boss gave the two of us, Jim and *me,* the day off. (You wouldn't say, "My boss gave *I* the day off.")

articles (see **adjectives**)

as (see also **like/as**)

Since *as* can mean so many things (*since, because, for, that, at that time, when, while,* and so on) and can be at least four **parts of speech (conjunction, preposition, adverb, pronoun),** it is often overused and misused, especially in speech. In writing, *as* is often weak or even ambiguous. Notice that the following sentence has two possible meanings:

EXAMPLE   *As* we were together, he revealed his plans.
MEANS   *Because* we were together, he revealed his plans.
OR   *While* we were together, he revealed his plans.

The word *as* can also contribute to **awkwardness** by appearing too many times in a sentence.

CHANGE   *As* we realized *as* soon *as* we began the project, the problem needed a solution.
TO   We realized the moment we began the project that the problem needed a solution.

(See also **because.**)

as much as/more than

These two **phrases** are sometimes incorrectly run together, especially when intervening phrases delay the completion of the phrase.

CHANGE The engineers had *as much,* if not *more,* influence in planning the program *than* the accountants.

TO The engineers had *as much* influence in planning the program *as* the accountants, if not *more.*

OR The engineers had *as much* influence *as* the accountants, if not *more,* in planning the program.

as regards/with regard to/in regard to/regarding

With regards to and *in regards to* are incorrect **idioms** for *with regard to* and *in regard to. As regards* and *regarding* both are acceptable variants.

CHANGE *With regards to* the building contract, this question is pertinent.

TO *With regard to* the building contract, this question is pertinent.

OR *In regard to* the building contract, this question is pertinent.

OR *As regards* the building contract, this question is pertinent.

OR *Regarding* the building contract, this question is pertinent.

as such

The **phrase** *as such* is seldom useful and should be omitted.

CHANGE The drafting department, *as such,* worked overtime to meet the deadline.

TO The drafting department worked overtime to meet the deadline.

CHANGE This program is poor. *As such,* it should be eliminated.

TO This program is poor and should be eliminated.

as to whether

The **phrase** *as to whether (as to when* or *as to where)* is clumsy and re-dundant. Either omit it altogether or use only *whether.*

CHANGE *As to whether* we will redesign the fuel-injection system, we are still undecided.

TO *Whether* we will redesign the fuel-injection system is still undecided.

Be wary of all phrases starting with *as to;* they are often redundant, vague, or indirect.

CHANGE *As to* his policy, I am in full agreement.

TO I am in full agreement with his policy.

as well as

Do not use *as well as* together with *both*. The two expressions have similar meanings; use one or the other.

> **CHANGE** *Both* General Motors, *as well as* Ford, are developing electric cars.
>
> **TO** *Both* General Motors *and* Ford are developing electric cars.
>
> **OR** General Motors, *as well as* Ford, is developing an electric car.

attribute/contribute

Attribute (with the accent on the second syllable) is a **verb** that means "point to a cause or source."

> **EXAMPLE** He *attributes* the plant's improved safety record to the new training program.

Attribute (with the first syllable accented) is a **noun** meaning a quality or characteristic belonging to someone or something.

> **EXAMPLE** His mathematical skill is his most valuable *attribute*.

Contribute means "give."

> **EXAMPLE** His mathematical skills will *contribute* much to the project.

audience (see **readers**)

augment/supplement

Augment means to increase or magnify in size, degree, or effect.

> **EXAMPLE** Many employees *augment* their incomes by working overtime.

Supplement means to add something to make up for a deficiency.

> **EXAMPLE** The physician told him to *supplement* his diet with vitamins.

average/mean/median

The *average*, or arithmetical *mean*, is determined by dividing a sum of two or more quantities by the number of quantities. For example, if one **report** is 10 pages, another is 30 pages, and a third is 20 pages, their *average* (or *mean*) length is 20 pages. It is incorrect, therefore, to say that "*each* averages 20 pages" because *each* report is a specific length.

> CHANGE   Each report *averages* 20 pages.
> TO   The three reports *average* 20 pages.

A *median* is the middle number in a sequence of numbers.

> EXAMPLE   The *median* of the series, 1, 3, 4, 7, 8 is 4.

awhile/a while

Awhile is an **adverb** meaning "for a short time." It is not preceded by *for* because the meaning of *for* is inherent in the meaning of *awhile*. *A while* is a **noun** phrase that means "a period of time."

> CHANGE   Wait for *awhile* before investing more heavily.
> TO   Wait for *a while* before investing more heavily.
> OR   Wait *awhile* before investing more heavily.

awkwardness

Any writing that strikes the **reader** as awkward—that is, as forced or unnatural—impedes the reader's understanding. Awkwardness has many causes, including overloaded sentences, overlapping **subordination,** grammatical errors, ambiguous statements, overuse of **expletives** or of the passive **voice,** faulty **logic,** unintentional **repetition, garbled sentences,** and jammed **modifiers.**

To avoid awkwardness, make your writing as direct and concise as possible. The following three guidelines will help you smooth out most awkward passages. (1) In general, you should keep your sentences uncomplicated. (2) Use the active **voice** unless you have a particular reason to use the passive voice. (3) Tighten up your writing by eliminating excess words.

bad/badly

Bad is the **adjective** form that follows such linking **verbs** as *feel* and *look.*

EXAMPLES   With the flu, you will feel *bad* for three days.
We don't want our department to look *bad* at the meeting.

Badly is an **adverb.**

EXAMPLE   The test model performed *badly* during the trial run.

To say "I feel badly" would mean, literally, that your sense of touch was impaired. (See also **good/well.**)

balance/remainder

One meaning of *balance* is "a state of equilibrium"; another meaning is "the amount remaining in a bank account after balancing deposits and withdrawals." *Remainder,* in all applications, is "what is left over."

EXAMPLES   The accounting department must attempt to maintain a *balance* between looking after the company's best financial interests and being sensitive to the company's research and development work.

The *balance* in the corporate account after the payroll has been met is a matter for concern.

Round off the fraction to its nearest whole number and drop the *remainder.*

Four of the speakers at the conference were from Germany, and the *remainder* were from the United States.

be sure and/be sure to

The **phrase** *be sure and* is colloquial and unidiomatic when used for *be sure to*. (See also **idioms.**)

CHANGE   When the claxon sounds, *be sure and* phone the guard desk immediately.

TO   When the claxon sounds, *be sure to* phone the guard desk immediately.

because

To express cause, *because* is the strongest and most specific connective (others are *for, since, as*). *Because* is unequivocal in stating causal relationship.

EXAMPLE   We didn't complete the project *because* the raw materials became too costly. (The use of *because* emphasizes the cause-and-effect relationship.)

For can express causal relationships, but it is weaker than *because,* and it allows the **clause** that follows to be a separate independent clause.

EXAMPLE   We didn't complete the project, *for* raw materials became too costly. (The use of *for* expresses but does not emphasize the cause-and-effect relationship.)

As a connective to express cause, *since* is also a weak substitute for *because.*

EXAMPLE   *Since* the computer is broken, paychecks will be delayed.

However, *since* is an appropriate connective when the **emphasis** is on circumstances, conditions, or time rather than on cause and effect.

EXAMPLES   *Since* I was in town anyway, I decided to visit the test site.
Since 1984, the company has earned a profit every year.

As is the least definite connective to indicate cause; its use for this purpose is better avoided.

CHANGE   I left the office early, *as* I had finished my work.
TO   I left the office early *because* I had finished my work.
OR   *Since* I had finished my work, I left the office early.

(See also **reason is because.**)

B   being as/being that

These **phrases** are nonstandard English and should not be used in writing. Use *because* or *since.*

beside/besides

Besides, meaning "in addition to" or "other than," should be carefully distinguished from *beside,* meaning "next to" or "apart from."

> EXAMPLE   *Besides* the two of us from the Systems Department, three people from Production were standing *beside* the president when he presented the award.

between/among

Between is normally used to relate two items or persons.

> EXAMPLE   The roll pin is located *between* the grommet and the knob.
>
> Preferred stock offers a buyer a middle ground *between* bonds and common stock.

Among is used to relate more than two.

> EXAMPLE   The subcontracting was distributed *among* the three firms.

between you and me

People sometimes use the incorrect expression *between you and I.* Because the **pronouns** are **objects** of the **preposition** *between,* the objective form of the personal pronoun *(me)* must be used. (See also **case [grammar].**)

> CHANGE   *Between you and I,* John should be taken off the job.
>   TO   *Between you and me,* John should be taken off the job.

bi/semi

When used with periods of time, *bi* means "two" or "every two." *Bimonthly* means "once in two months"; *biweekly* means "once in two weeks."

When used with periods of time, *semi* means "half of" or "occurring twice within a period of time." *Semimonthly* means "twice a month"; *semiweekly* means "twice a week."

B

Both *bi* and *semi* are normally joined with the following element without space or **hyphen.**

biannual/biennial

By conventional usage, *biannual* means "twice during the year," and *biennial* means "every other year." (See also **bi/semi.**)

bibliography

A *bibliography* is a **list** of the books, articles, and other source materials consulted in the preparation of a paper, **report,** or article. It provides a convenient alphabetical listing of these sources in a standardized form for **readers** interested in getting further information on the **topic** or in assessing the **scope** of the **research.** A bibliography is normally placed at the end of a document. (See **formal reports** for guidance on the placement of a bibliography in a report.)

Those works consulted for background information should be included in a bibliography in addition to those actually cited in the text. A bibliography is often appropriate as a supplement to lists of "Works Cited" or "Reference" sections, since such lists include only works actually referred to in the text. If a list of "References" or "Works Cited" is used, the bibliography should follow the same **format.**

The entries in a bibliography are listed alphabetically by the author's last name. If the author is unknown, the entry is alphabetized by the first word in the **title** (following *a, an,* or *the*). Entries may also be arranged into subject categories and then by alphabetical order within these categories.

An annotated bibliography is one that includes complete bibliographic information about a work (author, title, publisher) followed by a brief description or evaluation of what the work contains. (See also **literature reviews.**) The following annotation concisely summarizes and evaluates a book of historical interest:

> Rickard, T.A. *A Guide to Technical Writing.* San Francisco Mining and Scientific P, 1908.
> This book is of particular historical interest because it is the first published book on technical writing for the professional. The author comments, "It has been said that in this age the man of science appears to be the only one who has anything to say, and he is the one that least knows how to say it. . . . Write simply and clearly, be accurate and careful;

B

above all, put yourself in the other fellow's place. Remember the reader." Geared to the mining and metallurgical sciences, the 17 short, unnumbered chapters cover matters of language, usage, grammar, and mechanics slanted toward the needs of the technical writer. The book ends with a paper the author read before the American Association for the Advancement of Science, at Denver on August 28, 1901: "A Plea for Greater Simplicity in the Language of Science." "We must remember," the author suggests, "that the language in relation to ideas is a solvent, the purity and clearness of which effect what it bears in solution."

For explanations and samples of bibliographic and citation forms, see **documenting sources.**

blend words

A *blend word* is formed by combining part of one word with part of another.

> **EXAMPLES**   motor + hotel = motel
> breakfast + lunch = brunch
> smoke + fog = smog
> electric + execute = electrocute
> chuckle + snort = chortle

Although blend words (sometimes called "portmanteau words") may occasionally be created by a specialist to meet a specific need—such as *stagflation* (a stagnant economy coupled with inflation)—resist creating blend words in your writing unless an obvious need arises. If you must create a blend word, be sure to define it clearly for your **reader.** Otherwise, creating blend words could be confusing. (See also **new words.**)

both . . . and

Statements using the *both . . . and* construction should always be balanced both grammatically and logically.

> **EXAMPLE**   A successful photograph must be *both* clearly focused *and* adequately lighted.

Notice that *both* and *and* are followed logically by ideas of equal weight and grammatically by identical constructions.

CHANGE	For success in engineering, it is necessary *both* to develop writing skills *and* mastering calculus.
TO	For success in engineering, it is necessary *both* to develop writing skills *and* to master calculus.

Do not substitute *as well as* for *and* in this construction.

CHANGE	For success in engineering, it is necessary *both* to master calculus *as well as* to develop writing skills.
TO	For success in engineering, it is necessary *both* to master calculus *and* to develop writing skills.

(See also **parallel structure.**)

brackets

The primary use of brackets is to enclose a word or words inserted by an editor or writer into a **quotation** from another source.

EXAMPLE	The text stated, "Fissile and fertile nuclei spontaneously emit characteristic nuclear radiations *[such as neutrons and gamma rays]* that are sufficiently energetic to penetrate the container or cladding."

Brackets are used to set off a parenthetical item within **parentheses.**

EXAMPLE	We should be sure to give Emanuel Foose (and his brother Emilio [1812–1882] as well) credit for his role in founding the institute.

Brackets are also used in academic writing to insert the Latin word **sic,** which indicates that the writer has quoted material exactly as it appears in the original, even though it contains an obvious error.

EXAMPLE	Dr. Smith pointed out that "The earth does not revolve around the son *[sic]* at a constant rate."

bunch

Bunch refers to like things that grow or are fastened together. Do not use the word *bunch* to refer to people.

CHANGE	A *bunch* of trainees toured the site.
TO	A *group* of trainees toured the site.

can/may

In writing, *can* refers to capability, and *may* refers to possibility or permission.

EXAMPLES   I *can* have the project finished by January 1. (capability)
I *may* be in Boston on Thursday. (possibility)
I *can* be in Boston on Thursday. (capability)
May I have an extra week to finish the project? (permission)

cannot/can not

Cannot is one word.

CHANGE   We *can not* meet the deadline specified in the contract.
TO   We *cannot* meet the deadline specified in the contract.

cannot help but

Avoid the **phrase** *cannot help but* in writing. (See also **double negatives.**)

CHANGE   We *cannot help but* cut our staff.
TO   We cannot avoid cutting our staff.

canvas/canvass

Canvas is a **noun** meaning "heavy, coarse, closely woven cotton or hemp fabric." *Canvass* is a **verb** meaning "to solicit votes or opinions."

EXAMPLES   The maintenance crew spread the *canvas* over the equipment.
The executive committee decided to *canvass* the employees.

capital/capitol

Capital may refer either to financial assets or to the city that hosts the government of a state or a nation. *Capitol* refers to the building in which the state or national legislature meets. *Capitol* is often written with a small *c* when it refers to a state building, but it is always capitalized when it refers to the home of the United States Congress in Washington, D.C.

capital letters

The use of capital letters (or uppercase letters) is determined by custom and tradition. Capital letters are used to call attention to certain words, such as proper **nouns** and the first word of a sentence. Care must be exercised in using capital letters because they can affect the meaning of words (march/March, china/China, turkey/Turkey). Thus, capital letters can help eliminate **ambiguity.**

Proper Nouns

Proper nouns name a specific person, place, thing, concept, or quality and therefore are capitalized.

> EXAMPLES   Physics 101, General Electric, John Doe

Common Nouns

Common nouns name a general class or category of persons, places, things, concepts, or qualities rather than specific ones and therefore are not capitalized.

> EXAMPLES   a physics class, a company, a person

First Words

The first letter of the first word in a sentence is always capitalized.

> EXAMPLE   Of all the plans you mentioned, the first one seems the best.

The first word after a **colon** may be capitalized if the statement following is a complete sentence or if it introduces a formal resolution or question.

> EXAMPLE   Today's meeting will deal with only one issue: What is the firm's role in environmental protection?

If a subordinate element follows the colon, however, or if the thought is closely related, use a lowercase letter following the colon.

C

EXAMPLE	We had to keep working for one reason: our deadline was upon us.

The first word of a complete sentence in **quotation marks** is capitalized.

EXAMPLE	Dr. Vesely stated, "It is possible to postulate an imaginary world in which no decisions are made until all the relevant information is assembled."

Complete sentences contained as numbered items within a sentence may also be capitalized.

EXAMPLE	To make correct decisions, you must do three things: (1) Identify the information that would be pertinent to the decision anticipated, (2) Establish a systematic program for acquiring this pertinent information, and (3) Rationally assess the information so acquired.

The first word in the salutation and complimentary close of a letter is capitalized. (See also **correspondence.**)

EXAMPLES	Dear Mr. Smith:
Sincerely yours,
Best regards,

Specific People and Groups
Capitalize all personal names.

EXAMPLES	Walter Bunch, Mary Fortunato, Bill Krebs

Capitalize names of ethnic groups and nationalities.

EXAMPLES	American Indian, Italian, Jew, Cuban
Thus Italian immigrants contributed much to the industrialization of the United States.

Do not capitalize names of social and economic groups.

EXAMPLES	middle class, working class

Specific Places
Capitalize the names of all political divisions.

EXAMPLES	Chicago, Cook County, Illinois, Ontario, Iran, Ward Six

Capitalize the names of geographical divisions.

EXAMPLES   Europe, Asia, North America, the Middle East

Do not capitalize geographic features unless they are part of a proper name.

EXAMPLE   The mountains in some areas, such as the Great Smoky Mountains, make television transmission difficult.

The words *north, south, east,* and *west* are capitalized when they refer to sections of the country. They are not capitalized when they refer to directions.

EXAMPLES   I may travel south when I relocate to Delaware.
We may build a new plant in the South next year.
State Street runs east and west.

Capitalize the names of stars, constellations, and planets.

EXAMPLES   Saturn, Andromeda, Jupiter, Milky Way

Do not capitalize *earth, sun,* and *moon,* however, except when they are used with the names of other planets.

EXAMPLES   Although the sun rises in the east and sets in the west, the moon may appear in any part of the evening sky when darkness settles over the earth.
Mars, Pluto, and Earth were discussed at the symposium.

Specific Institutions, Events, and Concepts

Capitalize the names of institutions, organizations, and associations.

EXAMPLE   The American Society of Mechanical Engineers and the Department of Housing and Urban Development are cooperating in the project.

An organization usually capitalizes the names of its internal divisions and departments.

EXAMPLES   Faculty, Board of Directors, Engineering Department

Types of organizations are not capitalized unless they are part of an official name.

EXAMPLES   Our group decided to form a writers' association; we called it the American Association of Writers.
I attended Post High School. What high school did you attend?

Capitalize historical events.

EXAMPLE   Dr. Jellison discussed the Boston Tea Party at the last class.

Capitalize words that designate specific periods of time.

C

EXAMPLES   Labor Day, the Renaissance, the Enlightenment, January, Monday, the Great Depression, Lent

Do not, however, capitalize seasons of the year.

EXAMPLES   spring, autumn, winter, summer

Capitalize scientific names of classes, families, and orders, but do not capitalize species or English derivatives of scientific names.

EXAMPLES   Mammalia, Carnivora/mammal, carnivorous

Titles of Books, Articles, Plays, Films, Reports, and Memo Subject Lines

Capitalize the initial letters of the first and last words of a title of a book, article, play, or film, as well as all major words in the title. Do not capitalize articles *(a, an, the)*, **conjunctions** *(and, but, if)*, or short **prepositions** *(at, in, on, of)* unless they begin the title. Capitalize prepositions that contain more than four letters *(between, because, until, after)*.

EXAMPLES   The microbiologist greatly admired the book *The Lives of a Cell.*

Her favorite article is still "On the Universe Around Us."

The book *Year After Year* recounts the life story of a great scientist.

The report entitled "Alternative Sites for Plant Relocation" was submitted in February.

The memo "Analysis of Quality Assurance Procedures," dated November 13, explains the problem.

Some reference systems for scientific and technical publications do not follow these guidelines. See **documenting sources.**

Personal, Professional, and Job Titles

Titles preceding proper names are capitalized.

EXAMPLES   Miss March, Professor Galbraith

Appositives following proper names are not normally capitalized. (The word *President* is usually capitalized when it refers to the chief executive of a national government.)

EXAMPLE   Frank Jones, senator from New Mexico (but Senator Jones)

The only exception is an epithet that actually renames the person.

EXAMPLES   Alexander the Great, Solomon the Wise

Job titles used with personal names are capitalized.

    EXAMPLE   John Reems, Division Manager, will meet with us on Wednesday.

Job titles used without personal names are not capitalized.

    EXAMPLE   The division manager will meet with us on Wednesday.

Use capital letters to designate family relationships only when they occur before a name or substitute for a name.

    EXAMPLES   One of my favorite people is Uncle Fred.
                     Jim and Mother went along.
                     Jim and my mother went along.

Abbreviations

Capitalize **abbreviations** if the words they stand for would be capitalized.

    EXAMPLES   UCLA (University of California at Los Angeles)
                     p. (page)
                     Ph.D. (Doctor of Philosophy)

Letters

Capitalize letters that serve as names or indicate shapes.

    EXAMPLES   X-ray, vitamin B, T-square, U-turn, I-beam

Miscellaneous Capitalizations

The word *Bible* is capitalized when it refers to the Jewish or Christian Scriptures; otherwise, it is not capitalized.

    EXAMPLE   He quoted a verse from the *Bible,* then read from Blackstone, the lawyer's *bible.*

All references to deities (Allah, God, Jehovah, Yahweh) are capitalized.

    EXAMPLE   *God* is the *One* who sustains us.

A complete sentence enclosed in **dashes, brackets,** or **parentheses** is not capitalized when it appears as part of another sentence.

    EXAMPLES   We must make an extra effort in safety this year (accidents last year were up 10 percent).

                     Extra effort in safety should be made this year. (Accidents were up 10 percent.)

C

Certain units, such as parts and chapters of books and rooms in buildings, when specifically identified by number, are normally capitalized.

EXAMPLES   Chapter 5, Ch. 5; Room 72, Rm. 72

Minor divisions within such units are not capitalized unless they begin a sentence.

EXAMPLES   page 11, verse 14, seat 12

case (grammar)

Grammatically, *case* indicates the functional relationship of a **noun** or a **pronoun** to the other words in a sentence. Nouns change form only in the **possessive case;** pronouns may show change for the subjective, the objective, or the possessive case. The case of a noun or pronoun is always determined by its function in its **phrase, clause,** or sentence. If it is the subject of its phrase, clause, or sentence, it is in the subjective case; if it is an **object** within its phrase, clause, or sentence, it is in the objective case; if it reflects possession or ownership and modifies a noun, it is in the possessive case. The subjective case indicates the person or thing acting (*he* sued the vendor); the objective case indicates the thing acted upon (the vendor sued *him*); and the possessive case indicates the person or thing owning or possessing something (it was *his* company).

The different forms of a noun or pronoun indicate whether it is functioning as a subject (subjective case), as a **complement** (usually objective case), or as a **modifier** (possessive case).

Subjective Case	Objective Case	Possessive Case
I	me	my, mine
we	us	our, ours
he	him	his, his
she	her	her, hers
they	them	their, theirs
you	you	your, yours
who	whom	whose
it	it	its

Subjective Case

A pronoun is in the subjective case (also called the nominative case) when it represents the person or thing acting.

EXAMPLE   *I* wrote a letter to that company before *I* graduated.

A linking **verb** links a pronoun to its antecedent to show that they identify the same thing. Because they represent the same thing, the pronoun is in the subjective case even when it follows the verb, which makes it a subjective complement.

EXAMPLES   *He* is the head of the Quality Control Group. (subject)
The head of the Quality Control Group is *he*. (subjective complement)

Whether a pronoun is a subject or a subjective complement, it is in the subjective case.

The subjective case is used after the words *than* and *as* because of the understood (although unstated) portion of the clauses in which these words appear.

EXAMPLES   George is as good a designer as *I* [am].
Our subsidiary can do the job better than *we* [can].

Objective Case

A pronoun is in the objective case when it indicates the person or thing receiving the action expressed by the verb. (The objective case is also called the accusative case.)

EXAMPLE   They informed *me* by letter that they had received my résumé.

A pronoun is in the objective case when it is the object of a verb, gerund, or **preposition** and when it is the subject of an infinitive. Pronouns that follow prepositions must be in the objective case.

EXAMPLES   Between you and *me,* his facts are questionable.
Many of *us* attended the conference.

Pronouns that follow action verbs (which excludes all forms of the verb *be*) must be in the objective case. Don't be confused by an additional name.

EXAMPLES   The company promoted *me* in June.
The company promoted John and *me* in June.

Pronouns that follow gerunds must be in the objective case.

EXAMPLE   Training *him* was the best thing I could have done.

Subjects of infinitives must be in the objective case.

EXAMPLE   We asked *them* to recalibrate the instruments.

C

For determining the case of an object, English does not differentiate between direct objects and indirect objects; both require the objective form of the pronoun. (See also **complements.**)

> **EXAMPLES**   The interviewer seemed to like *me*. (direct object)
> They wrote *me* a letter. (indirect object)

Possessive Case

A noun or a pronoun is in the possessive case when it represents a person or thing owning or possessing something.

> **EXAMPLE**   Dr. *Peterson's* risk calculations appear in Appendix A of *his* report.

Nouns. Although exceptions are relatively common, it is a good rule of thumb to use the *'s* form of the possessive case with nouns referring to persons and living things and to use an *of* phrase for the possessive case of nouns referring to inanimate objects.

> **EXAMPLES**   The *chairman's* address was well received.
> The leaves *of the tree* look healthy.

If this rule leads to awkwardness or wordiness, however, be flexible.

> **EXAMPLES**   The *company's* pilot plants are doing well.
> The *plane's* landing gear failed.

Only the possessive form of a noun or pronoun should precede a gerund.

> **EXAMPLES**   *John's* working has not affected his grades.
> *His* working has not affected his grades.

The established **idiom** calls for the possessive case in many stock phrases.

> **EXAMPLES**   a day's journey, a day's work, a moment's notice, at his wit's end, the law's delay

In a few cases, the idiom even calls for a double possessive employing both the *of* and *'s* forms.

> **EXAMPLE**   That colleague *of George's* was at the conference.

Plural words ending in *s* need only add an **apostrophe** to form the possessive case.

> **EXAMPLE**   the *laborers'* union

When several words compose a single term, add the *'s* to the last word only.

EXAMPLES   The *Chairman of the Board's* statement was brief.

The *Department of Energy's* fiscal 19— budget shows increased revenues of $192 million for uranium enrichment.

To show individual possession with coordinate nouns, make both nouns possessive.

EXAMPLE   The *Senate's* and *House's* chambers were packed.

To show joint possession with coordinate nouns, make only the last possessive.

EXAMPLE   The *Senate and House's* joint declaration was read to the press.

Pronouns. The use of possessive pronouns does not normally cause problems except with gerunds and indefinite pronouns. Several indefinite pronouns *(all, any, each, few, most, none,* and *some)* require *of* phrases to form the possessive case.

EXAMPLE   Both dies were stored in the warehouse, but rust had ruined the surface *of each.*

Others, however, use the apostrophe.

EXAMPLE   *Anyone's* contribution is welcome.

Only the possessive form of a pronoun should be used with a gerund.

EXAMPLES   The safety officer insisted on *my* wearing a hardhat.
Our monitoring was not affected by changing weather conditions.

Pronouns in compound constructions should be in the same case.

EXAMPLES   This is just between *them* and *us.*
Both *they* and *we* must agree to the arrangement.

Appositives

An **appositive** is a noun or noun phrase that follows and amplifies another noun or noun phrase. Because it has the same grammatical function as the noun it complements, an appositive should be in the same case as the noun with which it is in apposition.

EXAMPLES   Two design engineers, Jim Knight and *I,* were asked to review the drawings. (subjective case)

The group leader selected two members to represent the department—Jim Knight and *me.* (objective case)

Tips on Determining the Case of Pronouns

C

One test to determine the proper case of a pronoun is to try it with some transitive verb such as *resembled* or *hit*. If the pronoun would logically precede the verb, use the subjective case; if it would logically follow the verb, use the objective case.

EXAMPLES   *She (he, they)* resembled her father. (subjective case)
Angela resembled *him (her, them)*. (objective case)

In the following type of sentence, try omitting the noun to determine the case of the pronoun.

EXAMPLES   *(We/Us)* pilots fly our own airplanes.

We [pilots] fly our own airplanes. (This correct usage sounds right.)

Us [pilots] fly our own airplanes. (This incorrect usage is obviously wrong.)

To determine the case of a pronoun that follows *as* or *than*, try mentally adding the words that are normally omitted.

EXAMPLES   The other operator is not paid as well *as she* [is paid]. (You would not write, "*Her* is paid.")

His partner was better informed than *he* [was informed]. (You would not write, "*Him* was informed.")

If compound pronouns cause problems, try using them singly to determine the proper case.

EXAMPLES   *(We/Us)* and the Johnsons are going to the Grand Canyon.

We are going to the Grand Canyon. (You would not write, "*Us* are going to the Grand Canyon.")

Who/Whom

Who and *whom* cause much trouble in determining case. *Who* is the subjective case form, whereas *whom* is the objective case form. When in doubt about which form to use, try substituting a personal pronoun to see which one fits. If *he* or *they* fits, use *who*.

EXAMPLES   *Who* is the congressman from the 45th district?
He is the congressman from the 45th district.

If *him* or *them* fits, use *whom*.

EXAMPLES   It depended on *them*.
It depended on *whom?*
It was they on *whom* it depended.

It is becoming common to use *who* for the objective case when it begins a clause or sentence, although some readers still object to such an "ungrammatical" construction, especially in formal contexts.

EXAMPLE *Who* should I call to report a fire?

The best advice is to know your **reader.** (See also **who/whom.**)

case (usage)

The word *case* is often merely filler. Be critical of the word, and eliminate it if it contributes nothing.

> CHANGE An exception was made in the *case* of those closely connected with the project.
>
> TO An exception was made for those closely connected with the project.

cause-and-effect method of development

When your purpose is to explain why something happened or why you think something will happen, the *cause-and-effect method of development* is a useful writing strategy.

The goal of the cause-and-effect **method of development** is to make as plausible as possible the relationship between a situation and either its cause or its effect. The conclusions you draw about the relationships should be based on the evidence you have gathered. Because not all evidence will be of equal value to you, keep some guidelines in mind for evaluating evidence.

Evaluating Evidence

The facts and arguments you gather should be pertinent to your **topic.** Be careful not to draw a conclusion that your evidence does not lead to or support. You may have researched some statistics, for example, showing that an increasing number of Americans are licensed to fly small airplanes. But you cannot use this information as evidence that there is a slowdown in interstate highway construction in the United States—the evidence does not lead to that conclusion. Statistics on the increase in small-plane licensing may be relevant to other conclusions, however. You could argue that the upswing has occurred because small planes save travel time, provide easy access to remote areas, and, once they are purchased, are economical to operate.

Your evidence should be adequate. Incomplete evidence can lead to false conclusions.

C

> EXAMPLE   Driver training classes do not help prevent auto accidents. Two people I know who completed driver training classes were involved in accidents.

Although the evidence cited to support the conclusion may be accurate, there is not enough of it. A thorough investigation of the usefulness of driver training classes in keeping down the accident rate would require many more than two examples. And it would require a comparison of the driving records of those who had completed driver training with drivers who had not.

Your evidence should be representative. If you conduct a survey to obtain your evidence, be sure that you do not solicit responses only from individuals or groups whose views are identical to yours; that is, be sure you obtain a representative sampling.

Your evidence should also be plausible. Two events that occur close to each other in time or place may or may not be causally related. Thunder and black clouds do not always signal rain, but they do so often enough that if we are outdoors and the sky darkens and we hear thunder, we seek shelter. If you sprain your ankle after walking under a ladder, however, you cannot conclude that a ladder brings bad luck. Merely to say that X caused Y (or will cause Y) is inadequate. You must demonstrate the relationship with pertinent facts and arguments.

Linking Causes to Effects

To show a true relationship between a cause and an effect, you must demonstrate that the existence of the one *requires* the existence of the other. It is often difficult to establish beyond any doubt that one event was *the* cause of another event. More often, a result will have more than one cause. As you research your subject, your task is to determine which cause or causes are most plausible.

When several probable causes are equally valid, report your findings accordingly, as in the following **paragraph** on the use of an energy-saving device called a furnace-vent damper. The damper is a metal plate fitted inside the flue or vent pipe of natural-gas or fuel-oil furnaces. When the furnace is on, the damper opens to allow the gases to escape up the flue. When the furnace shuts off, the damper closes, thus preventing warm air from escaping up the flue stack. The dampers are potentially dangerous, however. If the dampers fail to open at the proper time, they could allow poisonous furnace gases to back into the house and asphyxiate anyone in a matter of minutes.

Tests run on several dampers showed a number of probable causes for their malfunctioning.

> EXAMPLE   One damper was sold without proper installation instructions, and another was wired incorrectly. Two of the units had slow-opening dampers (15 seconds) that prevented the [furnace] burner from firing. And one damper jammed when exposed to a simulated fuel temperature of more than 700 degrees.

The investigator located all the causes of damper malfunctions and reported on them. Without such a thorough account, recommendations to prevent similar malfunctions would be based on incomplete evidence.

By substituting *problem* for "cause" and *solution* for "effect," you can use the same approach to develop a **report** dealing with a solution to a problem.

center on/center around

Substitute *on* or *in* for *around* in this phrase.

> CHANGE   The experiments *center around* the new discovery.
> TO   The experiments *center on* the new discovery.

Usually the idea intended by *center around* is best expressed by *revolve around*.

> EXAMPLE   The subcommittee hearings on computer security *revolved around* access codes.

chair/chairperson

The terms *chair, chairperson, chairman,* and *chairwoman* all are used to refer to a presiding officer. The titles *chair* and *chairperson*, however, avoid any sexual bias that might be implied by the other titles.

> EXAMPLE   Mary Roberts preceded John Stevens as *chair* (or *chairperson*) of the executive committee.

character

Character, used in the sense of "nature" or "quality," is often an unnecessary and inexact filler that should be omitted from your writing. Choose instead a word that conveys your meaning more specifically.

> CHANGE   The modifications changed the whole *character* of the engine.
> TO   The modifications changed the performance of the engine.

chronological method of development

C

The *chronological method of development* arranges the events under discussion in sequential order, beginning with the first event and continuing chronologically to the last event. **Trip reports, laboratory reports,** work schedules, some **minutes of meetings,** and certain **trouble reports** are among the types of writing in which information is organized chronologically.

Woodworking Plant Fire*

Exposed Building Destroyed   July 24, 19—
Notification Delayed   Burney, California

Setting

Wood bark, sawdust, and wood chips were stored in three piles about 100 feet south of one building at this lumber mill and 150 feet west of a second building. The second building, called the "panel plant," consisted of one story and a partial attic and was used in part as an electric shop, and in part for the storage of finished lumber. About six pallet loads of Class I flammable liquids in 55-gallon drums were stored in the western section. The building contained a sprinkler system.

Cause of fire

Fire, caused by spontaneous ignition of the piled bark, spread to sawdust and chip piles, then to the chip-loading facilities, and finally to the panel plant. A 40-mph wind was blowing in the direction of the panel plant.

Fire first noticed

A watchman first noticed the fire in the bark pile about 6 a.m. He notified the plant superintendent, who arrived more than an hour later, hosed down the smoldering bark pile, and set up several irrigation sprinklers to wet the area.

Fire department called

At about 1:20 p.m., smoke was seen at the farther end of the bark pile. The hose was not long enough to reach this area and the local fire department was called, nearly eight hours after the fire was originally discovered.

Start of pumping equipment delayed

The fire burned up into the hollow joisted roof of the panel plant. The sprinklers were on a dry system and, from accounts of witnesses, it is estimated that the fire pump was not started until after the fire had been burning in the panel plant for 15 to 30 minutes.

The plant was a $350,000 loss. . . . .

*"Bimonthly Fire Record," *Fire Journal* 72 (Mar. 1977): 24.

FIGURE 1 Chronological Method of Development

In the **report** shown in Figure 1, a firefighter describes a fire that took place at a lumber mill. After providing important background information, the writer presents the events as they occurred chronologically.

C

cite/site/sight

Cite means "acknowledge" or "quote an authority"; *site* is the place or plot of land where something is located; *sight* is the ability to see.

EXAMPLES   The speaker *cited* several famous economists to support his prediction about the stock market.

The *site* for the new factory is three miles from the middle of town.

After the accident, his vision was blurred, and he feared that he might lose his *sight*.

clarity

Clarity is essential. Strive to make all of your writing direct, orderly, and precise. Many factors contribute to clarity just as many other elements can defeat it. Logical development, unity, coherence, emphasis, subordination, pace, transition, an established point of view, conciseness, and word choice contribute to clarity. **Ambiguity, awkwardness,** vagueness, poor use of **idiom, clichés,** and inappropriate level of usage detract from clarity.

A logical **method of development** and a good **outline** are essential to clarity. Without a logical method of development and an outline, you may communicate only isolated thoughts to your **reader,** and you cannot achieve your **purpose** by a jumble of isolated thoughts. You must use a method of development that puts your thoughts together in a logical, meaningful sequence.

Proper **emphasis** and **subordination** are mandatory if you wish to achieve clarity. If you do not use these two complementary techniques wisely, your **clauses** and sentences all may appear to be of equal importance. Your reader will be forced to guess which are most important, which are least important, and which fall between the two extremes.

The **pace** at which you present your ideas is important to clarity because if the pace is not carefully adjusted to both the **topic** and the reader, your writing will appear cluttered and unclear.

Point of view establishes through whose eyes, or from what vantage point, the reader views the subject. A consistent point of view is essential to clarity; if you switch from the first person to the third person in midsentence, you are certain to confuse your reader.

Clear **transition** contributes to clarity by providing the smooth flow that enables the reader to connect your thoughts with one another without conscious effort.

That **conciseness** is a requirement of clearly written communication should be evident to anyone who has ever attempted to decipher an insurance policy or legal contract. Although words are our chief means of communication, too many of them can impede communication just as effectively as too many cars on a highway can impede traffic. For the sake of clarity, prune excess verbiage from your writing.

The selection of precise words over **vague words** advances clarity by defeating ambiguity and awkwardness.

clauses

A *clause* is a syntactical construction, or group of words, that contains a subject and a predicate and functions as part of a sentence.

> EXAMPLE   *The scaffolding fell* when the rope broke.

Every subject-predicate word group in a sentence is a clause. Unlike a **phrase,** a clause can make a complete statement because it contains a finite **verb** (as opposed to **verbals**) as well as a subject. Every sentence must contain at least one independent clause, with the obvious exception of minor sentences—**sentence fragments** that are acceptable because the missing part is clearly understood, such as "At last." or "So much for that."

A clause may be connected with the rest of its sentence by a coordinating **conjunction,** a subordinating conjunction, a relative **pronoun,** or a conjunctive **adverb.**

> EXAMPLES   It was 500 miles to the facility, *so* we made arrangements to fly. (coordinating conjunction)
>
> Mission control will have to be alert *because* at launch the space laboratory will contain a highly flammable fuel. (subordinating conjunction)

It was Robert M. Fano *who* designed and developed the earliest "Multiple Access Computer" system at M.I.T. (relative pronoun)

It was dark when we arrived; *nevertheless,* we began the tour of the factory. (conjunctive adverb)

Independent Clauses

An *independent clause* is a group of words containing a subject and a predicate that expresses a complete statement and that could stand alone as a separate sentence.

> **EXAMPLE** *Production was not as great as we had expected,* although we met our quota.

In the example, the independent clause is italicized and could clearly stand alone as a complete sentence. The second, unitalicized clause is a dependent clause; although it contains a subject and predicate, it does not express a complete thought and could not stand alone as a sentence.

Dependent Clauses

A *dependent (or subordinate) clause* is a group of words that has a subject and a predicate but nonetheless depends on a main clause to complete its meaning.

Dependent clauses are useful in making the relationship between thoughts more precise and succinct than if the ideas were presented in a series of simple sentences or compound sentences.

> **CHANGE** The sewage plant is located between Millville and Darrtown. Both villages use it. (two thoughts of approximately equal importance)
>
> **TO** The sewage plant, *which is located between Millville and Darrtown,* is used by both villages. (one thought subordinated to the other)
>
> **CHANGE** Title insurance is a policy issued by a title insurance company, and it protects against any title defects, such as outside claimants. (two thoughts of approximately equal importance)
>
> **TO** Title insurance, *which protects against any title defects such as outside claimants,* is a policy issued by a title insurance company. (one thought subordinated to the other)

Subordinate clauses are especially effective, therefore, for expressing thoughts that describe or explain another statement. Too much subordination, or a string of dependent clauses, however, may be worse than none at all.

C

> **CHANGE** He had selected teachers *who* taught classes *that* had a slant *that* was specifically directed toward students *who* intended to go into business.
>
> **TO** He had selected teachers *who* taught classes *that* were specifically directed to business students.

clichés

A *cliché* is an expression that has been used for so long that it is no longer fresh (although some clichés were, at one time, fresh **figures of speech**). Because they have been used continually over a long period of time, clichés come to mind easily. In addition to being stale, clichés are usually wordy and often vague. Each of the following clichés is followed by better, more direct words, or expressions:

EXAMPLES	quick as a flash	quickly, in five minutes
	straight from the shoulder	frank
	last but not least	last, finally
	as plain as day	clear, obvious
	abreast of the times	up to date, current
	the modern business world	business today

Clichés are often used in an attempt to make writing elegant or impressive (see also **affectation**). Because they are wordy and vague, however, they slow communication and can even irritate your **reader.** So, although clichés come to mind easily while you are **writing the draft,** they normally should be eliminated during the **revision** phase of the writing process.

> **CHANGE** Our new computer system will have a positive impact on the company *as a whole.* It will keep us *abreast of the times* and make our competition *green with envy.* The committee deserves *a pat on the back* for its *herculean efforts* in convincing management that it was *the thing to do.* I'm sure that their *untiring efforts* will not *go unrewarded.*
>
> **TO** Our new computer system will have a positive impact throughout the company. It will keep our operations up-to-date and make our competition envious. The committee deserves credit for their efforts in convincing management of the need for the computer. I'm sure that the value of their efforts will be recognized.

clipped forms of words

When the beginning or end of a word is cut off to create a shorter word, the result is called a *clipped form.*

C

EXAMPLES   dorm, lab, demo, phone, memo·

The word *specification,* for example, is often shortened to *spec.* Although acceptable in conversation, most clipped forms should not appear in writing unless they are commonly accepted as part of the special vocabulary of an occupational group.

Apostrophes are not normally used with clipped forms of words (not *'phone,* but *phone*). Since they are not strictly **abbreviations,** clipped forms are not followed by **periods** (not *lab.,* but *lab*).

Do not use clipped forms of spelling (*thru, nite,* and the like).

coherence

Writing is coherent when the relationships among ideas are made clear to the reader. Coherent writing moves logically and consistently from one point to another. Each idea should relate clearly to the others, with one idea flowing smoothly to the next. Many elements contribute to smooth and coherent writing; however, the major components are (1) a logical sequence of ideas and (2) clear **transitions** between ideas.

A logical sequence of presentation is the most important single requirement in achieving coherence, and the key to achieving the most logical sequence of presentation is the use of a good **outline.** The outline forces you to establish a beginning (**introduction**), a middle (body), and an end (**conclusion**), and this alone contributes greatly to coherence. The outline also enables you to lay out the most direct route to your **purpose**—without digressing into interesting but only loosely related side issues, a habit that inevitably defeats coherence. Drawing up an outline permits you to experiment with different sequences and choose the best one.

Thoughtful transition is also essential to coherence, for without transitions your writing cannot achieve the smooth flow from sentence to sentence and from **paragraph** to paragraph that is required for coherence. Notice the difference between the following two paragraphs; the first has no transition and the second has transition added.

C

CHANGE The moon has always been an object of interest to human beings. Until the 1960s, getting there was only a dream. Some thought that we were not meant to go to the moon. In 1969 Neil Armstrong stepped onto the lunar surface. Moon landings became routine to the general public.

TO The moon has always been an object of interest to human beings, *but* until the 1960s, getting there was only a dream. *In fact,* some thought that we were not meant to go to the moon. *However,* in 1969 Neil Armstrong stepped onto the lunar surface. *After that* moon landings became routine to the general public.

The transitional words and expressions of the second paragraph fit the ideas snugly together, making that paragraph read more smoothly than the first. Attention to transition in longer works is essential if your reader is to move smoothly from point to point in your writing.

Check your draft carefully for coherence during revision; if your writing is not coherent, you are not really communicating with your reader.

collaborative writing

Collaborative writing occurs when two or more writers work together to produce a single document for which they share responsibility and decision-making authority. The collaborating writers make approximately equal contributions, and communication among them is among equals, never superior to subordinate.

Collaborative writing teams are formed when (1) the size of a project or the time constraints imposed on it requires collaboration, (2) the project needs multiple areas of expertise, or (3) the project requires the melding of divergent views to form a single perspective that is acceptable to the whole collaborating team or another group.

The collaborative writing team is composed of peers, but its members recognize and take advantage of each member's expertise. Team members must respect one another's capabilities and be compatible enough to work together, although some conflict is natural.

The team must choose a leader, although that person does not have decision-making authority, just the extra responsibility of coordinating the team members' activities and organizing the project. Team leadership can be by mutual agreement of the team members, or it can be on a rotating basis if the team produces multiple documents.

Tasks of the Collaborative Writing Team

The collaborative writing team normally has four tasks: planning the document, researching and writing the draft, reviewing the drafts of other team members, and revising the drafts on the basis of the reviewers' comments. Collaborative authoring software is available to guide the collaborative writing team through some of these tasks.

Planning. The team, collectively, should identify the audience, purpose, and scope of the project, as well as its goals and the most effective organization for the whole document. The team analyzes the overall project, conceptualizes the document to be produced, creates a broad outline of the document, divides the document into segments, and assigns different segments to individual team members (often on the basis of expertise).

The team then plans, as a group, to the lowest level that is practical. Beyond a certain level, of course, the group does not have sufficient command of details to plan realistically and must leave any further planning to the individual team member who has the assigned responsibility for researching a specific section of the document.

In the planning stage, the team should produce a projected schedule and set any writing **style** standards that team members will be expected to meet. The agreed-on schedule should include the due dates for drafts, for reviews of the drafts by other team members, and for **revisions.** It is important that these deadlines be met, even if the drafts are not quite as polished as the individual author would like, because a missed deadline by one team member holds up the work of the entire team.

The team should not insist that the agreed-on broad **outline** be followed slavishly by individual team members. Once into the assignment, an individual writer may—and often will—find that the broad outline for a specific segment was based on insufficient knowledge and presents a plan that, in fact, is not possible or not desirable. The individual team member must be allowed the freedom to alter the broad outline on the basis of what is possible, more appropriate, or clearly more desirable.

Research and Writing. Planning is followed by research and writing, a period of intense independent activity by the individual members of the team. Each member researches his or her assigned segment of the document, fleshes out the broad outline with greater detail, and produces a draft from the detailed outline. The writers revise their drafts until they are as good as the individual writers can make them. Then, by the deadline established for drafts, the individual writers submit copies of their drafts to all other members of the team for review.

Reviewing. During the review stage, team members assume the role of the reading audience and try to clear up in advance any problems that

C

might arise for the reader. Each team member reviews the work of the other team members carefully and critically, but diplomatically. Team members review for things both large and small. They evaluate the organization of each segment, as well as each sentence and paragraph. They offer any advice or help that will enable the individual writer to improve his or her segment of the document.

The review stage may lead to renewed planning. If, for example, review makes it obvious that the original planning for a section was inadequate or incorrect, or if new information becomes available, the team must rethink the plan for that segment of the document on the basis of the newer understanding and knowledge.

Revising. The individual writers evaluate the reviews of all other team members and accept or reject their suggestions. Writers must be careful not to let ego get in the way of good judgment. They must evaluate each suggestion objectively—on the basis of its merit—rather than emotionally. The ability to accept criticism and use it to produce a better end product is one of the critical differences between an effective collaborator and an ineffective one.

Conflict. Members of a team never have exactly the same perspective on any subject, and differing perspectives can easily lead to conflict. A team that can tolerate some disharmony yet work through conflicting opinions to reach consensus produces better results. Although mutual respect among team members is necessary, too much deference can inhibit challenges—and that actually reduces the team's creativity. Writers must be willing to challenge one another—tactfully and diplomatically, however.

It is critically important to the quality of the document being produced that all viewpoints be considered. Under such circumstances, conflicts will occur, ranging from relatively mild differences over minor points to major conflicts over the basic approach. Regardless of the severity of the conflict, it must be worked through to a conclusion or a compromise that all team members can accept, even though all might not entirely agree.

The Advantages of Collaborative Writing

The work a collaborative writing team produces can be considerably better than the work any one of its members could have produced alone. This is true, at least in part, because team members lead each other to consider ideas different from those they would have explored individually.

Getting immediate feedback—even if it is sometimes contested and debated—is one of the great advantages of the peer writing team. Team members easily detect problems with **organization, clarity, logic,** and substance, and they point these things out during reviews.

Another advantage is that team members play devil's advocate for one another, taking contrary points of view to try to make certain that all important points are covered and that all potential problems have been exposed and resolved.

As a team member, you become more aware of, and involved in, the planning of a document than you would working alone, because of the team discussions that take place during the planning stage. The same is true of reviewing and revising.

collective nouns (see nouns)

colloquialism (see English, varieties of)

colons

The *colon* is a mark of anticipation and **introduction** that alerts the **reader** to the close connection between the first statement and the one that follows it.

A colon may be used to connect a list or series to the **clause,** word, or **phrase** with which it is in apposition.

> EXAMPLE   Three decontamination methods are under consideration: a zeolite-resin system, an evaporation and resin system, and a filtration and storage system.

Do not, however, place a colon between a **verb** and its **objects.**

> CHANGE   The three fluids for cleaning pipettes are: water, alcohol, and acetone.
> TO   The three fluids for cleaning pipettes are water, alcohol, and acetone.

One common exception is made when a verb is followed by a **list.**

> EXAMPLE   The corporations that manufacture computers include:
>
> NCR   IBM   Apple
> Unisys   Compaq   DEC

Do not use a colon between a **preposition** and its object.

> CHANGE I would like to be transferred to: Tucson, Boston, or Miami.
> TO I would like to be transferred to Tucson, Boston, or Miami.

A colon may be used to link one statement to another that develops, explains, amplifies, or illustrates the first.

> EXAMPLE Any large organization is confronted with two separate, though related, information problems: it must maintain an effective internal communication system, and it must see that an effective overall communication system is maintained.

A colon may be used to link an **appositive** phrase to its related statement if greater **emphasis** is needed.

> EXAMPLE There is only one thing that will satisfy Mr. Sturgess: our finished report.

Colons are used to link numbers signifying different identifying **nouns.**

> EXAMPLES Matthew 14:1 (chapter 14, verse 1)
> 9:30 a.m. (9 hours, 30 minutes)

In proportions, the colon indicates the ratio of one amount to another.

> EXAMPLE The cement is mixed with the water and sand at 7:5:14. (In this case, the colon replaces *to*.)

Colons are often used in mathematical ratios.

> EXAMPLE $7:3 = 14:x$

In **bibliography,** footnote, and reference citations, colons may link the place of publication with the publisher and perform other specialized functions.

> EXAMPLE Watson, R. L. *Statistics for Electrical Engineers.* Englewood, CA: EEE, 1997.

A colon follows the salutation in business letters, even when the salutation refers to a person by name.

> EXAMPLES Dear Ms. Jeffers:
> Dear Manager:
> Dear George:

The initial **capital letter** of a **quotation** is retained following a colon if the quoted material originally began with a capital letter.

> EXAMPLE The senator issued the following statement: "We are not concerned about the present. We are worried about the future."

A colon always goes outside **quotation marks.**

> EXAMPLE   This was the real meaning of his "suggestion": the division must show a profit by the end of the year.

When quoting material that ends in a colon, drop the colon and replace it with **ellipses.**

> CHANGE   "Any large corporation is confronted with two separate, though related, information problems:"
>
> TO   "Any large corporation is confronted with two separate, though related, information problems . . ."

The first word after a colon may be capitalized if (1) the statement following is a complete sentence or (2) it introduces a formal resolution or question.

> EXAMPLE   The members attending this year's conference passed a single resolution: Voting will be open to associate members next year.

If a subordinate element follows the colon, however, use a lowercase letter following the colon.

> EXAMPLE   There is only one way to stay within our present budget: to reduce expenditures for research and development.

comma splice

Do not attempt to join two independent **clauses** with only a comma; this is called a *comma splice.*

> EXAMPLE   It was five hundred miles to the facility, we made arrangements to fly.

Such a comma splice could be corrected in several ways.

1. Substitute a **semicolon,** or a semicolon and a conjunctive **adverb.**

> CHANGE   It was five hundred miles to the facility, we made arrangements to fly.
>
> TO   It was five hundred miles to the facility; we made arrangements to fly.
>
> OR   It was five hundred miles to the facility; *therefore,* we made arrangements to fly.
>
> When a conjunctive adverb connects two independent clauses, the conjunctive adverb must be preceded by a semicolon and followed by a comma.

C

CHANGE   It was five hundred miles to the facility, therefore, we made arrangements to fly.

TO   It was five hundred miles to the facility; therefore, we made arrangements to fly.

2. Add a coordinating **conjunction** following the comma.

EXAMPLE   It was five hundred miles to the facility, *so* we made arrangements to fly.

3. Create two sentences. (Be aware, however, that putting a **period** between two closely related and brief statements may result in two weak sentences.)

EXAMPLE   It was five hundred miles to the facility. We made arrangements to fly.

4. Subordinate one clause to the other.

EXAMPLE   *Because* it was five hundred miles to the facility, we made arrangements to fly.

commas

Like all **punctuation,** the comma helps **readers** understand the writer's meaning and prevents **ambiguity.** Notice how the comma helps make the meaning clear in the following examples:

CHANGE   To be successful managers with MBAs must continue to learn. (At first reading, the sentence seems to be about "successful managers with MBAs.")

TO   To be successful, managers with MBAs must continue to learn. (The comma makes clear where the main part of the sentence begins.)

CHANGE   When you see an airport fly over it at an altitude of 1,500 feet. (At first glance, this sentence seems to have airports flying.)

TO   When you see an airport, fly over it at an altitude of 1,500 feet. (The comma makes clear where the main part of the sentence begins.)

As these examples illustrate, effective use of the comma depends on your understanding of **sentence construction.**

To help you find the advice you need, the following entry is divided into the following uses of the comma:

Linking Independent Clauses
Enclosing Elements

Linking Independent Clauses

Use a comma between independent **clauses** that are linked by a coordinating conjunction *(and, but, or, nor,* and sometimes *so, yet,* and *for).* The comma precedes the **conjunction.**

> EXAMPLE   Human beings have always prided themselves on their unique capacity to create and manipulate symbols, but today computers are manipulating symbols.

Although many writers omit the comma when the clauses are short and closely related, the comma can never be wrong.

> EXAMPLES   The cable snapped and the power failed.
> The cable snapped, and the power failed.

Enclosing Elements

Commas are used to enclose nonrestrictive clauses and parenthetical elements. (For other means of punctuating parenthetical elements, see **dashes** and **parentheses.**)

> EXAMPLES   Our new Detroit factory, *which began operations last month,* should add 25 percent to total output. (nonrestrictive clause)
>
> We can, *of course,* expect their lawyer to call us. (parenthetical element)

(See also **restrictive and nonrestrictive elements.**)
    Yes and *no* are set off by commas in such uses as the following:

> EXAMPLES   I agree with you, *yes.*
> *No,* I do not think we can finish as soon as we would like.

A **direct address** should be enclosed in commas.

> EXAMPLE   You will note, *Mark,* that the surface of the brake shoe complies with the specifications.

C

Phrases in apposition (which identify another expression) are enclosed in commas.

> EXAMPLE   Our company, *The Blaylok Precision Company,* did well this year.

Commas enclose nonrestrictive participial **phrases.**

> EXAMPLE   The lathe operator, *working quickly and efficiently,* finished early.

Interrupting transitional words or phrases are usually set off with commas.

> EXAMPLE   We must wait for the written authorization to arrive, *however,* before we can begin work on the project.

Commas are omitted when the word or **phrase** does not interrupt the continuity of thought.

> EXAMPLE   I *therefore* suggest that we begin construction.

Introducing Elements

Clauses and Phrases.   It is generally a good rule of thumb to put a comma after an introductory clause or **phrase.** Identifying where the introductory element ends helps indicate where the main part of the sentence begins. Always place a comma after a long introductory clause.

> EXAMPLE   *Since many rare fossils seem never to occur free from their matrix,* it is wise to scan every slab with a hand lens.

When long modifying phrases precede the main clause, they should always be followed by a comma.

> EXAMPLE   *During the first series of field-performance tests last year at our Colorado proving ground,* the new motor failed to meet our expectations.

When an introductory phrase is short and closely related to the main clause, the comma may be omitted.

> EXAMPLE   *In two seconds* a 20°F temperature is created in the test tube.

A comma should always follow an introductory absolute phrase.

> EXAMPLE   *The tests completed,* we organized the data for the final report.

Words and Quotations.   Certain types of introductory words are followed by a comma. One such is a **noun** used in direct address.

> EXAMPLE   *Bill,* enclosed is the article you asked me to review.

An introductory **interjection** (such as *oh, well, why, indeed, yes,* and *no*) is followed by a comma.

EXAMPLES Yes, I will make sure your request is approved.
Indeed, I will be glad to send you further information.

A transitional word or phrase like *moreover* or *furthermore* is usually followed by a comma to connect the following thought with a preceding clause or sentence.

C

EXAMPLES *Moreover,* steel can withstand a humidity of 99 percent, provided that there is no chloride or sulphur dioxide in the atmosphere.

In addition, we can expect a better world market as a result of this move.

However, we should expect some shortages due to the overall economic climate.

When **adverbs** closely modify the **verb** or the entire sentence, however, they should not be followed by a comma.

EXAMPLE *Perhaps* we can still solve the environmental problem. Certainly we should try.

Use a comma to separate a direct **quotation** from its **introduction.**

EXAMPLE Morton and Lucia White said, "Men live in cities but dream of the countryside."

Do not use a comma, however, when giving an indirect **quotation.**

EXAMPLE Morton and Lucia White said that men dream of the countryside, even though they live in cities.

Separating Items in Series

Although the comma before the last word in a series is sometimes omitted, it is generally clearer to include it. The confusion that may result from omitting the comma is illustrated in the following sentence:

CHANGE Random House, Irwin, Doubleday and Dell are publishing companies. (Is "Doubleday and Dell" one company or two?)

TO Random House, Irwin, Doubleday, and Dell are publishing companies.

The presence of the comma removes the **ambiguity.**

Phrases and clauses in coordinate series, like words, are punctuated with commas.

EXAMPLE It is well known that plants absorb noxious gases, act as receptors of dirt particles, and cleanse the air of other impurities.

C

When **adjectives** modifying the same noun can be reversed and make sense, or when they can be separated by *and* or *or,* they should be separated by commas.

> **EXAMPLE**   The drawing was of a *modern, sleek, swept-wing* airplane.

When an adjective modifies a phrase, no comma is required.

> **EXAMPLE**   He was investigating his *damaged radar beacon system. (damaged* modifies the phrase *radar beacon system)*

Never separate a final adjective from its noun.

> **CHANGE**   He is a conscientious, honest, *reliable,* worker.
> **TO**   He is a conscientious, honest, *reliable* worker.

Clarifying and Contrasting

If you find you need a comma to separate the consecutive use of the same word to prevent misreading, rewrite the sentence.

> **CHANGE**   The assets we had, had surprised us.
> **TO**   We were surprised at the assets we had.

Use a comma to separate two contrasting thoughts or ideas.

> **EXAMPLES**   The project was finished on time, but not within the budget.
>
> The specifications call for 100-ohm resistors, not 1,000-ohm resistors.

Use a comma following an independent clause that is only loosely related to the dependent clause that follows it.

> **EXAMPLE**   The plan should be finished by July, even though I lost time because of illness.

Showing Omissions

A comma sometimes replaces a verb in certain elliptical constructions.

> **EXAMPLE**   Some were punctual; *others, late.* (replaces *were)*

However, it is better to avoid such constructions in on-the-job writing.

Using with Other Punctuation

Conjunctive adverbs *(however, nevertheless, consequently, for example, on the other hand)* joining independent clauses are preceded by a **semicolon** and followed by a comma. Such adverbs function as both **modifiers** and connectives.

EXAMPLES   Your idea is good; *however,* your format is poor.

He has held the project together; *moreover,* he has helped every-one's morale.

Use a semicolon to separate phrases or clauses in a series when one or more of the phrases or clauses contains commas.

EXAMPLE   Among those present were John Howard, president of the Howard Paper Company; Thomas Martin, president of Copco Corporation; and Larry Stanley, president of Stanley Papers.

A comma always goes inside **quotation marks.**

EXAMPLE   The operator placed the discharge bypass switch at "normal," which triggered a second discharge.

When an introductory phrase or clause ends with a **parenthesis,** the comma separating the introductory phrase or clause from the rest of the sentence always appears outside the parenthesis.

EXAMPLE   Although we left late (at 7:30 p.m.), we arrived in time for the keynote address.

Except with **abbreviations,** a comma should not be used with a **period, question mark, exclamation mark,** or **dash.**

CHANGE   "Have you finished the project?," I asked.
TO   "Have you finished the project?" I asked. (omit the comma)

Using with Numbers and Names

Commas are conventionally used to separate distinct items. Use commas between the elements of an address written on the same line.

EXAMPLE   Walter James, 4119 Mill Road, Dayton, Ohio 45401

A **date** can be written with or without a comma following the year if the date is in the month-day-year format.

EXAMPLES   October 26, 19—, was the date the project began.
October 26, 19— was the date the project began.

If the date is in the day-month-year format, do not set off the date with commas.

EXAMPLE   The date was 26 October 19— when the project began.

Use commas to separate the elements of Arabic **numbers.**

EXAMPLE   1,528,200

C

Use a space rather than a comma in metric values, because many countries use the comma as the decimal marker.

> **EXAMPLE**   1 528 200

A comma may be substituted for the **colon** in a personal letter. Do not use a comma in a business letter, however, even if you use the person's first name.

> **EXAMPLES**   Dear John, (personal letter)
> Dear John: (business letter)

Use commas to separate the elements of geographical names.

> **EXAMPLE**   Toronto, Ontario, Canada

Use a comma to separate names that are reversed or that are followed by an **abbreviation.**

> **EXAMPLES**   Smith, Alvin
> Jane Rogers, Ph.D.
> LMB, Inc.

Use commas to separate certain elements of footnote, reference, and **bibliography** entries.

> **EXAMPLES**   Bibliography—Garrett, Walter, ed. *Handbook of Engineering.* Westport, CT: Greenwood, 1997.
>
> Footnote—[1]Garrett, Walter, ed. *Handbook of Engineering.* (Westport, CT: Greenwood, 1997) 30.
>
> Reference—1. Garrett, Walter, ed. *Handbook of Engineering.* Westport, CT: Greenwood Press, 1997.

(See also **documenting sources.**)

Avoiding Unnecessary Commas

A number of common writing errors involve placing commas where they do not belong. These errors often occur because writers assume that a pause in a sentence should be indicated by a comma. It is true that commas usually signal pauses, but it is not true that pauses *necessarily* call for commas.

Be careful not to place a comma between a subject and verb or between a verb and its **object.**

> **CHANGE**   The cold conditions at the test site in the Arctic, made accurate readings difficult.
> **TO**   The cold conditions at the test site in the Arctic made accurate readings difficult.

CHANGE   He has often said, that one company's failure is another's opportunity.

TO   He has often said that one company's failure is another's opportunity.

Do not use a comma between the elements of a compound subject or a compound predicate consisting of only two elements.

CHANGE   The director of the engineering department, and the supervisor of the quality-control section both were opposed to the new schedules.

TO   The director of the engineering department and the supervisor of the quality-control section both were opposed to the new schedules.

CHANGE   The director of the engineering department listed five major objections, and asked that the new schedule be reconsidered.

TO   The director of the engineering department listed five major objections and asked that the new schedule be reconsidered.

An especially common error is the placing of a comma after a coordinating conjunction such as *and* or *but* (especially *but*).

CHANGE   The chairman formally adjourned the meeting, but, the members of the committee continued to argue.

TO   The chairman formally adjourned the meeting, but the members of the committee continued to argue.

CHANGE   I argued against the proposal. And, I gave good reasons for my position.

TO   I argued against the proposal. And I gave good reasons for my position.

Do not place a comma before the first item or after the last item of a series.

CHANGE   We are considering a number of new products, such as, calculators, scanners, and cameras.

TO   We are considering a number of new products, such as calculators, scanners, and cameras.

CHANGE   It was a fast, simple, inexpensive, process.

TO   It was a fast, simple, inexpensive process.

Do not unnecessarily separate a prepositional phrase from the rest of the sentence with a comma.

CHANGE   He met me, in the conference room, down the hall.

TO   He met me in the conference room down the hall.

CHANGE   We discussed the final report, on the new project.

TO   We discussed the final report on the new project.

committee

Committee is a collective **noun** that takes a singular **verb.**

> EXAMPLE   The *committee* is to meet at 3:30 p.m.

If you wish to emphasize the individuals on the committee, use *the members of the committee* with the plural verb form.

> EXAMPLE   The *members of the committee* were all in agreement.

common nouns (see **nouns**)

comparative degree (see **adjectives** and **adverbs**)

compare/contrast

When you *compare* things, you point out similarities or both similarities and differences. When you *contrast* things you point out only the differences. In either case, you compare or contrast only things that are part of a common category.

> EXAMPLES   He *compared* all the features of the two brands before buying.
> Their styles of selling *contrasted* sharply.

When *compare* is used to establish a general similarity, it is followed by *to.*

> EXAMPLE   *Compared to* the computer, the abacus is a primitive device.

When *compare* is used to indicate a close examination of similarities or differences, it is followed by *with* in formal usage.

> EXAMPLE   We *compared* the features of the new capacitor very carefully *with* those of the old one.

Contrast is normally followed by *with.*

> EXAMPLE   The new policy *contrasts* sharply *with* the earlier one in requiring that sealed bids be submitted.

When the **noun** form of *contrast* is used, one speaks of the *contrast between* two things or of one thing being in *contrast to* the other.

> EXAMPLES   There is a sharp *contrast between* the old and new policies.
> The new policy is in sharp *contrast to* the earlier one.

comparison

When making a comparison, make certain that both or all the elements being compared are clearly evident to your **reader.**

> CHANGE   The third-generation computer is *better.*
> TO   The third-generation computer is *better than the second-generation computer.*

The things being compared must be of the same kind.

> CHANGE   *Imitation alligator hide* is almost as tough as a *real alligator.*
> TO   *Imitation alligator hide* is almost as tough as *real alligator hide.*

Be sure to point out the parallels or differences between the things being compared. Don't assume your reader will know what you mean.

> CHANGE   Washington is farther from Boston *than Philadelphia.*
> TO   Washington is farther from Boston *than it is from Philadelphia.*
> OR   Washington is farther from Boston *than Philadelphia is.*

A double comparison in the same sentence requires that the first be completed before the second is stated.

> CHANGE   The discovery of electricity was *one of the great if not the greatest* scientific discoveries in history.
> TO   The discovery of electricity was *one of the great* scientific discoveries in history, *if not the greatest.*

Do not attempt to compare things that are not comparable.

> CHANGE   Farmers say that storage space is reduced by 40 percent compared with baled hay. *(Storage space* is not comparable to *baled hay.)*
> TO   Farmers say that baled hay requires 40 percent less storage space than loose hay requires.

comparison method of development

As a **method of development,** comparison points out similarities and differences between the elements of your subject. The *comparison method of development* can be especially effective because it can explain a difficult or unfamiliar subject by relating it to a simpler or more familiar one.

You must first determine the basis for the **comparison.** For example, if you were responsible for the purchase of chain saws for a logging company, you would have a number of factors to take into account in order to establish your bases of comparison. Because loggers use the equipment daily, you would have to select durable saws with the appropriate size engines, chain thicknesses, and bar lengths for the type of wood most frequently cut. Since chain saws produce noise and vibration, you would want to compare the quality and cost of the various silencers on the market. You would not include in your comparison such irrelevant factors as color or place of manufacture. Taking into account all of the important elements, however, you would establish a number of bases for choosing from among the available chain saws—engine size, chain thickness, bar length, and noise mufflers.

Once you have determined the basis (or bases) for comparison, you must decide how to present it. In the *whole-by-whole method,* all the relevant characteristics of one item are discussed before those of the next item are considered. In the *part-by-part method,* the relevant features of each item are compared one by one. The following discussion of typical woodworking glues, organized according to the whole-by-whole method, describes each type of glue and its characteristics before going on to the next type:

> *White glue* is the most useful all-purpose adhesive for light construction, but it cannot be used on projects that will be exposed to moisture, high temperature, or great stress. Wood that is being joined with white glue must remain in a clamp until the glue dries, which will take about 30 minutes.
>
> *Aliphatic resin glue* has a stronger and more moisture-resistant bond than white glue. It must be used at temperatures above 50°F. The wood should be clamped for about 30 minutes. . . .
>
> *Plastic resin glue* is the strongest of the common wood adhesives. It is highly moisture resistant—though not completely waterproof. Sold in powdered form, this glue must be mixed with water and used at temperatures above 70°F. It is slow setting and the joint should be clamped for four to six hours. . . .
>
> *Contact cement* is a very strong adhesive that bonds so quickly it must be used with great care. It is ideal for mounting sheets of plastic laminate on wood. It is also useful for attaching strips of veneer to the edges of plywood. Since this adhesive bonds immediately when two pieces are pressed together, clamping is not necessary, but the parts to be joined must be very carefully aligned before being placed together. Most brands are quite flammable and the fumes can be harmful if inhaled. To meet current safety standards, this type of glue must be used in a well-ventilated area, away from flames or heat.

As is often the case when the whole-by-whole method is used, the purpose of this comparison is to weigh the advantages and disadvantages of each glue for certain kinds of woodworking. The comparison could be expanded, of course, by the addition of other types of glue.

If, on the other hand, your purpose were to consider, one at a time, the various characteristics of all the glues, the information might be arranged according to the part-by-part method, as in the following example:

> Woodworking adhesives are rated primarily according to their bonding strength, moisture resistance, and setting times.
>
> *Bonding strengths* are categorized as very strong, moderately strong, or adequate for use with little stress. Contact cement and plastic resin glue bond very strongly, while aliphatic resin glue bonds moderately strongly. White glue provides a bond least resistant to stress.
>
> The *moisture resistance* of woodworking glues is rated as high, moderate, and low. Plastic resin glues are highly moisture resistant. Aliphatic resin glues are moderately moisture resistant; white glue is least moisture resistant.
>
> *Setting times* for these glues vary from an immediate bond to a four-to-six-hour bond. Contact cement bonds immediately and requires no clamping. Because the bond is immediate, surfaces being joined must be carefully aligned before being placed together. White glue and aliphatic resin glue set in thirty minutes; both require clamping to secure the bond. Plastic resin, the strongest wood glue, sets in four to six hours and also requires clamping.

The part-by-part method could accommodate further comparison. Comparisons might be made according to temperature ranges, special warnings, common uses, and so on.

complaint letters

Businesses sometimes err when providing goods and services to customers, and so customers write *complaint letters* (or *claim letters*) asking that such situations be corrected. The **tone** of such a letter is important; the most effective complaint letters do not sound angry. Do not use a complaint letter to vent your anger; remember that the **reader** of your letter probably had nothing to do with whatever went wrong, and berating that person is not likely to achieve anything positive. In most cases, you need only state your claim, support it with all the pertinent facts, and then ask for the desired adjustment. Most companies are very willing to correct whatever went wrong.

The **opening** of your complaint letter should include all identifying data concerning the transaction: item, date of purchase, place of purchase if pertinent, cost, invoice number, and so on.

C

BAKER MEMORIAL HOSPITAL | *Television Services*
501 Main Street
Springfield, OH 45321
(513) 683-8100 (Telephone)
(513) 683-8110 (Fax)

September 23, 19—

General Television, Inc.
5521 West 23rd Street
New York, NY 10062

Attention: Customer Relations Manager

On July 9th I ordered nine TV tuners for your model MX-15 color receiver.
The tuner part number is TR-5771-3.

On August 2nd I received from your Newark, New Jersey, parts warehouse
seven tuners, labeled TR-413-7. I immediately returned these tuners with a note
indicating the mistake that had been made. However, not only have I failed to
receive the tuners I ordered, but I have also been billed repeatedly.

Please either send me the tuners I ordered or cancel my order. I have enclosed a
copy of my original order and the most recent bill.

Sincerely,

Paul Denlinger
Manager

PD:sj
Enclosure

FIGURE 1 Complaint Letter

The body of your letter should explain logically and clearly what
happened. You should present any facts that prove the validity of your
claim. Be sure of your facts, and present them concisely and objectively, carefully avoiding any overtones of accusation or threat. You
may, however, wish to state any inconvenience or loss created by the
problem, such as a broken machine stopping an entire assembly line.

Your **conclusion** should be friendly, and it should request action. State what you would like your reader to do to solve your problem.

Large organizations often have special departments to handle complaints. If you address your letter to one of these departments—for example, to Customer Relations or Consumer Affairs—it should reach someone who can respond to your claim. In smaller organizations you might write to a vice-president in charge of sales or service. For a very small business, write directly to the owner.

Figure 1 is an example of a typical complaint letter. (See also **adjustment letters** and **refusal letters.**)

complement/compliment

Complement means "anything that completes a whole." It is used as either a **noun** or a **verb.**

> EXAMPLES   A *complement* of four employees would bring our staff up to its normal strength. (noun)
>
> The two programs *complement* one another perfectly. (verb)

Compliment means "praise." It too is used as either a noun or a verb.

> EXAMPLES   The manager *complimented* the staff on its efficient job. (verb)
> The manager's *compliment* boosted staff morale. (noun)

complements

A complement is a word, **phrase,** or **clause** used in the predicate of a sentence to complete the meaning of the sentence.

> EXAMPLES   Pilots fly *airplanes*. (word)
> To live is *to risk death*. (phrase)
> John knew *that he would be late*. (clause)

Four kinds of complements are generally recognized: direct **object,** indirect object, objective complement, and subjective complement.

A *direct object* is a **noun** or noun equivalent that receives the action of a transitive **verb;** it answers the question *what* or *whom* after the verb.

> EXAMPLES   John built *an antenna*. (noun)
> I like *to work*. (verbal)
> I like *it*. (pronoun)
> I like *what I saw*. (noun clause)

An *indirect object* is a noun or noun equivalent that occurs with a direct object after certain kinds of transitive verbs such as *give, wish, cause,* and *tell.* It answers the question *to whom* or *for whom* (or *to what* or *for what*).

> **EXAMPLE**   We should *buy the Milwaukee office* a color copier. (*color copier* is the direct object)

An *objective complement* is a noun or adjective that completes the meaning of a sentence by revealing something about the direct object of a transitive verb; it may either describe (**adjective**) or rename (**noun**) the direct object.

> **EXAMPLES**   We painted the house *white.* (adjective)
> I like my coffee *hot.* (adjective)
> They call her a *genius.* (noun)

A *subjective complement* is a noun or adjective in the predicate of a sentence; it completes the meaning of a linking verb by describing or renaming the subject of the verb.

> **EXAMPLES**   The project director seems *confident.* (adjective)
> Acme Corp. is *our major competitor.* (noun phrase)
> His excuse was *that he had been sick.* (noun clause)
> He is an *engineer.* (noun)

The subjective complement is also known as a predicate nominative (noun) or a predicate adjective (adjective).

complex sentences (see **sentence construction**)

compound sentences (see **sentence construction**)

compound-complex sentences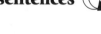
(see **sentence construction**)

compound words

A compound word is made from two or more words that are either hyphenated or written as one word. (If you are not certain whether a compound word should be hyphenated, check a **dictionary.**)

> **EXAMPLES**   nevertheless, mother-in-law, courthouse, run-of-the-mill, low-level, high-energy

Be careful to distinguish between compound words and words that frequently appear together but do not constitute compound words, such as *high school* and *post office*. Also be careful to distinguish between compound words and word pairs that mean different things, such as *greenhouse* and *green house*.

Plurals of compound words are usually formed by adding an *s* to the last letter.

EXAMPLES   bedrooms, masterminds, overcoats, cupfuls

When the first word of the compound is more important to its meaning than the last, however, the first word takes the *s* (when in doubt, check your dictionary).

EXAMPLES   editors-in-chief, fathers-in-law

Possessives are formed by adding *'s* to the end of the compound word.

EXAMPLES   the *vice-president's* speech, his *brother-in-law's* car, the *pipeline's* diameter, the *antibody's* action

comprise/compose

Comprise means "include," "contain," or "consist of." The whole *comprises* the parts.

EXAMPLE   The mechanism *comprises* 13 moving parts.

Compose means "create" or "make up the whole." The parts *compose* the whole.

EXAMPLE   The 13 moving parts *compose* the mechanism.
OR   The mechanism is *composed* of 13 moving parts.

computer graphics

Computer graphics are among the fastest developing areas of applications software for personal computers. Both new software and novel combinations of existing software are continually being developed. Accordingly, rather than attempting a definitive classification of this field, this entry provides a concise, basic introduction to the most common types of graphics software and their capabilities. The discussion focuses on graphics produced primarily for use in documents rather than for use in oral or multimedia presentations.

Graphics software for personal computers can be divided into two broad categories: *bitmapped graphics* and *vector graphics*.

Bitmapped Graphics

C

Bitmapped graphics are composed of an array of evenly spaced "pixel elements." Pixels are similar to the dots that make up the images on a television screen. Because bitmapped graphics are composed of a finite number of pixel elements, they tend to produce images with a distinctive jagged-edge appearance (see Figure 1). Bitmapped graphics tend to require large amounts of computer memory to manipulate and store.

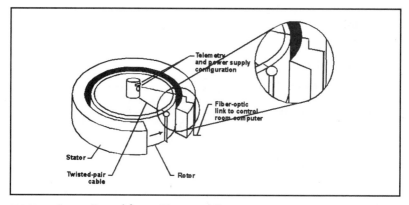

FIGURE 1   Image Created from a Bitmapped Program

Bitmapped graphics can be obtained in several ways: by *scanning* an existing image, by using *bitmapped clipart,* or by using a *bitmap editor.*

Scanned images are created by using a computer device called a scanner to convert an existing paper copy graphic (**graph,** diagram, **photograph,** etc.) into a bitmapped image on a computer screen through a process called *digitizing.* Once digitized in bitmap form, the scanned graphic can be modified on screen by using a bitmap editor and then printed for use in a document. If you scan a **copyrighted** image for use in your document, of course, you must obtain prior written permission to do so.

A *bitmap editor* is a computer software application that permits the editing and manipulation of bitmapped graphics. In their simplest form, bitmap editors allow you to modify an image by adding and removing pixels (or dots) from it. More sophisticated bitmapped editors allow you to rotate, reduce, and enlarge images. Bitmap editors also allow you to create freehand images from scratch, much as an artist wields pencil, brush, charcoal, and other tools to create images on paper and canvas.

Bitmapped clip art is computer artwork created by professional graphic artists and sold on diskette or CD-ROM. Clip art graphics are frequently grouped and sold by topics, such as vehicles, animals, symbols, children, insignia, and many others. Figure 2 shows a grouping of typical clip art images. Since clip art images are not copyrighted, you need not obtain prior approval to use these images after you purchase the software.

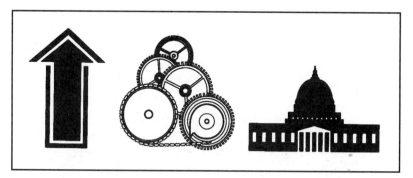

FIGURE 2   Sample Clip Art Images

Vector Graphics

Vector graphics are composed of predefined shapes, called graphic primitives (boxes, triangles, arcs, circles, lines, and so forth), instead of pixels (Figure 3). Vector graphics are far easier to manipulate than pixel graphics and they produce images that are of higher resolution (better visual quality) than bitmap images. Note the crispness of the image in Figure 4 compared with the bitmapped image in Figure 1.

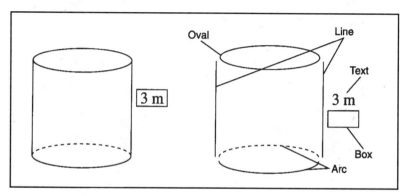

FIGURE 3   Graphics Primitives

C

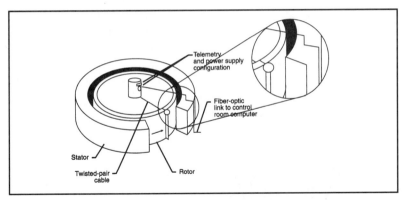

FIGURE 4   Image Created from a Vector Graphic Program

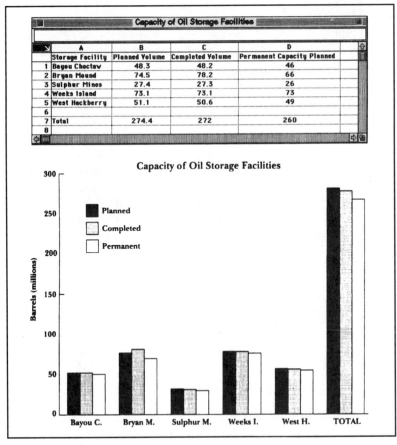

FIGURE 5   Spreadsheet Table to Graph Image

Vector graphics can be obtained either by vector drawing programs or by vector clip art.

Vector drawing programs are a class of computer software applications that allows you to create complex graphic images. These images are manipulated by using the coordinates, or vectors, of the beginning and end points of the primitives that form them. By modifying the primitives, you can quickly modify the image. In general, vector graphics are ideal for creating graphs, charts, and diagrams.

Vector clip art is like that described for bitmapped clip art, except for the technology by which the images are created.

Uses of Computer Graphics

Business graphics programs, also called *charting* programs, produce preformatted charts, graphs, and **tables** based on numerical data taken either from a computer spreadsheet program, as in Figure 5, or from text and data entered from the computer keyboard. Figure 6 shows a range of typical bar and pie charts created by business graph-

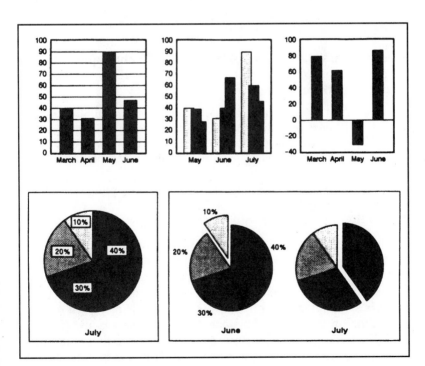

FIGURE 6 Typical Computer-Generated Bar and Pie Charts

C

ics software. After creating your graphics, follow the guidelines listed in **illustrations** for integrating them into your document.

Other computer graphics programs are available for creating **maps,** statistical analyses, animation, and computer-aided design (CAD) images used in engineering and manufacturing. They offer specialized applications for the graphics, engineering, architectural, or scientific professional that are beyond the scope of this text.

concept/conception

A *concept* is a thought or an idea. A *conception* is the sum of a person's ideas, or concepts, on a subject.

EXAMPLES   This final *conception* of the whole process evolved from many smaller *concepts.*

From the *concept* of combustion evolved the *conception* of the internal combustion engine.

conciseness/wordiness

Effective writers make all words, sentences, and **paragraphs** count by eliminating unnecessary words and **phrases.** Wordiness results from needless **repetition** of the same idea in different words.

CHANGE   Modern students *of today* are more technologically sophisticated than their parents. (The phrase *of today* repeats the thought already expressed by the adjective *modern.*)

TO   Modern students are more technologically sophisticated than their parents.

CHANGE   The walls were sky-blue *in color.* (The phrase *in color* is redundant.)

TO   The walls were sky-blue.

Careful writers remove every word, **phrase, clause,** or **sentence** they can without sacrificing **clarity.** In doing so, they are striving to be as concise as clarity permits—but note that conciseness is not a synonym for brevity. In fact, when brevity is overdone it can become **telegraphic style.** Brevity may or may not be desirable in a given passage depending on the writer's **purpose,** but conciseness is always desirable. The writer must distinguish between language that is used for effect and mere wordiness that stems from lack of care or judgment, as often happens with **nominalizations.**

C

A concise sentence is not guaranteed to be effective, but a wordy sentence always loses some of its readability and **coherence** because of the extra load it must carry. Wordiness is to be expected in a first draft, but it should never survive **revision.** "I would have written a shorter letter if I'd had more time" is a truism.

Causes of Wordiness

Modifiers that repeat an idea already implicit or present in the word being modified contribute to wordiness by being redundant.

EXAMPLES   *active* consideration   *present* status
           final outcome   *past* history
           personal opinion   descended *down*
           completely finished   circle *around*
           basic essentials   worthy *of merit*
           advance planning   the reason *is because*

Another cause of wordiness is the "it . . . that . . ." construction. Not only is this construction wordy, but often it forces you to use the passive voice (a major cause of wordiness in its own right).

CHANGE   *It* is agreed *that* our new design will strive for simplicity.
TO   We agree that our new design will strive for simplicity.

Coordinating **synonyms** that merely repeat one another contribute to wordiness.

EXAMPLES   any and all   each and every
           finally and for good basic and fundamental
           first and foremost

Excess qualification also contributes to wordiness, as the following examples demonstrate:

EXAMPLES   utterly rejected—rejected
           perfectly clear—clear
           completely compatible—compatible
           completely accurate—accurate
           radically new—new

The use of **expletives,** relative **pronouns,** and relative **adjectives,** although they have legitimate purposes, often results in wordiness.

CHANGE   *There are* (expletive) many supervisors in the area *who* (relative pronoun) are planning to attend the workshop *that* (relative adjective) is scheduled for Friday.
TO   Many supervisors in the area plan to attend the workshop scheduled for Friday.

C

Circumlocution (a long, indirect way of expressing things) is a leading cause of wordiness.

> CHANGE The payment to which a subcontractor is entitled should be made promptly so that in the event of a subsequent contractual dispute we, as general contractors, may not be held in default of our contract by virtue of nonpayment.
>
> TO Pay subcontractors promptly. Then if a contractual dispute should occur, we cannot be held in default of our contract because of nonpayment.

Conciseness can be overdone. If you respond to a written request that you cannot understand with "Your request was unclear" or "I don't understand," you will probably offend your reader. Instead of attacking the writer's ability to phrase a request, consider that what you are really doing is asking for more information. Say so.

> EXAMPLE I will need more information before I can answer your request. Specifically, can you give me the title and the date of the report you are looking for?

This version is a little longer than the others, but it is both more polite and more helpful.

How to Achieve Conciseness

Conciseness can be achieved by effective use of **subordination.** This is, in fact, the best means of tightening wordy writing.

> CHANGE The chemist's report was carefully illustrated, and *it covered five pages.*
>
> TO The chemist's *five-page* report was carefully illustrated.

Conciseness can be achieved by using simple words and phrases.

> CHANGE It is the policy of the company to provide the proper equipment to enable each employee to conduct the telephonic communication necessary to discharge his responsibilities; such should not be utilized for personal communications.
>
> TO Your telephone is provided for company business; do not use it for personal calls.

Conciseness can be achieved by eliminating undesirable **repetition.**

> CHANGE Postinstallation testing, which is offered to all our customers at no further cost to them whatsoever, is available with each Line Scan System One purchased from this company.
>
> TO Free postinstallation testing is offered with each Line Scan System One.

Conciseness can sometimes be achieved by changing a sentence from the passive to the active **voice** or from the indicative to the imperative **mood.** The following example does both:

CHANGE   Codes are normally used when it is known that the records are to be processed by a computer, and controls are normally used when it is known that the records are designed to be processed at a tab installation.

TO   Use codes when you process the cards on a computer, and use controls when you process them at a tab installation.

Eliminate wordy introductory **phrases** or pretentious words and phrases of any kind.

EXAMPLES   As you may recall   In view of the foregoing
            As you know   In the case of
            Needless to say   It may be said that
            In view of the fact that   It is interesting to note that

CHANGE   in order to, so as to, so as to be able to, with a view to
TO   to

CHANGE   due to the fact that, for the reason that, owing to the fact that, the reason for
TO   because

CHANGE   by means of, by using, utilizing, through the use of
TO   by or with

CHANGE   at this time, at this point in time, at present, at the present
TO   now

CHANGE   at that time, at that point, at that point in time, as of that date, during that period
TO   then

Overuse of **intensifiers** (such as *very, more, most, best, quite*) contributes to wordiness; conciseness can be achieved by eliminating them. The same is true of excessive use of **adjectives** and **adverbs.** (See also **nominalizations** and **pace.**)

concluding

Concluding a piece of writing not only ties together all the main ideas but can do so emphatically by making a final significant point. This final point may be to recommend a course of action, to make a prediction, to offer a judgment, to speculate on the implications of your ideas, or merely to summarize your main points.

C

The way you conclude depends on both the purpose of your writing and the needs of your **reader.** For example, a committee **report** about possible locations for a new manufacturing plant could end with a recommendation. A report on a company's annual sales figures might conclude with a judgment about why sales are up or down. A letter about consumer trends could end by speculating on the implications of these trends. A document that is particularly lengthy will often end with a summary of its main points. Study the following examples:

Recommendation   These results indicate that you need to alter your testing procedure to eliminate the impurities we found in specimens A through E.

Prediction   Although my original estimate on equipment ($20,000) has been exceeded by $2,300, my original labor estimate ($60,000) has been reduced by $3,500; therefore, I will easily stay within the limits of my original bid. In addition, I see no difficulty in having the arena finished for the December 23 Christmas program.

Judgment   Although our estimate calls for a substantially higher budget than in the three previous years, we believe that it is justified by our planned expansion.

Summary   As this letter has indicated, we would attract more recent graduates by (1) increasing our advertising in local student newspapers, (2) resuming our co-op program, (3) sending a representative to career day programs at local colleges, (4) inviting local college instructors to teach in-house courses here at the plant, and (5) encouraging our employees to attend evening classes at local colleges.

The concluding statement may merely present ideas for consideration, but also it may call for action or deliberately provoke thought.

Ideas for consideration   The new prices become effective the first of the year. Price adjustments are routine for the company, but some of your customers will not consider them so. Please consider the needs of both your customers and the company as you implement these new prices.

Call for action   Send us a check for $250 now if you wish to keep your account active. If you have not responded to our previous letters because of some special hardship, I will be glad personally to work out a solution with you.

Thought-provoking statement	Can we continue to accept the losses incurred by careless workmanship? Must we accept it as inevitable? Or should we consider steps to control it firmly now?

C

Because this statement appears in a position of great **emphasis,** be careful to avoid closing with a **cliché.** Some such expressions do say what you want to say so precisely that they are considered acceptable closings, however.

EXAMPLE   If I can provide further information, please let me know.

Be especially careful not to introduce a new **topic** when you conclude. A concluding thought should always relate to and reinforce the ideas presented in your writing. (See also **conclusions.**)

conclusions

The *conclusion* section of a **report** or article pulls together the results or findings and interprets them in the light of the study's purpose and the methods by which it was conducted. The evidence for these findings makes up the discussions in the body of the report or article; the conclusions must grow out of the information discussed there. Moreover, the conclusions must be consistent with what the **introduction** stated that the report would examine (its purpose) and how it would do so (its method). If the introduction stated that the report had as its **purpose** to learn the economic costs of relocating a plant from one city to another, the conclusion should not discuss the social or aesthetic impacts the new plant could have on the new location. Of course, if "costs" are defined in the introduction to include social and aesthetic concerns, then these also must be accounted for in the report and discussed in the conclusions.

The sample conclusion on page 114 comes from a report discussing the effects of hazardous waste sites on housing values. (The introduction for this report appears in the entry on introductions, page 297.) The author restates the purpose of the study in the first sentence as a finding and summarizes the other findings in the first paragraph. In the second paragraph, the author elaborates on several ways that the findings may be of value for public officials in planning the cleanup of other hazardous waste sites.

Not all reports or articles require a separate conclusions section. Periodic **progress reports** for a long-term project, for example, may simply indicate what was done during the last month or quarter. The final report for such a project, however, would require a conclusions

C

CONCLUSIONS

Finding

The results from this study show that information on the toxicity of sites does influence housing prices. It appears that information from sources including the Environment Protection Agency (EPA) and local community groups does impact sales prices; thus at least in this case the EPA announcements seem to provide some additional information to the residents. Official announcements that

Finding

sites will be cleaned up are not necessarily believed by residents, or the agency may not be perceived as able to eliminate the externality effectively. It is possible that after the cleanup is completed and residents have an opportunity to assess the success of the process, prices would respond. Regardless, studies which have looked at prices for brief periods before and after official announcements

Finding

have potentially missed the source of the information as well as the timing of the movement in house prices.

The results from this study could be used to provide dollar estimates of the costs of a toxic waste site to nearby homeowners from

Implication of finding

the damage to house values. It appears that the EPA does contribute to the market adjustment process during the announcement phase, and that perhaps there is a role for federal agencies in disseminating information in the siting of facilities such as incinerators. However, it does not appear that the EPA affects the market process in the cleaning phases. If prices do respond to cleanup efforts in the

Implication of finding

future, benefits would give government officials at all levels a more complete framework within which they can allocate scarce financial resources for cleaning up toxic sites in a more efficient and equitable way. It would also allow for the design of a more complete compensation program to help those who have been harmed by the existence of these sites.

K. A. Kiel, "Measuring the Impact of the Discovery and Cleaning of Identified Hazardous Waste Sites on House Values," *Land Economics* 71 (4) Nov. 1995 428–35.

FIGURE 1   Sample Conclusion

section in which the findings and their implications were discussed. Nor do many other types of writing require a fully developed conclusion. For examples of less formal conclusions, see **concluding.**

(For guidance about the location of the conclusions section in a report, see **formal reports.**)

conjunctions

A conjunction connects words, **phrases,** or **clauses** and can also indicate the relationship between the two elements it connects.

Types of Conjunctions

A *coordinating conjunction* is a word that joins two sentence elements that have the same functions. The coordinating conjunctions are *and, but, or, for, nor, yet,* and *so.*

EXAMPLES Nature *and* technology are only two conditions that affect petroleum operations around the world. (joining two nouns)

To hear *and* to obey are two different things. (joining two phrases)

He would like to include the test results, *but* that would make the report too long. (joining two clauses)

Coordinating conjunctions can join words, **phrases,** or **clauses** of equal rank.

EXAMPLES Only crabs *and* sponges managed to escape the red tide. (words)

Our twin objectives are to increase power *and* to decrease noise levels. (phrases)

The average city dweller in the United States now has 0.17 parts per million of lead in his blood, *and* the amount is increasing yearly. (clauses)

Correlative conjunctions are coordinating conjunctions that are used in pairs. The correlative conjunctions are *either . . . or, neither . . . nor, not only . . . but also, both . . . and,* and *whether . . . or.*

EXAMPLES The shipment contains the parts *not only* for the seven machines scheduled for delivery this month *but also* for those scheduled for delivery next month.

The shipment will arrive *either* on Wednesday *or* on Thursday.

To add not only symmetry but also logic to your writing, follow correlative conjunctions with parallel sentence elements (such as *on Wednesday* and *on Thursday* in the previous example).

A problem can develop with correlative conjunctions joining a singular **noun** to a plural noun. The problem is with the **verb.** If you use this construction, make the verb agree with the nearest noun. In the following example, you would make the verb *fail* agree with the noun *coils.*

EXAMPLE Ordinarily, neither the pressure switch nor the relay *coils fail* the test.

(See also **parallel structure.**)

Subordinating conjunctions connect sentence elements of varying importance, normally independent clauses and dependent clauses; they usually introduce the dependent clause. The most frequently used subordinating conjunctions are *so, although, after, because, if, where, than, since, as, unless, before, though, when,* and *whereas.* They are distinguished from coordinating conjunctions, which connect elements of equal importance.

EXAMPLES   He finished his report, *and* he left the office. (coordinating conjunction)

He left the office *after* he finished his report. (subordinating conjunction)

A clause introduced with a subordinating conjunction is a dependent clause.

EXAMPLE   *Because* we did not oil the crankcase, the engine was ruined.

The word *because* is a subordinating conjunction. Do not use the coordinating conjunction *and* as a substitute for it.

CHANGE   We didn't add oil to the crankcase, *and* it ruined the engine.
TO   *Because* we didn't add oil to the crankcase, the engine was ruined.

When the subordinating conjunction begins the sentence, a comma should follow the dependent clause in which it appears.

EXAMPLE   *Because* we were late, the client was angry.

When the dependent clause beginning with a subordinating conjunction comes at the end of the sentence, do not separate the independent clause and the dependent clause with a comma.

EXAMPLE   The client was angry *because* we were late.

Be aware that words may serve multiple functions. *When, where, how,* and *why* serve as both interrogative adverbs and subordinating conjunctions.

EXAMPLES   *When* will we go? (interrogative adverb)
We will go *when* he arrives. (subordinating conjunction)

Since, until, before, and *after* are used as both subordinating conjunctions and **prepositions.**

EXAMPLES   *Since* we all had arrived, we decided to begin the meeting early. (subordinating conjunction)

I have worked on this project *since* May. (preposition)

A *conjunctive adverb* is an adverb because it modifies the clause that it introduces; it operates as a conjunction because it joins two independent clauses. The most common conjunctive adverbs are *however, nev-*

ertheless, moreover, therefore, further, then, consequently, besides, accordingly, also, and *thus.*

> **EXAMPLE**   The engine performed well in the laboratory; *however,* it failed under road conditions.

Punctuating Conjunctions

Two independent clauses that are joined by a conjunctive **adverb** require a **semicolon** before and a comma after the conjunctive adverb because the conjunctive adverb is introducing the independent clause that follows it and is indicating its relationship with the independent clause preceding it. The conjunctive adverb connects ideas but does not grammatically connect the clause stating the ideas; the semicolon must do that. The conjunctive adverb both connects and modifies. As a **modifier,** it is part of one of the two clauses that it connects. The use of the semicolon makes it a part of the clause it modifies.

> **EXAMPLES**   The building was not finished on the scheduled date; *nevertheless,* Rogers moved in and began to conduct the business of the department from the unfinished offices.
>
> The new project is almost completed and is already over cost estimates; *however,* we must emphasize the importance of adequate drainage.

Two independent clauses separated by a coordinating conjunction should have a comma immediately preceding the coordinating conjunction if the clauses are relatively long.

> **EXAMPLE**   The Alpha project was her third assignment, *and* it was the one that made her reputation as a project leader.

If two independent clauses that are joined by a coordinating conjunction have commas within them, a semicolon may precede the conjunction.

> **EXAMPLE**   Even though the schedule is tight, we must meet our deadline; *and* all the staff, including programmers, will have to work overtime this week.

Conjunctions in Titles

Conjunctions in the titles of books, articles, plays, movies, and so on should not be capitalized unless they are the first or last word in the title. (See also **capital letters.**)

> **EXAMPLE**   The book *Technical and Professional Writing* was edited by Herman Estrin.

Conjunction at Beginning of Sentence

Occasionally, a conjunction may begin a sentence and even a **paragraph,** as in the following example:

> EXAMPLE The executive is impressed by the marvels of computer technology before him; he has difficulty in understanding the programming endeavor that makes the computer run. He walks away feeling that the annual report will look better because of these machines—and the data processing manager dreams of a larger paycheck.
>
> *But* the balloon bursts the first time the system gets into trouble and requires a modification, and the lead programmer is no longer with the company or cannot recall the details of the program due to the passage of time. . . .
>
> —William L. Harper, *Data Processing Documentation*

conjunctive adverbs (see **adverbs**)

connected with/in connection with

Connected with and *in connection with* are wordy **phrases** that can usually be replaced by *in* or *with*.

> CHANGE He is *connected with* the TFT Corporation.
> TO He is *with* the TFT Corporation.
> OR He is employed by the TFT Corporation.

> CHANGE The fringe benefits *in connection with* (or *connected with*) the job are quite good.
> TO The fringe benefits *with* the job are quite good.

connotation/denotation

The terms *connotation* and *denotation* refer to two different ways of interpreting a word's meaning. The denotation of a word is its literal and objective meaning. Defined this way, a *bureaucrat* is an unelected public official who administers government policy. The word *bureaucrat* frequently conjures up visions of someone who insists on rigid adherence to arbitrary rules and regulations. The secondary meanings are a word's connotations.

In your writing you must be sure to consider the connotations of your words. You know what you mean to say, but you must be careful to say it in words that will not be misinterpreted by your reader. For example, you might not want to refer to an item as *cheap*, meaning inexpensive, for the word *cheap* also implies shoddy workmanship or poor materials. The usage notes in a reputable dictionary can help you make these distinctions. Technical writing in particular requires an objective and impartial presentation of information, and so you should always be as careful as possible to use words without misleading or unwanted connotations.

(See also **defining terms.**)

consensus of opinion

Since *consensus* normally means "harmony of opinion," the **phrase** *consensus of opinion* is redundant. The word *consensus* can be used only to refer to a group, never one or two people.

CHANGE   The *consensus of opinion* of the members was that the committee should change its name.

TO   The *consensus* of the members was that the committee should change its name.

continual/continuous

Continual means "happening over and over" or "frequently repeated."

EXAMPLE   Writing well requires *continual* practice.

Continuous means "occurring without interruption" or "unbroken."

EXAMPLE   The *continuous* roar of the machines was deafening.

contractions

A *contraction* is a shortened **spelling** of a word or **phrase** with an **apostrophe** substituting for the missing letters.

EXAMPLES   cannot   can't
will not   won't
have not   haven't
it is   it's

Contractions are often used in speech but should be used discriminatingly in **reports,** formal letters, and most technical writing. (See also **style.**)

coordinating conjunctions (see conjunctions)

copyright ⟲

Copyright is the right of exclusive ownership by an author of the benefits resulting from the creation of his or her work. This right gives authors, or others to whom they transfer ownership of copyright, control over the reproduction and dissemination of their works. Once copyrighted, a work cannot be indiscriminately reproduced unless the copyright owner gives permission, usually in exchange for royalties or other compensation.

The copyright law provides a "fair use" provision that allows teachers, librarians, reviewers, and others to reproduce a limited portion of copyrighted materials for educational and certain other purposes without compensation to the copyright owners. The law refers to such purposes as "criticism, comment, news reporting, teaching (including multiple copies for classroom use), scholarship, or research." The law includes guidelines for the fair use of copyrighted materials in the classroom. These guidelines set three standards: brevity, spontaneity, and cumulative effect. How a copying situation adheres to these standards determines whether or not the instructor needs permission to use multiple copies. Although the law does not exactly define "fair use," it sets out four criteria for determining whether a given use is fair: (1) the purpose and nature of the use, (2) the amount of material used in relation to the copyrighted work as a whole, (3) the nature of the copyrighted work, and (4) the effect of the use on the potential market for or value of the copyrighted work. Publications that explain the copyright law in detail are available from the Copyright Office, Library of Congress, Washington, DC 20559.

If you use copyrighted materials in your written work, be aware of the following guidelines:

1. In many situations, a small amount of material from a copyrighted source may be used in your written work without permission or payment as long as the use satisfies the fair use criteria. However, you should give credit to the source from which the material was taken.

2. A work first published after March 1, 1989, receives copyright protection whether or not it bears a notice of copyright, although all published materials generally contain such a notice, usually on the back of the title page. (For an example of a copyright notice, turn to the back of the title page of this book.)

3. All publications created by U.S. government agencies are in the public domain—that is, they are not copyrighted.

With the advent of the **Internet,** many copyrighted works formerly available solely in print are now distributed around the world in cyberspace. Copyright law applies to cyberspace works just as it does to their print counterparts. If you plan to reproduce or further distribute copyrighted works available on the Internet, you must first obtain permission from the copyright holder, unless the fair use provision of copyright law applies to your intended use. As with printed works, document the source of material (text, graphics, tables) obtained on the Internet.

(See also **documenting sources** and **plagiarism.**)

correlative conjunctions (see conjunctions)

correspondence

The process of writing letters, **memorandums,** or **email** messages involves many of the same steps that go into most other on-the-job tasks. The following list summarizes these steps:

1. Establish your **purpose,** and determine your **reader's** attitude and needs.
2. Prepare an **outline,** even if it is only a list of points to be covered in the order you wish to cover them.
3. Write the first draft (see also **writing the draft**).
4. Allow a "cooling" period (time for weaknesses to become obvious).
5. Revise the draft (see also **revision**).
6. Use **proofreading** techniques.

These guidelines will help you write a clear, well-organized message. You might also find it valuable to look at the "Five Steps to Successful Writing" at the front of this book. Keep in mind, however, that one very important element in business letters is the impression they leave on the reader. To convey the right impression—of yourself as well as of your company or organization—you must take particular care with both the **tone** and the **style** of your writing.

Tone

Letters are generally written directly to another person who is identified by name. You may or may not know the person, but never forget that you are writing to another human being. For this reason, letters are always more personal than are **reports** or other forms of business writing. Successful writers find that it helps to imagine their reader sitting across the desk from them as they write; they then write to the reader as if they were talking to him or her in person. This technique helps them keep their language natural.

As a letter writer addressing yourself directly to your reader, you are in a good position to take into account your reader's needs. If you ask yourself, "How might I feel if I were the recipient of such a letter?" you can gain some insight into the likely needs and feelings of your reader—and then tailor your message to fit those needs and feelings. Furthermore, you have a chance to build goodwill for your business or organization. Many companies spend millions of dollars to create a favorable public image. A letter to a customer that sounds impersonal and unfriendly can quickly tarnish that image, but a thoughtful letter that communicates sincerity can greatly enhance it.

Suppose, for example, you are a department-store manager who receives a request for a refund from a customer who forgot to enclose the receipt with the request. In a letter to that customer, you might write the following:

> **EXAMPLE**   The sales receipt must be enclosed with the merchandise before we can process a refund.

But if you consider how you might keep the customer's goodwill, you might word the request this way:

> **EXAMPLE**   Please enclose the sales receipt with the merchandise so that we can send you your refund promptly.

However, you can go one step further as a business-letter writer. You can put the reader's needs and interests at the center of the letter by writing from the reader's perspective. Often, although not always, doing so means using the words *you* and *your* rather than *we, our, I,* and *mine.* That is why the technique has been referred to as the "you" viewpoint. Consider the following revision, which is written with the "you" viewpoint:

> **EXAMPLE**   So that you can receive your refund promptly, please enclose the sales receipt with the merchandise.

In this example, the reader's benefit and interest is stressed: to get a refund as quickly as possible. By emphasizing the reader's needs, the writer will be more likely to accomplish his or her objective, to get the reader to act.

Obviously, both goodwill and the "you" viewpoint can be overdone. Used thoughtlessly, both techniques can produce a fawning, insincere tone—what might be called "plastic goodwill," as in the following:

CHANGE	You're the sort of forward-thinking person whose outstanding good judgment is obvious from your selection of the Model K-50 copier.
TO	Congratulations on selecting the Model K-50 copier. We believe the K-50 is one of the finest on the market.

Direct and Indirect Pattern: Good News and Bad News Letters

It is generally more effective to present good news directly and bad news indirectly, because readers form their impressions and attitudes very early in letters. In fact, some readers do not finish reading a letter when bad news is presented first. Other readers may finish a letter that presents bad news at the outset, but they tend to read what follows with a predetermined opinion. They may be skeptical about an explanation, or they may reject a reasonable alternative presented by the writer. Furthermore, even if you must refuse a person's request or say no to someone, you still may wish to work with him or her in the future, and an abruptly phrased rejection early in the letter may prevent you from reestablishing an amicable relationship in the future.

Consider the thoughtlessness of the rejection in Figure 1. Although the letter is direct and uses the pronouns *you* and *your,* the writer has apparently not considered how the reader will feel as she reads the letter. There is no expression of regret that Mrs. Mauer is being rejected for the position nor any appreciation of her efforts in applying for the job. The letter is, in short, rude. The pattern of this letter is (1) bad news, (2) explanation, (3) close.

A better general pattern for "bad news" letters is as follows:

1. Buffer
2. Bad news
3. Goodwill

The "buffer" may be either neutral information or an explanation that makes the bad news *understandable.* Bad news is never pleasant; however, information that either puts the bad news in perspective or makes the bad news seem reasonable maintains goodwill between the writer and the reader. Consider, for example, a revision of the rejection letter, as shown in Figure 2. This letter carries the same disappointing news as the first one does, but the writer is careful to thank the reader for her time and effort, to explain why she was not accepted for the job, and to offer her encouragement in finding a position in another office. Presenting good news is, of course, easier. Present good news

C

Southtown Dental Center
3221 Ryan Road   San Diego, CA 92217
(714) 321-1579 (Telephone)
(714) 321-1580 (Fax)

November 11, 19—

Mrs. Barbara L. Mauer
157 Beach Drive
San Diego, CA 92113

Dear Mrs. Mauer:

Your application for the position of dental receptionist at
Southtown Dental Center has been rejected. We have found
someone more qualified than you.

Sincerely,

Mary Hernandez

Mary Hernandez
Office Manager

MH/bt

FIGURE 1   Poor "Bad News" Letter

early—at the outset, if at all possible. The pattern for "good news" letters should be as follows:

1. Good news
2. Explanation or facts
3. Goodwill

By presenting the good news first, you will increase the likelihood that
the reader will pay careful attention to details, and you will achieve

C

Southtown Dental Center
3221 Ryan Road   San Diego, CA 92217
(714) 321-1579 (Telephone)
(714) 321-1580 (Fax)

November 11, 19—

Mrs. Barbara L. Mauer
157 Beach Drive
San Diego, CA 92113

Dear Mrs. Mauer:

Buffer   Thank you for your time and effort in applying for the position of dental receptionist at Southtown Dental Center.

Bad news   Since we need someone who can assume the duties here with a minimum of training, we have selected an applicant with over ten years of experience.

Goodwill   I am sure that with your excellent college record you will find a position in another office.

Sincerely,

Mary Hernandez

Mary Hernandez
Office Manager

MH/bt

FIGURE 2   Courteous and Effective "Bad News" Letter

goodwill from the start. Figure 3 is a good example of a "good news" letter.

Writing Style

Letter-writing style may vary from informal, in a letter to a close business associate, to formal, or restrained, in a letter to someone you do not know. (Even if you are writing a business letter to a close associate, you should always follow the rules of standard grammar, spelling, and punctuation.)

C

Southtown Dental Center
3221 Ryan Road   San Diego, CA 92217
(714) 321-1579 (Telephone)
(714) 321-1580 (Fax)

November 11, 19—

Mrs. Barbara L. Mauer
157 Beach Drive
San Diego, CA 92113

Dear Mrs. Mauer:

Good news   Please accept our offer for the position of dental receptionist at Southtown Dental Center.

Explanation   If the terms we discussed in the interview are acceptable to you, please come in at 9:30 a.m. on November 15. At that time we will ask you to complete our personnel form, in addition to. . . .

Goodwill   Everyone here at Southtown is looking forward to working with you. We all were very favorably impressed with you during your interview.

Sincerely,

Mary Hernandez

Mary Hernandez
Office Manager

MH/bt

FIGURE 3   Effective "Good News" Letter

INFORMAL   It worked! The new process is better than we had dreamed.
RESTRAINED   You will be pleased to know that the new process is more effective than we had expected.

You will probably find yourself relying on the restrained style more frequently than on the informal one, since an obvious attempt to sound casual, like overdone goodwill, may strike the reader as insincere. Do not adopt such a formal style, however, that your letters read like legal contracts. Using legalistic-sounding words in an effort to impress your reader will make your writing seem stuffy and pompous—and may well irritate your reader.

CHANGE   In response to your query, I wish to state that we no longer have an original copy of the brochure requested. Be advised that a photographic copy is enclosed herewith. Address further correspondence to this office for assistance as required.

TO   Because we are currently out of original copies of our brochure, I am sending you a photocopy of it. If I can help further, please let me know.

The excessively formal writing style in the original version is full of largely out-of-date business language; expressions like *query, I wish to state, be advised that,* and *herewith* are both old-fashioned and pretentious. Good business letters today have a more personal, down-to-earth style, as the revision illustrates.

The revised version is not only less stuffy but also slightly more concise. Being concise in writing is important, but don't be so concise that you become blunt (or lapse into **telegraphic style**). If you respond to a written request that you cannot understand with "Your request was unclear" or "I don't understand," you will probably offend your reader. Instead of attacking the writer's ability to phrase a request, consider that what you are really doing is asking for more information. Say so.

EXAMPLE   I will need more information before I can answer your request. Specifically, can you give me the title and the date of the report you are looking for?

This version is a bit longer, but it is both more polite and more helpful.

Accuracy

Since a letter is a written record, it must be accurate. Facts, figures, dates, and explanations that are incorrect or misleading may cost time, money, and goodwill. Remember that when you sign a letter, you are responsible for its contents. Always allow yourself time to review a letter before mailing it. When time permits, ask someone who is familiar with its contents to review an important letter. Listen with an open mind to the criticism of others about what you have said, and make any changes you believe necessary.

A second kind of accuracy to check for is the mechanics of writing—**punctuation, grammar,** and **spelling.** When you sign a letter, you are responsible for its contents and form.

Appearance

The appearance of a business letter may be crucial in influencing a recipient who has never seen you. The rules for preparing a neat,

C

attractive letter are not difficult to master, and they are important. Use unruled white bond paper of standard size, and use envelopes of the same quality. Center the letter on the page vertically and horizontally—a "picture frame" of white space should surround the contents. When you use your company's letterhead stationery, consider the bottom of the letterhead as the top edge of the paper.

Parts of a Business Letter

The two most common formats of business letters are the full block style (shown in Figure 5 on page 134) and the modified block style (shown in Figure 6 on page 135). The full block style, though easier to type because every line begins at the left margin, is suitable only with business letterhead stationery. In the modified block style, the return address, the date, and the complimentary closing all begin at the center of the page, and the other elements are aligned at the left margin. All other letter styles are variations of these two styles. Again, if your employer recommends or requires a particular style, follow it carefully. Otherwise, choose a style and follow it consistently. Figure 6 is an example of the modified block style.

If your employer recommends or requires a particular format and typing style, use it. You may also wish to consult a secretarial guide, such as *The Gregg Reference Manual* by William A. Sabin (McGraw-Hill). Otherwise, follow the guidelines provided here, and review the illustrations for placement with full- and modified-block style (pp. 134 and 135).

Heading. The writer's full address (street, city, state, and ZIP code) and the date are given in the heading. Because the writer's name appears at the end of the letter, it need not be included in the heading. Words like *street, avenue, first,* or *west* should be spelled out rather than abbreviated. You may either spell out the name of the state in full or use the U.S. Postal Service abbreviations. The date usually goes directly beneath the last line of the address. Do not abbreviate the name of the month.

EXAMPLE   1638 Parkhill Drive East
Great Falls, MT 59407
April 8, 19—

For modified block style, align the heading on the page at the center line. If you are using company letterhead that gives the address, type in only the date, two spaces below the last line of printed copy.

The Inside Address. The recipient's full name and address are given in the inside address. You can begin the inside address two spaces below the date if the letter is long, or four spaces below if the letter is quite short. The inside address should be aligned with the left margin—and the left margin should be at least one inch wide. Include the reader's full name and title (if you know them) and his or her full address, including ZIP code.

EXAMPLE   Ms. Gail Silver
Production Manager
Quicksilver Printing Company
14 President Street
Sarasota, FL 33546

The Salutation. Place the salutation, or greeting, two spaces below the inside address, also aligned with the left margin. In most business letters, the salutation contains the recipient's title (*Mr., Ms., Dr.,* and so on) and last name, followed by a **colon.** If you are on a first-name basis with the recipient, include his or her title and full name on the inside address, but use only the first name in the salutation. Notice that the titles *Mr., Ms., Mrs.,* and *Dr.* may be abbreviated but that other titles, such as *Captain* or *Professor,* should always be spelled out.

EXAMPLES   Dear Ms. Silver:
Dear Dr. Lee:
Dear Captain Ortiz:
Dear Professor Murphy:
Dear George:

Address women without a professional title as *Ms.,* whether they are married or unmarried. If a woman has expressed a preference for *Miss* or *Mrs.,* however, honor her preference. If you do not know whether the recipient is a man or a woman, you may use a title appropriate to the context of the letter. The following are examples of the kinds of titles you may find suitable:

EXAMPLES   Dear Customer: (Letter from a department store)
Dear Homeowner: (Letter from an insurance agent soliciting business)
Dear Parts Manager: (Letter to an auto-parts dealer)

In the past, correspondence to large companies or organizations was customarily addressed to "Gentlemen." Today, however, writers who do not know the name or title of the recipient often address the letter to an appropriate department or identify the subject in a "subject line" and use no salutation.

C

EXAMPLE   National Business Systems
501 West National Avenue
Minneapolis, MN 55407

Attention: Customer Relations Department

I am returning three calculators that failed to operate. . . .

EXAMPLE   National Business Systems
501 West National Avenue
Minneapolis, MN 55407

Subject: Defective Parts for SL-100 Calculators

I am returning three calculators that failed to operate. . . .

When a person's first name could be either feminine or masculine, one solution is to use both the first and the last name in the salutation.

EXAMPLE   Dear Pat Smith:

Avoid "To Whom It May Concern" because it is impersonal and dated.

The Body. The body of the letter should begin two spaces below the salutation (or directly below the heading if there is no salutation). Single-space within paragraphs, and double-space between paragraphs. If a letter is very short and you want to suggest a fuller appearance, you may instead double-space throughout and indicate paragraphs by indenting the first line of each paragraph five spaces from the left. The right margin should be approximately as wide as the left margin. (In very short letters, you may increase both margins to about an inch and a half.)

Two very important elements in the body are the **opening** and **closing.** One effective way to arrange your letter is to open with a short **paragraph,** followed by one or more longer paragraphs for the message and another short paragraph for concluding. This is called a diamond arrangement, and it can help focus your letter whether you are using the direct or indirect pattern. Never underestimate the importance of the opening and closing; in fact, their positions of **emphasis** make them particularly significant.

In your opening you should identify your subject so as to focus its relevance for the reader. Remember that your reader may not immediately recognize or see the importance of your topic—indeed, he or she may be preoccupied with some other business. Therefore, it is important to focus his or her attention on the subject at hand. Be particularly careful to get directly to the point; leave out any less important details.

EXAMPLE Yesterday, I received your letter and the defective tuner, number AJ 50172, that you described. I sent the tuner to our laboratory for tests.

Carol Moore, our lead technician, reports that preliminary tests indicate. . . .

Your closing should let the reader know what he or she should do next or establish goodwill—or often both.

EXAMPLE Thanks again for the report, and let me know if you want me to send you a printout of the tests.

Because a closing is in a position of emphasis, be especially careful to avoid **clichés.** Of course, some very commonly used closings are so precise that they are hard to replace.

EXAMPLES Thank you for your advice.
If you have further questions, please let me know.

The Complimentary Closing. Type the complimentary closing two spaces below the body. Use a standard expression like *Yours truly, Sincerely,* or *Sincerely yours.* (If the recipient is a friend as well as a business associate, you can use a less formal closing, such as *Best wishes* or *Best regards.*) Capitalize only the initial letter of the first word, and follow the expression with a **comma.** Type your full name four spaces below, aligned with the closing at the left. On the next line you may type in your business title, if it is appropriate to do so. Write your signature in the space between the complimentary closing and your typed name. If you are writing to someone with whom you are on a first-name basis, it is acceptable to sign only your given name; otherwise, sign your full name.

EXAMPLE Sincerely,

Thomas R. Castle

Thomas R. Castle
Treasurer

Sometimes the complimentary closing is followed by the name of the firm. Then comes the actual signature, between the name of the firm and the typed name.

C

EXAMPLE   Sincerely yours,
VIKING SUPPLY COMPANY, INC.

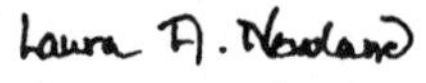

Laura A. Newland

Laura A. Newland
Controller

A Second Page. If a letter requires a second page, always carry at least two lines of the body text over to that page; do not use a continuation page to type only the letter's closing. The second page should be typed on plain paper of quality equivalent to that of the letterhead stationery. It should have a heading with the recipient's name, the page number, and the date. The heading may go in the upper left-hand corner or across the page, as shown in Figure 4.

Additional Information.   Business letters sometimes require the typist's initials, an enclosure notation, or a notation that a copy of the letter is being sent to one or more people. Place any such information at the

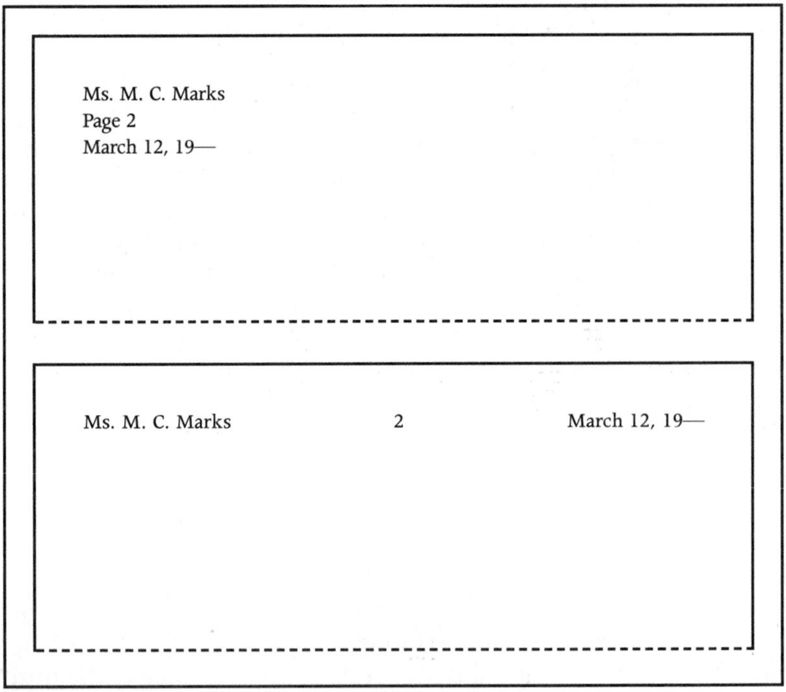

FIGURE 4   Headings for the Second Page of a Letter

left margin, two spaces below the last line of the complimentary closing in a long letter, four spaces below in a short letter.

The *typist's initials* should follow the letter writer's initials, and the two sets of initials should be separated by either a colon or a slash. The writer's initials should be in capital letters, and the typist's initials should be in lowercase letters. (When the writer is also the typist, no initials are needed.)

EXAMPLES   CBG:pbg
APM/sjl

Enclosure notations indicate that the writer is sending material along with the letter (an invoice, an article, and so on). They may take several forms, as illustrated below; choose the form that seems most helpful to your readers.

EXAMPLES   Enclosure: Preliminary report invoice
Enclosures (2)
Enc. (Encs.)

Enclosure notations are included in long, formal letters or in any letters where the enclosed items would not be obvious to the reader. Remember, though, that an enclosure notation cannot stand alone. You must mention the enclosed material in the body of the letter.

Copy notations tell the reader that a copy of the letter is being sent to one or more named individuals. You should add a copy notation for both carbon copies and photocopies.

EXAMPLE   cc: Ms. Marlene Brier
Mr. David Williams
Ms. Bonnie Ng
Ms. Robin Horton

A business letter may, of course, contain all three items of additional information.

EXAMPLE   Sincerely yours,

Jane T. Rogers

Jane T. Rogers
JTR/pst
Enclosure: Preliminary report invoice
cc: Ms. Marlene Brier
Mr. David Williams
Ms. Bonnie Ng
Ms. Robin Horton

C

Letterhead	**Parkside Office Machines Co.** **123 Oceanview Drive** **Seattle, WA 98002**
Date	April 2, 19—
Inside address	Ms. Judith Sparks Technical Director Components Division Parkside Office Machines Co. Pines, NJ 04113
Salutation	Dear Judy:
Body	We've discovered a problem with our current block size restriction for the Decade 2000 computer. Many of our internal and customer publications on utility routines state that the Decade 2000 computer is normally restricted to a maximum block length of 4023 characters, but that this restriction can be lifted by setting Flag 20. This statement is misleading, and I am beginning to wonder if we should take a different approach. As I understand it, the problem is a hardware restriction in that the processor was wired to accommodate a maximum block length of 4023 characters. This restriction created a probelm for FRAN users because a FRAN track could accept a block of 4623 characters. The solution was to make a wiring change to make use of the first bit in the second "TA" character of control words, thereby doubling the maximum block length to 8046. This change was made about two years ago, but existing systems were not modified. Consequently, if a user who has one of the old systems sets Flag 20, the restriction is not going to be lifted as we say it will. A couple of solutions quickly come to mind. We could instruct the user to set the Flag only if his system can accommodate an 8046-character block (if we assume that he has this information). We could give him the change number of the update to the processor and instruct him to use Flag 20 only if his processor has that number or greater. I propose that we use the second solution because the user is more likely to know his change number than whether he can accommodate an 8046-character block. Since the customer is directly involved, the solution to the problem should probably be approved by your department. Please let me know as quickly as possible whether you approve of this solution because some of our customer manuals are presently being revised.
Complimentary close	Sincerely yours,
Signature	Gerald Stein
Typed name	Gerald Stein
Title	
Additional information	GS/bb

FIGURE 5 Business Letter (Full Block) with Standard Elements

Letter Formats

The two most common formats of business letters are the full block style (shown in Figure 5) and the modified block style (shown in Figure 6). The full block style, though easier to type because every line begins at the left margin, is suitable only with business letterhead stationery. In the modified block style, the return address, the date, and the complimentary closing all begin at the center of the page, and the

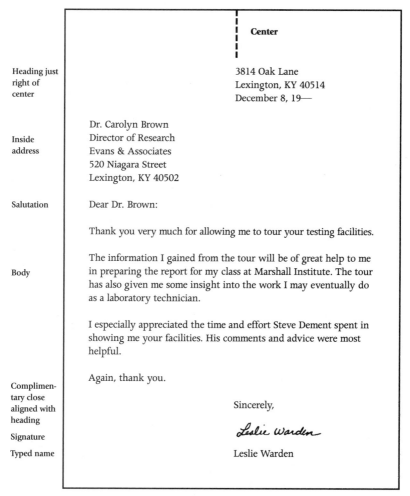

Center

Heading just right of center

3814 Oak Lane
Lexington, KY 40514
December 8, 19—

Inside address

Dr. Carolyn Brown
Director of Research
Evans & Associates
520 Niagara Street
Lexington, KY 40502

Salutation

Dear Dr. Brown:

Thank you very much for allowing me to tour your testing facilities.

Body

The information I gained from the tour will be of great help to me in preparing the report for my class at Marshall Institute. The tour has also given me some insight into the work I may eventually do as a laboratory technician.

I especially appreciated the time and effort Steve Dement spent in showing me your facilities. His comments and advice were most helpful.

Again, thank you.

Complimentary close aligned with heading

Sincerely,

Signature

Leslie Warden

Typed name

Leslie Warden

FIGURE 6 Modified Block Style

other elements are aligned at the left margin. All other letter styles are variations of these two styles. Again, if your employer recommends or requires a particular style, follow it carefully. Otherwise, choose a style and follow it consistently.

For information on specific types of correspondence, see **acceptance letters, acknowledgment letters, adjustment letters, application letters, complaint letters, inquiry letters, reference letters, refusal letters,** and **resignation letter or memorandum.**

cover letters

C

The *cover letter* identifies the item being sent, the person to whom it is being sent, and the reason for sending it. A cover letter provides a permanent record of the transmittal for both the writer and the **reader.**

Keep your remarks brief in a cover letter. Open with a **paragraph** explaining what is being sent and why. In an optional second para-

---

WATERFORD PAPER PRODUCTS

P.O. BOX 413
WATERFORD, WI 53474

(414) 738-2191

November 12, 19—

Ms. Nancy Melcher
2021 State Street
Racine, WI 53307

Dear Ms. Melcher:

Thank you for your interest in Waterford Paper Products. The enclosed brochure describes our product line and our current prices.

If I can be of future help, please call me at extension 2349.

Sincerely,

Lawrence Smith

Lawrence Smith
Customer Relations

LS/ik
Enclosure

FIGURE 1 Brief Cover Letter

C

WATERFORD PAPER PRODUCTS

P.O. BOX 413
WATERFORD, WI 53474

(414) 738-2191

January 16, 19—

Mr. Roger Hammersmith
Ecology Systems, Inc.
1015 Clarke Street
Chicago, IL 60615

Dear Mr. Hammersmith:

Enclosed is the report estimating our power consumption for the year as requested by John Brenan, Vice President, on September 4.

The report is a result of several meetings with the Manager of Plant Operations and her staff and an extensive survey of all our employees. The survey was delayed by the temporary layoff of key personnel in Building "A" from October 1 to December 5. We believe, however, that the report will provide the information you need to furnish us with a cost estimate for the installation of your Mark II Energy Saving System.

We would like to thank Diana Biel of ESI for her assistance in preparing the survey. If you need any more information, please let me know.

Sincerely,

James G. Evans
New Projects Office

JGE/fst
Enclosure

FIGURE 2   Cover Letter

graph, you might include a summary of any information you're sending. A letter accompanying a **proposal,** for example, might point out any sections in the proposal of particular interest to the reader. The letter could then go on to present evidence that the writer's firm is the best one to do the job. This paragraph could also mention the conditions under which the material was prepared, such as limitations of time or budget. The closing paragraph should contain acknowledgments, offer additional assistance, or express the hope that the material will fulfill its purpose.

The first of the above examples (Figure 1) is brief and to the point. Figure 2 is a bit more detailed, for it touches on the manner in which the information was gathered.

credible/creditable

Something is *credible* if it is believable.

> EXAMPLE   The statistics in this report are *credible*.

Something is *creditable* if it is worthy of praise or credit.

> EXAMPLE   The chief engineer did a *creditable* job.

criterion/criteria

Criterion means "an established standard for judging or testing." *Criteria* and *criterions* both are acceptable plural forms of *criterion*, but *criteria* is generally preferred.

> EXAMPLE   In evaluating this job, we must use three *criteria*. The most important *criterion* is quality of workmanship.

critique

A *critique* **(noun)** is a written or oral evaluation of something. Avoid using *critique* as a **verb** meaning "criticize."

> CHANGE   Please *critique* his job description.
>   TO   Please prepare a *critique* of his job description.

dangling modifiers

Phrases that do not clearly and logically refer to the proper **noun** or **pronoun** are called *dangling modifiers*. Dangling modifiers usually appear at the beginning of a **sentence** as an introductory phrase.

> CHANGE *While eating lunch in the cafeteria,* the computer malfunctioned. (The problem, of course, is that the sentence neglects to mention who was eating lunch in the cafeteria.)
>
> TO While *the operator* was eating lunch in the cafeteria, the computer malfunctioned.

They can, however, appear at the end of the sentence as well.

> CHANGE The program gains in efficiency *by eliminating the superfluous instructions.* (An action is stated, but no one is identified who could perform the stated action.)
>
> TO The program gains in efficiency *when you* eliminate the superfluous instructions.

Dangling modifiers are often caused by overuse of the passive **voice.** In an active-voice clause, the doer of the action appears in front of the verb; since the doer of the action is normally the intended subject of the introductory phrase, the phrase does not dangle. If the clause is in the passive voice, however, the doer of the action appears after the verb or is not identified, thereby causing the phrase to dangle.

> CHANGE To evaluate the feasibility of the project, the centralized plan will be compared with the present system of dispersing facility sites. (The main clause is in the passive voice, and the doer of the action expressed by the **verb** *compared* is not identified; therefore, the introductory phrase "To evaluate the feasibility of the project" has nothing that it can logically modify because there is no one in the main clause who could possibly evaluate anything.)

TO	To evaluate the feasibility of the project, *the committee* will compare the centralized plan with the present system of dispersing facility sites.

To test whether a phrase is a dangling modifier, turn it into a **clause** with a subject and a verb. If the expanded phrase and the independent clause do not have the same subject, the phrase is dangling.

CHANGE	After finishing the research, the paper was easy to write. (The implied subject of the phrase is intended to be *I*, and the subject of the independent clause is *paper*; therefore, the sentence contains a dangling modifier.)
TO	After finishing the research, *I found that* the paper was easy to write. (Changing the subject of the independent clause to agree with the implied subject of the introductory phrase eliminates the dangling modifier.)
OR	After I finished the research, the paper was easy to write. (This version changes the phrase to a dependent clause with an explicit subject.)

dashes

The *dash* is a versatile, yet limited, mark of **punctuation.** It is versatile because it can perform all the duties of punctuation (to link, to separate, and to enclose). It is limited because it is an especially emphatic mark that is easily overused. Use the dash cautiously, therefore, to indicate more informality or **emphasis** (a dash gives an impression of abruptness) than would be achieved by the conventional punctuation marks. In some situations, a dash is required; in others, a dash is a forceful substitute for other marks.

A dash can emphasize a sharp turn in thought.

EXAMPLE	The project will end January 15—unless the company provides additional funds.

A dash can indicate an emphatic pause.

EXAMPLES	Consider the potential danger of a household item that contains mercury—a very toxic substance.
	The job will be done—after we are under contract.

Sometimes, to emphasize contrast, a dash is also used with *but*.

EXAMPLE	We may have produced work more quickly—but the result was not as good.

A dash can be used before a final summarizing statement or before **repetition** that has the effect of an afterthought.

EXAMPLE   It was hot near the ovens—steaming hot.

Such a thought may also complete the meaning.

EXAMPLE   We try to speak as we write—or so we believe.

A dash can be used to set off an explanatory or **appositive** series.

EXAMPLE   Three of the applicants—John Evans, Mary Fontana, and Thomas Lopez—seem well qualified for the job.

Dashes set off parenthetical elements more sharply and emphatically than do **commas.** Unlike dashes, **parentheses** tend to reduce the importance of what they enclose. Compare the following sentences:

EXAMPLES   Only one person—the president—can authorize such activity.
Only one person, the president, can authorize such activity.
Only one person (the president) can authorize such activity.

Use dashes for **clarity** when commas appear within a parenthetical element; this avoids the confusion of too many commas.

EXAMPLE   Retinal images are patterns in the eye—made up of light and dark shapes in addition to areas of color—but we do not see patterns; we see objects.

The first word after a dash is never capitalized unless it is a proper noun.

data/datum

In much technical writing, *data* is considered a collective singular. In formal scientific and scholarly writing, however, *data* is generally used as a plural, with *datum* as the singular form.

EXAMPLES   The *data are* voluminous in support of a link between smoking and lung cancer. (formal)
The *data is* now ready to be evaluated. (less formal)

dates

In business and industry, *dates* have traditionally been indicated by the month, day, and year, with a **comma** separating the figures.

EXAMPLE   October 26, 19—

The day–month–year system used by the military does not require commas.

EXAMPLE   26 October 19—

A date can be written with or without a comma following the year if the date is in the month–day–year format.

> EXAMPLES   October 26, 19—, was the date the project began.
> October 26, 19— was the date the project began.

D

If the date is in the day–month–year format, do not set off the date with commas.

> EXAMPLE   The project began on 26 October 19—.

The strictly numerical form for dates (10/26/97) should be used sparingly, and never on business letters or formal documents, since it is less immediately clear. When this form is used, the order in American usage is always month/day/year. For example, 5/7/97 is May 7, 1997.

Centuries

Confusion often occurs because the spelled-out names of centuries do not correspond to the numbers of the years. (See also **numbers.**)

> EXAMPLE   The twentieth century is the 1900s (1900–1999).

When the century is written as a **noun,** do not use a **hyphen.**

> EXAMPLE   The nineteenth century produced many great inventions.

When the century is written as an **adjective,** however, use a hyphen.

> EXAMPLE   Twenty-first-century technology must rely on clear technical communications.

decided/decisive

When something is *decided*, it is "clear-cut," "without doubt," "unmistakable." When something is *decisive*, it is "conclusive" or "has the power to settle a dispute."

> EXAMPLE   The executive committee's *decisive* action gave our firm a *decided* advantage over the competition.

decreasing-order-of-importance method of development

Decreasing order of importance is a **method of development** in which the writer's major points are arranged in descending order of importance. It begins with the most important fact or point, then goes

INTEROFFICE MEMORANDUM

To: Mary Vincenti, Chief, Personnel Department
From: Frank W. Russo, Chief, Claims Department *FWR*
Date: November 13, 19—
Subject: Selection of Chief of the Claims Processing Section

I feel there are three possible candidates for the position of Chief of the Claims Processing Section.

The most qualified candidate for Chief of the Claims Processing Section is Mildred Bryand, who is at present Acting Chief of the Claims Processing Section. In her twelve years in the Claims Department, Ms. Bryand has gained wide experience in all facets of the department's operations. She has maintained a consistently high production record and has demonstrated the skills and knowledge that are required for the supervisory duties she is now handling in an acting capacity. A number of additional qualifications also make her the best candidate: her valuable contributions to and cooperation during the automation of the department's claims records; her performance on several occasions as acting chief in other sections of the Claims Department; the continual rating of "outstanding" she has received in all categories of her job-performance appraisals.

Michael Bastik, Claims Coordinator, our second choice, also has strong potential for the position. An able administrator, he has been with the company for seven years. For the past year he has been enrolled in several management training courses at the university. He is ranked second, however, because he lacks supervisory experience and because his most recent work has been with the department's maintenance and supply components. He would be the best person to fill many of Mildred Bryand's responsibilities if she should be made full-time Chief of the Claims Processing Section.

Jane Fine, our third-ranking candidate, has shown herself a skilled administrator in her three years with the Claims Consideration Section. Despite her obvious potential, my main objection to her is that compared with the other top candidates, she lacks the breadth of experience in claims processing that would be required of someone responsible for managing the Claims Processing Section. Jane Fine also lacks on-the-job supervisory experience.

Please let me have your recommendations on filling this position at your earliest convenience.

mo

FIGURE 1   Decreasing Order of Importance

to the next most important, and so on, ending with the least important. It is an especially appropriate method of development for a **report** addressed to a busy decision maker, who may be able to reach a decision after considering only the most important points, or for a report addressed to various **readers,** some of whom may be interested only in the major points and others in all of the points.

The advantages of this method of development are that (1) it gets the reader's attention immediately by presenting the most important point first, (2) it makes a strong initial impression, and (3) it ensures that the hurried reader will not miss the most important point.

The example shown in Figure 1 uses decreasing order of importance.

D

de facto/de jure

De facto means that something "exists" or is "a fact" and therefore is accepted for practical purposes. *De jure* means that something "legally" exists.

> **EXAMPLE** The law states that no signs should be erected along Highway 127. Store owners have disregarded this law, and many signs exist along Highway 127. The presence of the signs along Highway 127 is *de facto* but not *de jure*—their presence is "a fact," but it is not "lawful."

defective/deficient

If something is *defective,* it is faulty.

> **EXAMPLE** The wiring was *defective.*

If something is *deficient,* it lacks a necessary ingredient.

> **EXAMPLE** The compound was found to be *deficient* in calcium.

defining terms

One of the most basic rules of good writing is to be sure that your **readers** understand and follow what they are reading. You must always keep in mind their familiarity with the topic and be careful to define any terms that they may not understand.

Terms can be defined either formally or informally, depending on your purpose and on your readers. (For techniques on expanding a definition to explain the meaning of a term, see **definition method of development.**)

A *formal definition* is a form of classification. A term is defined by placing it in a category and then identifying what features distinguish it from other members of the same category.

Term	Category	Distinguishing Features
An auction is	a public sale	in which property passes to the highest bidder through successively increased offers.
A contract is	a binding agreement	between two or more persons or parties.
A lease is	a contract	that conveys real estate for a term of years at a specified rent.

D

An *informal definition* explains a term by giving a familiar word or **phrase** as a **synonym.**

EXAMPLES An *invoice* is a bill.

Many states have set up wildlife *habitats* (or living spaces).

Plants have a *symbiotic,* or mutually beneficial, relationship with certain kinds of bacteria.

Definitions should normally be stated positively; focus on what the term is rather than on what it is not. For example, "In a legal transaction, real property is not personal property," although a true statement, does not tell the reader exactly what real property is. The definition could just as easily be stated positively: "Real property is legal terminology for the right or interest a person has in land and the permanent structures on that land." (For a discussion of when negative definitions are appropriate, see **definition method of development.**)

Avoid circular definitions, which merely restate the term to be defined and therefore fail to clarify it.

CIRCULAR *Spontaneous combustion* is fire that begins spontaneously.
REVISED *Spontaneous combustion* is the self-ignition of a flammable material through a chemical reaction like oxidation and temperature buildup.

Avoid "is when" and "is where" definitions. Such definitions fail to include the category and are too indirect: They say when or where rather than what.

"IS WHEN" A *contract* is when two or more people agree to something.
REVISED A *contract* is a binding agreement between two or more people.
"IS WHERE" A *day-care center* is where working parents can leave their preschool children during the day.
REVISED A *day-care center* is a facility at which working parents can leave their preschool children during the day.

Avoid definitions that include terms that may be unfamiliar to your readers. Even informal writing will occasionally require terms used in a special sense that your readers may not understand; such terms should always be defined.

> **EXAMPLE** In these specifications the term *safety can* refers to an approved container of not more than five-gallon capacity having a spring-closing spout cover designed to relieve internal pressure when the can is exposed to fire.

definite/definitive

Definite and *definitive* both apply to what is "precisely defined," but definitive more often refers to what is complete and authoritative.

> **EXAMPLE** When the committee took a *definite* stand, the president made it the *definitive* company policy.

definition method of development

To define something is basically to identify precisely its fundamental qualities. In technical writing, as in all writing, clear and accurate definitions are critical. Sometimes simple, dictionary-like definitions suffice (see **defining terms**), but at other times definitions need to be expanded with additional details, examples, comparisons, or other explanatory devices.

When more than a **phrase** or a sentence or two is needed to explain an idea, use an extended definition, which explores a number of qualities of the item being defined. How an extended definition is developed depends on your reader's needs and on the complexity of the subject. A **reader** familiar with a **topic** or an area might be able to handle a long, fairly complex definition, whereas one less familiar with a topic would require simpler language and more basic information.

Probably the easiest way to give an extended definition is with specific examples. Listing examples gives your reader easy-to-picture details that help him or her to see and thus to understand the term being defined.

> **EXAMPLE** Form, which is the shape of landscape features, can best be represented by both small-scale features, such as *trees and shrubs,* and by large-scale elements, such as *mountains and mountain ranges.*

Another useful way to define a difficult concept, especially when you are writing for nonspecialists, is to use an **analogy** to link the unfamiliar concept with a simpler or more familiar one. An analogy is a resemblance in certain aspects between things otherwise unalike, and it can help the reader understand an unfamiliar term by showing its similarities with a more familiar one. Defining radio waves in terms of the length (long) and frequency (low), a writer develops an analogy to show why a low frequency is advantageous.

> EXAMPLE The low frequency makes it relatively easy to produce a wave having virtually all its power concentrated at one frequency. Think, for example, of a group of people lost in a forest. If they hear sounds of a search party in the distance, they all will begin to shout for help in different directions. Not a very efficient process, is it? But suppose that all the energy which went into the production of this noise could be concentrated into a single shout or whistle. Clearly the chances that the group will be found would be much greater.

Some terms are best defined by an explanation of their causes. Writing in a professional journal, a nurse describes an apparatus used to monitor blood pressure in severely ill patients. Called an in-dwelling catheter, the device displays blood-pressure readings on an oscilloscope and on a numbered scale. Users of the device, the writer explains, must understand what a *dampened wave form* is.

> EXAMPLE The *dampened wave form,* the smoothing out or flattening of the pressure wave form on the oscilloscope, is *usually caused by an obstruction* that prevents blood pressure from being freely transmitted to the monitor. The obstruction may be a *small clot or bit of fibrin* at the catheter tip. More likely, *the catheter tip* has become *positioned against the artery wall* and is preventing the blood from flowing freely.

Sometimes a formal definition of a concept can be made simpler by breaking the concept into its component parts. In the following example, the formal definition of *fire* is given in the first paragraph, and the component parts are given in the second:

> EXAMPLE Fire is the visible heat energy released from the rapid oxidation of a fuel. A substance is "on fire" when the release of heat energy from the oxidation process reaches visible light levels.
>
>   The classic fire triangle illustrates the elements necessary to create fire: *oxygen, heat,* and *burnable material* or *fuel.* Air provides sufficient oxygen for combustion; the intensity of the heat needed to start a fire depends on the characteristics of the burnable material or fuel. A burnable substance is one that

will sustain combustion after an initial application of heat to start it.

Under certain circumstances, the meaning of a term can be clarified and made easier to remember by an exploration of its origin. Scientific and medical terms, because of their sometimes unfamiliar Greek and Latin roots, benefit especially from an explanation of this type. Tracing the derivation of a word can also be useful when you want to explain why a word has favorable or unfavorable associations—particularly if your goal is to influence your reader's attitude toward an idea or an activity.

EXAMPLE   Efforts to influence legislation generally fall under the head of *lobbying*, a term that once referred to people who prowl the lobbies of houses of government, buttonholing lawmakers and trying to get them to take certain positions. Lobbying today is all of this, and much more, too. It is a respected—and necessary—activity. It tells the legislator which way the winds of public opinion are blowing, and it helps inform him of the implications of certain bills, debates, and resolutions he must contend with.
—Bill Vogt, *How to Build a Better Outdoors*

Sometimes it is useful to point out what something is not in order to clarify what it is. A negative definition is effective only when the reader is familiar with the item with which the defined item is contrasted. If you say x is not y, your readers must understand the meaning of y for the explanation to make sense. In a crane operators' manual, for instance, a negative definition is used to show that for safety reasons, a hydraulic crane cannot be operated in the same manner as a lattice boom crane.

EXAMPLE   A hydraulic crane is *not* like a lattice boom crane in one very important way. In most cases, the safe lifting capacity of a lattice boom crane is based on the *weight needed to tip the machine.* Therefore, operators of friction machines sometimes depend on signs that the machine might tip to warn them of impending danger.
    This practice is very dangerous with a hydraulic crane. . . .
—*Operator's Manual* (Model W-180), Harnishfeger Corporation.

demonstrative adjectives (see adjectives)

demonstrative pronouns (see pronouns)

dependent clauses (see **clauses**)

description

Effective *description* uses words to transfer a mental image from the writer's mind to the reader's. The key to effective description is the accurate presentation of all pertinent details.

In describing a mechanical device, it is best to describe the whole device before getting into a detailed description of how the parts work together. The description should conclude with an explanation of the way the parts work together to get their particular jobs done.

The following **paragraph** describes a mechanical assembly. Intended for the assembler/repairman, this description includes an illustration of the assembly mechanism (Figure 1).

Number of
features and
relative size
differences
noted

The Die Block Assembly consists of two machined block sections, eight Code Pins, and a Feed Pin. The larger section, called the Die Block, is fashioned of a hard, noncorrosive beryllium-copper alloy. It houses the eight Code Pins and the smaller Feed Pin in nine finely machined guide holes. The guide holes

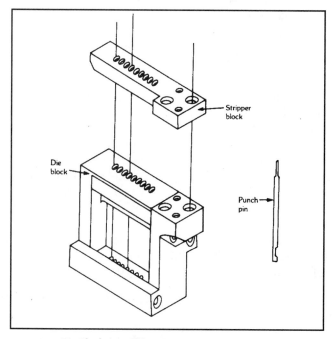

FIGURE 1 Die Block Assembly

D

Use of
analogy

Definition

at the upper part of the Die Block are made smaller to conform to the thinner tips of the Feed Pins. Extending over the top of the Die Block and secured to it at one end is a smaller, arm-like block called the Stripper Block. The Stripper Block is made from hardened tool steel, and it also has been drilled through with nine finely machined guide holes. It is carefully fitted to the Die Block at the factory so that its holes will be precisely above those in the Die Block and so that the space left between the blocks will measure .015" (plus or minus .003"). The residue (chad) from the operation is pushed out through the top of the Stripper Block and guided out of the assembly by means of a plastic Residue Collector and Residue Collector Extender.

Note that this description concentrates on the number of pieces—their sizes, shapes, dimensions—and on their relationship to one another to perform their function. It also specifies the materials of which the hardware is made. Because the description is illustrated (with identifying captions) and is intended for technicians who have been trained on the equipment, it does not require the use of "bridging devices" to explain the unfamiliar in relation to the familiar. Even so, an important term is defined (chad), and crucial alignment dimensions are specified (plus or minus .003").

For descriptions intended for readers unfamiliar with a topic, more details are needed. This type of description benefits from showing or demonstrating (as opposed to telling), primarily through the use of images and details.

This type of description can use analogy to explain unfamiliar concepts in terms of familiar concepts. For example, the plastic device that holds the letters and symbols for computer printers is a wheel-shaped piece of plastic, with each letter and symbol positioned at the circumference of the wheel and attached to the hub at the center of the wheel by thin plastic strips. Using analogy, you can think of the strips as petals radiating from the center of a flower. Because the strips are uniform in length and width, similar to the petals of a daisy, the print elements are commonly referred to, in an extension of the flower analogy, as daisy wheel print elements.

The following example, which was written when computers still used punched cards, skillfully uses images and details from everyday life (the familiar) as a bridge to the unfamiliar.

Familiar
(print/page)

When information appears in print, as on this page, people like you and me are able to read and understand it. However, if information is to be processed by a machine, like a computer, then other ways must be found to put these same letters and

Unfamiliar (machine)	words into the machine. Computers, of course, do not have eyes like humans—but they do have electrical sensing equipment that in certain ways does almost the same thing.
Familiar (doors at market)	For example, most of us have walked into supermarkets through doors that open by themselves. These doors are controlled by electrical sensing equipment known as photoelectric cells, which act like eyes. Each door is controlled by two photoelectric cells that shine a beam of light to each other. As you walk through the door, the light beam is broken causing the
Unfamiliar (photoelectric cell)	photoelectric cells or "eyes" to sense that someone is beginning to enter. The electric eye reacts in a split second by sending a burst of electrical energy to the door's mechanical hinge, and this automatically swings the door open.
Transitional ideas (beam of light/hole in card)	The photoelectric sensing idea can also be used to detect the presence of a hole in a card or piece of paper.
New concept (Machine "reading" text)	Thus, if holes are punched in paper to represent certain letters of the alphabet or words, it is possible for a machine to electrically sense or read this information.

—Joseph Becker, *The First Book of Information Science*

design and layout

Good visual design is crucial to the success of a document, whether it is a simple letter or an elaborate product of desktop publishing. A well-designed document should make even complex information look accessible and give **readers** a favorable impression of the writer and the organization. To accomplish those goals, a design should

- be visually simple and uncluttered
- highlight structure, hierarchy, and order
- help readers find information they need
- help reinforce an organization's image

Good design achieves visual simplicity by establishing compatible or harmonious relationships, such as using the same family of type in a document or the same highlighting device for similar items. Useful design reduces the complexity of prose, for example, by creating "chunks" of information that can be quickly noticed and absorbed. Effective design also reveals hierarchy by signaling the difference between **topics** and subtopics, between primary and secondary information, and between general points and examples. Visual cues make information easy to find the first and subsequent times it is sought. They also make information accessible to different readers within an audience by allowing them to locate the information they need. Finally, the design of a document should project the appropriate image of an organization. Good design can build both goodwill for the writer and the organization and respect for a company's products or services.

You can achieve effective design and layout through your selection of type for the text, your choice of devices that highlight information, and your arrangement of text and visual components on a page.

Using Typography

A complete set of all the words, **numbers,** and **symbols** available in one typeface (or style) is called a font.

The letters comprising a font have a number of distinctive characteristics, some of which are shown in Figure 1.

There are key elements that give a typeface both its style and its legibility.

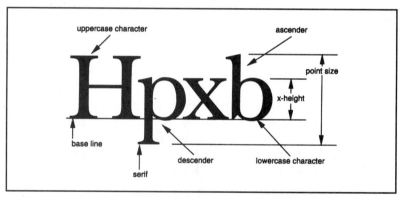

FIGURE 1   Primary Components of Letter Characters

Selecting a Typeface.   For most on-the-job writing, select a typeface primarily for its legibility, which is the speed with which readers can recognize each letter and word. Avoid typefaces that may distract readers with contrasts in thickness or with odd features, as is often the case with script, or cursive, typefaces. In addition, avoid typefaces that fade when printed or copied. Choose popular typefaces with which readers are familiar, such as the following:

- Baskerville
- Bodoni
- Caslon
- Century
- Futura
- Garamond
- Gill Sans
- Helvetica
- Times New Roman
- Univers

Avoid more than two typeface styles in a document, even though you may have access to many more. To create a dramatic contrast between **headings** and text, as in a newsletter, use a typeface that is distinctively different from the text. You may also use a noticeably different typeface inside a graphic element. In any case, experiment before making final decisions, and keep in mind that not all fonts have the same assortment of symbols and other characters.

One basic distinction of typefaces is *serif* and *sans serif* type. Serifs, shown in Figure 1, are the small projections at the end of each stroke in a letter. Serif type styles have these lines; sans (French for "without") serif styles do not. Although sans serif type has a modern look, serif type is easier to read, especially in the smaller sizes. Sans serif, however, does work well for headings. If you do choose sans serif typefaces for text, pick one with large "counters," the fully or partially enclosed white spaces in letters like *c* and *b*.

Do not use a type that is too small for text because it will cause eye strain and make the text look crammed and uninviting. Six-point type is the smallest that can be read without a magnifying glass (see Figure 2). Type that is too large may use more space than necessary, can make reading difficult and inefficient, and can make readers perceive words in parts rather than wholes.

6 pt.	Type size can determine legibility.
8 pt.	Type size can determine legibility.
10 pt.	Type size can determine legibility.
12 pt.	Type size can determine legibility.
14 pt.	Type size can determine legibility.

FIGURE 2   Samples of 6- to 14-Point Type

Ideal point sizes for text range from 8 to 12 points (see Figure 2); 10-point type is most commonly used. Print samples of text in various type sizes and typefaces—and trust your reaction to them.

The distance from which a document will be read should help determine type size. For example, choose a larger typeface for a set of **instructions** that will be placed on a table while the user of the instructions stands. Consider the audience's age too. Visually impaired readers and some older adults may need large type sizes.

Adjusting Line Length and Leading. Designers sometimes use the following rule of thumb for determining optimal line length in typeset copy:

D

Double the point size of the typeface for line length in picas. (There are six picas to an inch.) Thus, if the type size is 12 points, the line length should be 24 picas (or 4 inches). Other designers suggest that a maximum of 10 to 12 words on a line is acceptable for many typefaces, depending on the amount of white space on the page, as well as other elements discussed later in this entry.

Leading refers to the space between lines. Leading should be proportional to line length and point size—about 20 percent of the point size. The table in Figure 3 shows standard type sizes with recommended proportional line lengths and leading used in commercial publishing. Ranges are given (in **parentheses**) because some typefaces can affect the leading you should use. You may also use extra leading between **paragraphs** as well as indention for paragraphs to achieve maximum readability.

Type Size (in points)	Line Length		Leading (in points)
	(in picas)	(in inches)	
6	12	2	1(0–1)
8	16	$2^2/_3$	1½ (0–2)
9	18	3	2(0–3)
10	20	$3^1/_3$	2(0–3)
11	22	$3^2/_3$	2(1–3)
12	24	4	3(2–4)

FIGURE 3   Optimal Type Sizes, Line Lengths, and Leading

Choosing Justified or Ragged Right Margins. Ragged right margins are generally easier to read than justified right margins because the uneven contour of the right margin provides the eyes with more landmarks to identify. Ragged right may also be preferable if justification or proportional spacing in your **word processing** software inserts irregular-sized spaces between words, producing unwanted white space or unevenness in blocks of text. Do not justify short lines; doing so will leave huge gaps between words.

Because ragged right margins look slightly informal, justified text may be more appropriate for publications aimed at a broad readership that expects a more formal, polished appearance. Further, justification is often useful with multiple-column formats because the spaces between the columns (called *alleys*) need the definition that justification provides.

Using Highlighting Devices

Writers use a number of means to emphasize important words, passages, and sections within documents:

D

- typographical devices
- headings and captions
- **headers and footers**
- rules and boxes
- icons and pictograms
- color and screening

When used thoughtfully, such highlighting and finding devices give a document a visible sense of **logic** and **organization.** For example, they can set off steps or **illustrations** from surrounding explanations.

Keep in mind also that typographical devices and special graphic effects should be used in moderation. Just because a feature is available on your program does not mean that you should use it. In fact, too many design devices clutter a page and interfere with comprehension. Check for clutter by simply counting the number of visual elements in a text and their frequency.

Consistency is important. When you choose a highlighting technique to designate a particular feature, always use the same technique for that feature throughout your document.

Typographical Emphasis.   One method of typographical emphasis is the use of uppercase or **capital letters.** BUT ALL UPPERCASE LETTERS ARE DIFFICULT TO READ AT LONG STRETCHES BECAUSE THEIR UNIFORMITY OF SIZE AND SHAPE DEPRIVES READERS OF IMPORTANT VISUAL CLUES AND THUS SLOWS READING. Letters in lowercase have ascenders and descenders (see Figure 1) that make the letters distinctive and easy to identify; therefore, a mixture of uppercase and lowercase is most readable. Use all uppercase letters only in short spans—three or four words, as in headings.

As with all uppercase letters, use **italics** sparingly. *Continuous italic type reduces legibility and thus slows readers because they must expend extra effort.* Of course, italics may be useful where you wish to slow readers, as in cautions and warnings. Boldface, used to cross-reference entries in this handbook, may be the best cuing device because it is visually different yet retains the customary shape of the letters and numbers.

Headings and Captions. Headings function as signposts that give readers a sense of what is covered in a section of a document. **Headings** also reveal the organization of the document and indicate hierarchy within

D

it. Many readers scan a document's headings before reading any text. Headings should therefore help readers decide which sections they need to read. Using too few headings forces readers to work to find their way; conversely, using too many headings can confuse readers and make a document look like an **outline.** Captions are key words that highlight or describe **illustrations** or blocks of text. Captions often appear in the left or right margins of textbooks to summarize passages. Keep in mind that they can be easily overused and do not emphasize hierarchy.

Numerous heading formats are possible, and organizations often set standards to ensure a consistent look for headings in their documents. Headings may appear in many typeface variations (boldface being most common) and often use sans serif styles. The most common positions for headings and subheadings are centered, flush left, indented, or by themselves in a wide left margin. Major section or chapter headings normally appear at the top of a new page. Never leave a heading on the final line of a page. Instead, carry it over to the start of the next page. Insert one additional line of space or extra leading above a heading to emphasize that it marks a logical division.

Headers and Footers. A *header* contains the identifying information carried at the top of each page; a *footer* contains a similar information at the bottom of each page. Pages may have one or the other or both. Although practices vary, **headers and footers** carry such information as the **topic** or subtopic of a section, identifying **numbers,** a document date, page numbers, document name, or a section **title.** Although headers and footers are important reference devices, too much information in them can create visual clutter.

Rules, Icons, and Color. *Rules* are vertical or horizontal lines used to divide one area of the page from another, such as setting off headers and footers from the rest of the page. Rules can also be combined to box off elements on the page. Do not overuse rules; too many rules and boxed elements can create a cluttered look.

An *icon* is a pictorial representation of an idea; examples include the **symbols** used to designate men's and women's restrooms, wheelchair accessibility, and parking for the handicapped.

In documents, icons are useful to identify sections or actions and to warn readers of danger. To be effective, icons must be appropriate, simple, and intuitively recognizable—or at least easy to define. Icons can be placed in headers, in footers, next to headings, or in the open left column of a page.

Color and screening can distinguish one part of a document from another or unify a series of documents. (Screening refers to shaded areas on a page.) Color and screening can set off sections within a document, highlight examples, or emphasize warnings.

D

Designing the Page

Page design is the process of combining the various design elements on a page. The flexibility and lavishness of your design will be based on the capabilities of your word-processing or desktop publishing system, how the document will be reproduced, and the budget available. (Color printing, for example, is far more expensive than black-and-white printing.)

Thumbnail Sketch.   Before you put actual text and visuals on the screen, you may wish to create a *thumbnail sketch*. To create a thumbnail sketch, use a page that is the actual size of the finished document and sketch blocks that indicate the placement of elements. You can go further by laying out a rough assembly of all the thumbnail pages showing size, shape, form, and general style of a publication. This mock-up, called a *dummy,* allows you to see how a publication will look. As you work with elements on the page, experiment with different designs. Often what seems a useful concept in principle turns out not to work in practice.

Defining Columns.   As you design pages, consider the size and number of columns. Figure 4 shows eight ways of placing text on a page. Pattern *A* provides maximum text; patterns *B* and *C* provide more white space; patterns *D* and *E* combine maximum text and readability; and patterns *F, G,* and *H* provide ample illustrative space. For traditionally typewritten material, such as **reports,** the maximum block (pattern *A*) is acceptable if double-spaced. For fairly solid prose set in type, the traditional two-column structure (pattern *D*) enhances legibility by keeping text columns narrow enough so that readers need not scan back and forth across the width of the entire page for every line.

Avoid both widows and orphans. A word at the end of a column is called a *widow.* This term is also used for carried-over letters of hyphenated words. If a widow is carried over to the top of the next column or page, it is called an *orphan.*

Using White Space.   White space makes difficult subjects seem easier to comprehend by visually framing information and breaking it into manageable chunks. Even white space between paragraphs helps readers

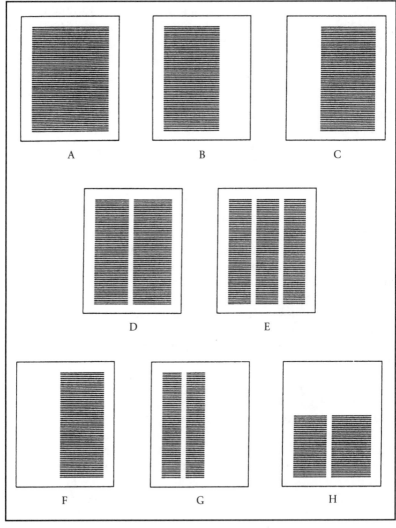

FIGURE 4 Eight Ways of Using Columns

see the information in that paragraph as a unit. Use extra white space between sections to signal to the reader that one section is ending and another is beginning—a visual cue indicating the organization of a document. You need not have access to sophisticated equipment to make use of white space. You can easily indent and skip lines for paragraphs, **lists,** and other blocks of material.

Using Lists. *Lists* provide an effective way to highlight words, phrases, and short sentences. Lists are particularly useful for certain types of information:

- steps in sequence
- materials or parts needed
- items to remember
- criteria for evaluation
- **concluding** points
- recommendations

Follow the advice given in the entry lists, such as avoiding both too many lists and too many items in lists.

Using Illustrations. Readers notice *illustrations* before text and large illustrations before small ones. Thus, the size of an illustration is the reader's gauge of its importance. Consider the proportion of the illustration to the text. Proportion often means employing the three-fifths rule: page layout is more dramatic and appealing when the major element (photo, drawing, and so on) occupies three-fifths rather than half of the available space.

Illustrations can be gathered in one place (as at the end of a report), but placing them within the text makes them more effective by putting them closer to their accompanying explanations, as well as providing visual relief.

desktop publishing (see **design and layout**)

despite/in spite of

Although there is no literal difference between *despite* and *in spite of,* *despite* suggests an effort to avoid blame.

EXAMPLES *Despite* our best efforts, the plan failed. (We are not to blame for the failure.)

In spite of our best efforts, the plan failed. (We did everything possible, but failure overcame us.)

Despite and *in spite of* (both meaning "notwithstanding") should not be blended into *despite of.*

CHANGE *Despite of* our best efforts, the plan failed.
TO *In spite of* (or *despite*) our best efforts, the plan failed.

diacritical marks

Diacritical marks are **symbols** added to letters to indicate their specific sounds. They include the phonetic symbols used by dictionaries as well as the marks used with foreign words. Some **dictionaries** place a reference list of common marks and sounds at the bottom of each page. The following is a list of common diacritical marks with their equivalent sound values. (See also **foreign words in English.**)

Name	Symbol	Example	Meaning
macron	¯	căke	a "long" sound that signifies the standard pronunciation of the letter.
breve	˘	brăcket	a "short" sound, in contrast to the standard pronunciation of the letter.
dieresis	¨	coöperate	indicates that the second of two consecutive vowels is to be pronounced separately.
acute accent	´	cliché	indicates a primary vocal stress on the indicated letter.
grave accent	`	crèche	indicates a deep sound articulated toward the back of the mouth.
circumflex	^	crêpe	indicates a very soft sound.
tilde	~	cañon	a Spanish diacritical mark identifying the palatal nasal *ny* sound.
cedilla	ç	garçon	a mark placed beneath the letter c in French, Portuguese, and Spanish to indicate an *s* sound.

diagnosis/prognosis

Because they sound somewhat alike, these words are often confused with each other. *Diagnosis* means "an analysis of the nature of something" or "the conclusions reached by such analysis."

> EXAMPLE   The chairman of the board *diagnosed* the problem as the allocation of too few dollars for research.

Prognosis means "a forecast or prediction."

> EXAMPLE   He offered his *prognosis* that the problem would be solved next year.

diction

The term *diction* is often misunderstood because it means both "the choice of words used in writing and speech" *and* "the degree of distinctness of enunciation of speech." In discussions of writing, diction applies to choice of words. (See also **word choice.**)

dictionaries

A *dictionary* lists, in alphabetical order, a selection of the words in a language. It defines them, gives their **spelling** and pronunciation, and indicates their function as a **part of speech.** In addition, it provides information about a word's origin and historical development. Often it provides a list of **synonyms** for a word and, where pertinent, an **illustration** to help clarify the meaning of the word.

The explanation of a word's meaning makes up the bulk of a dictionary entry. A dictionary gives primarily denotative rather than connotative meanings. It also labels the field of knowledge to which a specific meaning applies (grid, *electricity;* merger, *law*). Some specialized dictionaries list such terms exclusively. (See **library research** for examples.) The order in which a word's meanings are given varies. Some dictionaries give the most widely accepted current meaning first; others list the meaning in historical order, with the oldest meaning first and the current meaning last. If a word has two or more fundamental meanings, they are listed in separate paragraphs, with secondary meanings numbered consecutively within each paragraph. A dictionary's preface normally indicates whether current or historical meanings are listed in an entry.

A dictionary entry also includes a word's spelling. It lists any variant spellings (align/aline, catalog/catalogue) and indicates which is most commonly used. Entries show if and where a word ought to be hyphenated—information especially helpful for **compound words**—and how the word can be abbreviated.

A word's pronunciation, indicated by phonetic symbols, is also given. The **symbols** are explained in a key in the front pages of the dictionary, and an abbreviated symbol key is often found at the bottom of each page as well. (See also **diacritical marks.**)

Information about the origin and history of a word (its etymology) is also given, usually in **brackets** at the end of the entry. This information can help clarify a word's meaning.

Entries also include information about a word's part of speech and its inflected forms (how it forms plurals or how it changes form to

D

express comparative and superlative degree). Many entries include a list of **synonyms** in a separate paragraph following the main **paragraph.** Some dictionaries label words according to **usage** levels to indicate whether a word is considered standard English, nonstandard English, slang, and so on. Many dictionaries supplement a word's definition by providing **illustrations** in the form of **maps, photographs, tables, graphs,** and the like.

Types of Dictionaries

Desk dictionaries are often abridged versions of larger dictionaries. There is no single "best" dictionary, but there are several guidelines for selecting a good desk dictionary. Choose a recent edition. The older the dictionary, the less likely it is to have the up-to-date information you need. Select a dictionary with upward of 125,000 entries. Pocket dictionaries are convenient for checking spelling, but for detailed information the larger range of a desk dictionary is necessary. Many of the following dictionaries are available in CD-ROM and include a human-voice pronunciation of each word. Many general and specialized dictionaries are also available at a variety of **Internet** sites. The following are considered good desk dictionaries:

1. *The Random House College Dictionary.* Rev. ed. New York: Random.
2. *The American Heritage Dictionary of the English Language.* Boston: Houghton.
3. *Funk & Wagnalls Standard College Dictionary.* New, updated ed. New York: Funk.
4. *Webster's New World Dictionary of the American Language.* Cleveland: World.
5. *Merriam Webster's Collegiate Dictionary,* 10th ed. Springfield, MA: Merriam-Webster.

Unabridged dictionaries provide complete and authoritative linguistic information, but they are impractical for desk use because of their size and expense.

1. *Webster's Third New International Dictionary of the English Language, Unabridged* (Springfield, MA: Merriam, 1986) contains over 450,000 entries. Since word meanings are listed in historical order, the current meaning is given last. This dictionary does not list personal and geographical names, nor does it include usage information.
2. *The Random House Dictionary of the English Language,* 2nd ed. (New York: Random, 1987) contains nearly 300,000 entries, many illustrations, a color atlas, a manual of style, a chart of the chemical ele-

ments, and quick translating dictionaries of French, German, Spanish, and Italian. It gives a word's widely used current meaning first, and includes personal and geographic names.

3. *The Oxford English Dictionary,* 2nd ed. (Oxford, England: Clarendon, 1991) is the standard historical dictionary of the English language, with definitions of over 500,000 terms. It follows the chronological development of the terms defined, beginning about the year 1000, providing numerous examples of usage and sources. The 20-volume second edition integrates the original dictionary, published in 1928, with the four supplemental volumes, published between 1972 and 1986, into one alphabetical sequence.

differ from/differ with

Differ from is used to suggest that two things are not alike.

> **EXAMPLE**   The vice-president's background *differs from* the general manager's background.

Differ with is used to indicate disagreement between persons.

> **EXAMPLE**   Our attorney *differed with* theirs over the selection of the last jury member.

different from/different than

In formal writing, the **preposition** *from* is used with *different.*

> **EXAMPLE**   The fourth-generation computer is *different from* the third-generation computer.

Different than is acceptable when it is followed by a **clause.**

> **EXAMPLE**   The job cost was *different than* we had estimated it.

direct address

Direct address refers to a sentence or **phrase** in which the person being spoken or written to is explicitly named.

> **EXAMPLE**   *John,* call me as soon as you arrive at the airport.
> Call me, *John,* as soon as you arrive at the airport.

Notice that the person's name in a direct address is set off by **commas.**

discreet/discrete

Discreet means "having or showing prudent or careful behavior."

> **EXAMPLE**   Since the matter was personal, he asked Bob to be *discreet* about it at the office.

Discrete means something is "separate, distinct, or individual."

> **EXAMPLE**   The publications department was a *discrete* unit of the marketing division.

disinterested/uninterested

Disinterested means "impartial, objective, unbiased."

> **EXAMPLE**   Like good judges, scientists should be passionately interested in the problems they tackle but completely *disinterested* when they seek to solve these problems.

Uninterested means simply "not interested."

> **EXAMPLE**   Despite Jane's enthusiasm, her manager remained *uninterested* in the project.

division-and-classification method of development

One effective way to develop a complex subject is to divide it into logical parts and then go on to explain each part separately. This technique is especially well suited to subjects that can be readily broken down into units, and it can be used as the **method of development** in much technical writing. You might use this approach to describe a physical object, like a machine; to examine an idea, like the terms of a new labor-management contract; to explain a process, like the stages of an illness; or to give **instructions,** like the steps necessary to prepare a rusty metal surface for painting.

To explain the different types of printing processes currently in use, for example, you could divide the field into its major components, and where a fuller explanation is required, subdivide those (Figure 1). The emphasis in division is on breaking down a complex whole into a number of like units, because it is easier to consider smaller units and to examine the relationship of each to the other.

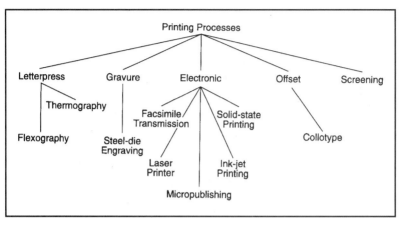

FIGURE 1   Printing Processes

The process by which a subject is divided is similar to the process by which a subject is classified. Division involves the separation of a whole into its component parts; classification is the grouping of a number of units (people, objects, ideas, and so on) into related categories. Consider the following list:

Triangular file	Steel tape ruler
Needle-nose pliers	Vise
Pipe wrench	Keyhole saw
Mallet	Tin snips
C clamps	Rasp
Hacksaw	Plane
Glass cutter	Ball-peen hammer
Steel square	Spring clamp
Claw hammer	Utility knife
Crescent wrench	Folding extension ruler
Slip-joint pliers	Crosscut saw
Tack hammer	Utility scissors

To group the items in the list, you would first determine what they have in common. The most obvious characteristic they share is that they all belong in a carpenter's tool chest. With that observation as a starting point, you can begin to group the tools into related categories. Pipe wrenches belong with slip-joint pliers because both tools grip objects. The rasp and the plane belong with the triangular file because all three tools smooth rough surfaces. By applying this kind of think-

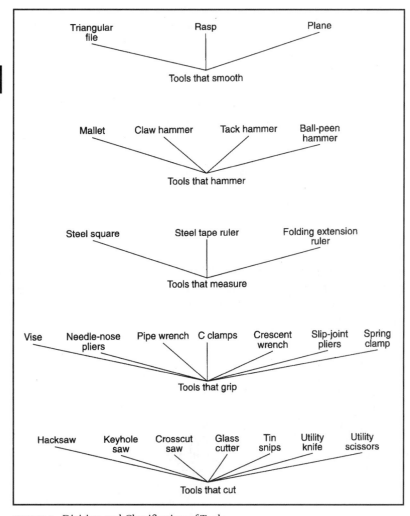

FIGURE 2   Division and Classification of Tools

ing to all the items in the list, you can group each tool according to function (Figure 2).

To divide or classify a subject, you must first divide the subject into its largest number of equal units. The basis for division depends, of course, on your subject and your purpose. For explaining the functions of tools, the classifications in Figure 2 (smoothing, hammering, measuring, gripping, and cutting) are excellent. For recommending which tools a new homeowner should buy first, however, the classifications are not helpful—they all contain tools that a new homeowner might want to purchase right away. To give homeowners advice on purchasing tools, you should clarify the types of repairs they are most likely to have to do. That classification can serve as a guide to tool purchase.

D

Once you have established the basis for the division or classification, apply it consistently, putting each item in only one category. For example, it might seem logical to classify needle-nose pliers both as "tools that cut" and as "tools that grip" because most needle-nose pliers have a small section for cutting wires. However, the primary function of needle-nose pliers is to grip. So listing them only under "tools that grip" would be most consistent with the basis used for listing the other tools.

In the example given in Figure 3, two Canadian park rangers classify typical park users according to four categories; the rangers then go on to discuss how to deal with potential rule breaking by members of each group. The rangers could have classified the visitors in a variety of other ways, of course: as city and country residents, backpackers and drivers of recreational vehicles, United States and Canadian citizens

---

Dealing with Groups

First, recognize the various types of campers. They can be broken down as follows:

1. Family for weekend stay
2. Small groups ("a few of the boys")
3. Large groups or conventions
4. Hostile gangs

Persons in groups A and B can probably be dealt with on a one-to-one basis. For example, suppose a member of the group is picking wild flowers, which is an offense in most park areas. Two courses of action are open. You could either issue a warning or charge the person with the offense. In this situation, a warning is preferable to a charge. First, advise the person that this action is an offense but, more importantly, explain why. Point out that the flowers are for all to enjoy and that most wild flowers are delicate and die quickly when picked.

For large groups, other approaches may be necessary. Every group has a leader. The leader may be official or, in informal or hostile groups, unofficial. If the group is organized, seek the official leader and hold this person responsible for the group's behavior. For informal groups, seek the person who assumes command and try to deal with this person.

—A. W. Moore and J. Mitchell, "Vandalism and Law Enforcement in Wildland Areas"

FIGURE 3 Division-and-Classification Method of Development

and so on. But for law enforcement agents in public parklands, the size of a group and the relationships among its members were the most significant factors.

D

documentation

The word *documentation* in the academic world refers to the giving of formal credit to sources used or quoted in a research paper or **report**— that is, the form used in a **bibliography,** footnote, or a reference. In the business and industrial world, *documentation* may refer also to information that is recorded, or "documented," on paper. The **technical manuals** provided to customers by manufacturers are examples of such documentation. Even **flowcharts** and engineering **drawings** are considered documentation.

documenting sources

By documenting sources, writers identify where they obtained the facts, ideas, **quotations,** and **paraphrases** used in preparing a **report, trade journal article,** book, or other document. The information can come from books, manuals, **correspondence, interviews,** software documentation, reference works, the **Internet** and other sources. Full and accurate documentation allows readers to locate and consult the information given and to find further information on the subject. It also gives proper credit to others so that the writer avoids **plagiarism.** The sources of all facts and ideas that are not common knowledge to the intended audience should be documented, as well as the sources of all direct quotations. The documentation must be complete, accurate, and consistent in format.

This entry describes and provides examples of the three principal methods of documenting sources: (1) *parenthetical documentation*— putting brief citations in parentheses in the text and providing full information in a list of "Works Cited"; (2) *reference documentation*— referring to sources with numbers in parentheses or by superscripts in the text and providing full information in a "References" section where the entries are listed numerically in the order of their first citation in the text; and (3) *notational documentation*—using superscript numbers in the text to refer to notes either at the bottom of the page (footnotes) or at the end of the paper, article, or chapter (endnotes).

Many professional organizations and journals publish style manuals that describe their own **formats** for documenting sources. A list of such style manuals appears at the end of this entry. If you are writing for publication in a professional field, consult the manual for that field or the style sheet for the journal to which you are submitting your article, and follow it exactly. Whatever format you choose, be sure to follow it consistently, in every detail of order, **punctuation,** and **capitalization.**

Use the following contents guide to this entry to locate information quickly. Parenthetical samples begin on page 169; and notational samples begin on page 175.

Parenthetical Documentation

The method that follows is recommended by the Modern Language Association of America (MLA) in the *MLA Handbook for Writers of Research Papers,* 4th ed. This method gives an abbreviated reference to a source parenthetically in the text and lists full information about the source in a separate section called "Works Cited."

When documenting sources in text, include only the author and page number in parentheses. If the author's name is mentioned in the text, include only the page number of the source. The parenthetical citation should include no more information than is necessary to enable the reader to relate it to the corresponding entry in the list of works cited. When referring to an entire work rather than to a particular page in a work, mention the author's name in the text, and omit the parenthetical citation.

The following passages contain sample parenthetical citations.

EXAMPLES

Preparing a videotape of measurement methods is cost effective and can expedite training (Peterson 151).

Peterson (183–191) summarized the results of these measurements in a series of tables.

When placing parenthetical citations in text, insert them between the closing quotation mark or the last word of the sentence (or clause) and the period or other punctuation. Use the spacing shown above. If the parenthetical citation follows an extended quotation or paraphrase, however, place it outside the last sentence of the quotation or paraphrase, with one space between the period and the first parenthesis. Within the citation itself, allow one space (no punctuation) between the name of the author and the page number. Don't use the word *page* or its abbreviation.

If you are citing a page or pages of a multivolume work, give the volume number, followed by a **colon,** space, and page number: (Jones 2: 53). If the entire volume is being cited, follow the punctuation in the two samples below:

Nonstellar celestial objects include clusters, nebulae, galaxies, quasi-stellar objects, radio sources, and X-ray sources (Hirshfeld, Sinnott, and Ochsenbein, vol. 2).

Although dwarf elliptical systems are the most common type of galaxy in the universe, their low luminosity causes them to be outnumbered by spiral galaxies in any catalogue of galaxies above even a relatively low apparent brightness (Hirshfeld, Sinnott, and Ochsenbein 2: xxxi).

If your list of works cited includes more than one work by the same author, include the title of the work (or a shortened version if the title is long) in the parenthetical citation, unless you mention it in the text. If, for example, your list of works cited included more than one work by John Roebuck, a proper parenthetical citation for his book *Anthropometric Methods: Designing to Fit the Human Body* would appear as (Roebuck, *Anthropometric Methods* 112).

Use only one space between the title and the page number. If two or more authors have the same last names, include their first names or initials to avoid confusion.

The following section explains the content and format of entries in a list of works cited. The format style described is that recommended in the *MLA Handbook for Writers of Research Papers,* 4th ed. The guidelines and models for citing computer software and online sources are based on the *MLA Style Manual,* 2nd ed.

D

Citation Format for Works Cited
The list of works cited should begin on the first new page following the end of the text. Each new entry should begin at the left margin, with the second and subsequent lines within an entry indented five spaces. Double-space within and between entries.

AUTHOR

Entries appear in alphabetical order by the author's last name (by the last name of the first author if the work has more than one author). Works by the same author should be alphabetized by the first major word of the title (following *a, an,* or *the*). If no author is given, the entry should begin with the title and be alphabetized by the first significant word in the title. (Articles in reference works, like encyclopedias, are sometimes signed with initials; you will find a list of the contributors' initials and full names elsewhere in the work, probably near the introductory material or the index.) If the author is a corporation, the entry should be alphabetized by the name of the corporation; if the author is a government agency, entries are alphabetized by the government, followed by the agency (for example, "United States. Dept. of Health and Human Services."). Some sources require more than one agency name (for example, "United States. Dept. of Labor. Bureau of Labor Statistics.").

After the first listing for an author, substitute three hyphens and a period in place of the name for subsequent entries by the same author. Hyphens used to represent an editor's name are followed by a comma and the abbreviation "ed."

McNeill, William H. <u>The Pursuit of Power</u>. Chicago: U of Chicago P, 1982.
---. <u>The Rise of the West</u>. Chicago: U of Chicago P, 1963.

TITLE

The title of the work is usually the second element. Capitalize the first and last words and each significant word in between. Underline the title of a book or pamphlet. Place quotation marks around the title of an article in a periodical, an essay in a collection, and a paper in a proceedings. Each title should be followed by a period, but do not underline the period. If the edition used is not the first, the edition should be specified immediately after the title.

D

PERIODICALS

For an article in a periodical (journal, magazine, or newspaper), the volume number, date (for a magazine or newspaper, simply the date), and the page numbers should immediately follow the title of the periodical, which itself directly follows the article title. Underline the journal title. See the sample entries below for the proper punctuation.

SERIES OR MULTIVOLUME WORKS

To cite works in a series or two or more volumes of a multivolume work, you should follow the title (or edition) with either the name of the series and the series number of the work in question or the number of volumes.

PUBLISHING INFORMATION

The final elements of the entry for a book, pamphlet, or conference proceedings are the place of publication, publisher, and date of publication. Use a shortened form of the publisher's name (for example, *St. Martin's* for *St. Martin's Press, Inc., Random* for *Random House, Oxford UP* for *Oxford University Press*). If any of these cannot be found in the work, use the abbreviations n.p. (no publication place), n.p. (no publisher), and n.d. (no date), respectively. For familiar reference works, list only the edition and year of publication.

Sample Entries (MLA Style)

BOOK, ONE AUTHOR

Roebuck, John A., Jr. Anthropometric Methods: Designing to Fit the Human Body. Santa Monica: Human Factors and Ergonomic Society, 1995.

BOOK, TWO OR MORE AUTHORS

Peixoto, José P., and Abraham H. Oort. Physics of Climate. New York: American Institute of Physics, 1992.

BOOK, CORPORATE AUTHOR

National Fire Protection Association. NFPA Inspection Manual. 7th ed. Quincy: National Fire Protection Association, 1994.

WORK IN AN EDITED COLLECTION

Klatzky, Robert L., and M. M. Ayoub. "Health Care." Emerging Needs and Opportunities for Human Factors Research. Ed. Raymond S. Nickerson. Washington: National Academy, 1995. 131–57.

BOOK EDITION, IF NOT THE FIRST

Cember, Herman. Introduction to Health Physics, 3rd ed. New York: McGraw, 1996.

TRANSLATED WORK

Texereau, Jean. How to Make a Telescope. Trans. Allen Strickler. 2nd English ed. Richmond: Willmann, 1984.

MULTIVOLUME WORK

Hirshfeld, Alan, Roger W. Sinnott, and Francois Ochsenbein, eds. 2nd ed. <u>Sky Catalogue</u> <u>2000.0</u>. 2 vols. Cambridge: Sky, 1991.

WORK IN A SERIES

Ruskin, Arnold M., and W. Eugene Estes. <u>What Every Engineer Should Know about Project</u> <u>Management</u>. What Every Engineer Should Know, Vol. 33. New York: M. Dekker, 1995.

GOVERNMENT PUBLICATION

United States. General Accounting Office. <u>Job Training: Small Business Participation in</u> <u>Selected Training Programs</u>. GAO/HEHS-96-106. Washington: General Accounting Office, 1996.

UNPUBLISHED THESIS OR DISSERTATION

White, Elizabeth L. "Control Integration in Heterogeneous Distributed Software Applications." Diss. U of Maryland, 1995.

ENCYCLOPEDIA ARTICLE

Kinsler, Lawrence E. "Vibration." <u>McGraw-Hill Encyclopedia of Science and Technology</u>. 7th ed. 1992.

PROCEEDINGS

El-Genk, M. S., and M. D. Hoover, eds. <u>Proceedings of the Ninth Symposium on Space</u> <u>Nuclear Power Systems</u>. Institute for Space Nuclear Power Studies. 12–16 Jan. 1992, Albuquerque, NM. New York: American Institute of Physics, 1992.

PAPER IN A PROCEEDINGS

Whitfield, R. P., and B. M. Kleiner. "Environmental Benchmarking as a Management Tool for Performance Improvement." <u>Working Towards a Cleaner Environment: Waste Pro-</u> <u>cessing, Transportation, Storage and Disposal, Technical Programs and Public Educa-</u> <u>tion</u>. Ed. R. G. Post. Vol 1. Proc. of the Symposium on Waste Management, 27 Feb.–3 Mar. 1994. Tucson: Laser Optics, 1994. 239–44.

TRADE JOURNAL ARTICLE

Thorton, W. A. "A Rational Approach to Design of Tee Shear Connections." <u>Engineering</u> <u>Journal</u> 33 (1996): 34–37.

MAGAZINE ARTICLE

Livingston, James D. "Magnets on the Rise." <u>Technology Review</u> May-June 1996: 32–40.

ANONYMOUS ARTICLE (IN WEEKLY PERIODICAL)

"Skinny Monitor Doesn't Eat Up Power." <u>Government Computer News</u> 15 April 1996: 36.

NEWSPAPER ARTICLE

Pepper, Jonathan. "Computer as Private Photo Lab." <u>New York Times</u> 9 May 1996: C2.

LETTER FROM ONE OFFICIAL TO OTHERS

Brown, Charles L. Letter to retired members of Bell System Presidents' Conference. 8 Jan. 1996.

LETTER PERSONALLY RECEIVED

Harris, Robert S. Letter to the author. 3 Dec. 1996.

PERSONAL INTERVIEW
Denlinger, Virgil. Personal interview. 15 Mar. 1996.

COMPUTER SOFTWARE
As-Sabil for Windows. Montreal: Alis Technologies, 1996.

ONLINE SOURCES

When citing online sources, include all information as for sources published in print: author, title, publication information, and page numbers or number of pages (if available). In addition, provide the date you accessed the source and the URL in angle brackets. The following citation is to an article in an online journal about architecture, called *Architronic*. (For more information on citing sources from the World Wide Web, refer to the *MLA Style Manual,* 2nd ed., or visit the MLA's Web site at *www.mla.org.*)

Harris, Elizabeth L. "Housing the War-Time Workers." Architronic 4.1 (1995): 11 pp. 20 Apr. 1998. <http://www.saed.kent.edu//Architronic/v4nl/v4n1.02.html>.

Reference Documentation

The reference system of documenting sources lists the numbered documentation references at the end of the work. The references on the list are identified by numbers in the text that correspond to the numbered items in the list. The reference list cites only those works referred to in text.

The styles for references vary. Consult the style manual of the publication for which you are writing for their preferred reference style.

The citations in a reference list are arranged in numerical sequence (1, 2, 3), according to the order in which they are first mentioned in the text. The most common method for directing readers from your text to specific references is to place the reference number in parentheses after the work cited. Thus in the text, the number one in parentheses (1) after a reference to a book, article, or other work refers the reader to the first citation in the reference list. The number five in parentheses (5) refers the reader to the fifth citation in the reference list. A second number in the parentheses, separated from the first by a colon (3:27), refers to the page number of the source from which the information was taken. There are several other common methods for directing readers from your text to specific references. You can place the word *Reference,* or the abbreviation *Ref.,* within parentheses with the numbers: (Reference 1) or (Ref. 1). Or you can write the reference number as a superscript: text[1]. Sometimes you may cite your sources within a sentence in the text; then you should spell out the word *Reference:* "The data in Reference 3 include . . ." For subsequent references in the text, simply repeat the reference number of the first citation.

Notational Documentation

D

Notes in publications have two uses: (1) to provide background information in publications or explanations that would interrupt the flow of thought in the text and (2) to provide documentation references. (For guidance on footnotes used with tables, see **tables.**)

Explanatory Footnotes

Explanatory or content notes are useful when the basis for an assumption should be made explicit but spelling it out in the text might make readers lose the flow of an argument. Because explanatory or content notes can be distracting, they should be kept to a minimum. If you cannot work the explanatory or background material into your text, it may not belong there. Lengthy explanations should be placed in an **appendix.** (See **appendix/appendixes/appendices.**)

Documentation Notes

Notes that document sources can appear as either endnotes or footnotes. Endnotes are placed in a separate section at the end of a report, article, chapter, or book; footnotes (including explanatory notes) are placed at the bottom of a text page. The only drawback to using endnotes is that readers may find them inconvenient to locate and difficult to correlate with the text. Footnotes are easier to locate.

Use superscript numbers in the text to refer readers to notes, and number them consecutively from the beginning of the report, article, or chapter to the end. Place each superscript number at the end of a sentence, clause, or phrase, at a natural pause point like this,[1] right after the period, comma, or other punctuation mark (except a dash, which the number should precede[2]—as here). Use lower case letter (a, b) or **symbols** (#, +, *) to identify footnotes in table and figures.

Footnote Format

To type footnotes, leave four lines below the text on a page, and indent the first line of each footnote five spaces. Begin each note with the appropriate superscript number, and skip one space between the number and the text of the footnote. If the footnote runs longer than one line, the second and subsequent lines should begin at the left margin. Single-space within each footnote, and double-space between footnotes.

EXAMPLE   Assume that this is the last line of text on a page.

¹ Begin the first footnote at this position. When it runs longer than one line, begin the second and all following lines at the left margin.

² The second footnote follows the same spacing as the first. Single-space within footnotes, and double-space between them.

Endnote Format
Endnotes should begin on a separate page after the end of the text entitled "Works Cited," "References," or "Notes." The individual notes should be typed as for footnotes. Double-space within and between endnotes.

Documentation Note Style
The following guidelines for documentation note style are based on those provided in the Modern Language Association's *MLA Handbook for Writers of Research Papers,* 4th ed.

Documentation notes give the author's full name, title, and publication information for the source, including the page or pages from which material was taken. This information is given in full in the first note for the source, and the second and subsequent notes give abbreviated author and title information.

In a footnote or endnote, the information is given in a somewhat different order and uses somewhat different punctuation than in a bibliography or a list of references or works cited. The author's name is given in normal order (Jane F. Smith) rather than last name first; the author and title are separated by a comma rather than a period; and the publication information for a book is enclosed in parentheses. The first reference to a book appears as follows:

¹ David S. Landes, Revolution in Time: Clocks and the Making of the Modern World (Cambridge: Belknap-Harvard UP, 1983) 340.

There is no punctuation mark between the title and the opening parenthesis or between the closing parenthesis and the page number, and the page number is not preceded by the word *page* or its abbreviation. Compare this sample note with how the same book would be listed in a bibliography or list of works cited:

Landes, David S. Revolution in Time: Clocks and the Making of the Modern World. Cambridge: Belknap-Harvard UP, 1983.

Sample Entries
The following sample notes for first references are based on the format style provided in the *MLA Handbook*:

BOOK, ONE AUTHOR
   [1] John A. Roebuck Jr., <u>Anthropometric Methods: Designing to Fit the Human Body</u> (Santa Monica: Human Factors and Ergonomics Society, 1995) 14.

BOOK, TWO OR MORE AUTHORS
   [2] José P. Peixoto and Abraham H. Oort, <u>Physics of Climate</u> (New York: American Institute of Physics, 1992) 41.

BOOK WITH A CORPORATE AUTHOR
   [3] National Fire Protection Association, <u>NFPA Inspection Manual</u>, 7th ed. (Quincy: National Fire Protection Association, 1994) 130.

WORK IN AN EDITED COLLECTION
   [4] Roberta L. Klatzky and M. M. Ayoub, "Health Care," <u>Emerging Needs and Opportunities for Human Factors Research</u>, ed. Raymond S. Nickerson (Washington: National Academy, 1995) 135.

GOVERNMENT PUBLICATION
   [5] United States, General Accounting Office, <u>Job Training: Small Business Participation in Selected Training Programs</u>, GAO/HEHS-96-106 (Washington: General Accounting Office, 1996).

TRADE JOURNAL ARTICLE
   [6] W. A. Thorton, "A Rational Approach to Design of Tee Shear Connections," <u>Engineering Journal</u> 33 (1996): 36.

MAGAZINE ARTICLE
   [7] James D. Livingston, "Magnets on the Rise," <u>Technology Review</u> May-June 1996: 35.

NEWSPAPER ARTICLE
   [8] Jonathan Pepper, "Computer as Private Photo Lab," <u>New York Times</u> 9 May 1996: C2.

UNPUBLISHED THESIS OR DISSERTATION
   [9] Elizabeth L. White, "Control Integration in Heterogeneous Distributed Software Applications," diss., U of Maryland, 1995, 27.

For periodicals, the date (enclosed in parentheses if the periodical is a journal) is followed by a colon.

Second and Subsequent References

Notes for second and subsequent references to works should contain only enough information to allow the reader to relate it to the note for the first reference. Generally, the author's name and page number will suffice:

   [7] Landes 343.

Even if the second reference immediately follows the first, put the author's last name and the page number as shown. (The abbreviations *ibid.* and *op. cit.*, in older books and articles, are no longer used as ways to refer to previous citations.)

If two or more works by the same author are being cited, include a shortened version of the appropriate title in the second and subsequent notes. Insert a comma between the author's name and the title, with no punctuation between the title and the page number.

Bibliography Format
If you are using the documentation note style and your instructor also requires a separate list of works cited or a bibliography, use the style for works-cited entries described on pages 171–174.

Style Manuals

Many professional societies, publishing companies, and other organizations publish manuals that prescribe bibliographic reference formats for their publications or publications in their fields. For an annotated bibliography of style manuals issued by commercial publishers, government agencies, and university presses in fields ranging from agriculture to zoology, see John Bruce Howell, *Style Manuals of the English-speaking World* (Phoenix: Oryx, 1983).

Following is a selected list of style manuals for various fields:

Biology
Council of Biology Editors. *Scientific Style and Format: The CBE Manual for Authors, Editors, and Publishers in the Biological Sciences.* 6th ed. New York: Cambridge UP, 1994.

Chemistry
American Chemical Society. *ACS Style Guide: A Manual for Authors and Editors.* Washington, DC: American Chemical Soc., 1986.

Geology
U. S. Geological Survey. *Suggestions to Authors of the Reports of the United States Geological Survey.* 7th ed. Washington, DC: U.S. GPO, 1991.

Mathematics
American Mathematical Society. *A Manual for Authors of Mathematical Papers.* Providence, RI: American Mathematical Soc., 1990.

Medicine
American Medical Association. *American Medical Association Manual of Style.* 8th ed. Baltimore: Williams, 1989.

Physics
American Institute of Physics, Publications Board. *AIP Style Annual.* 4th ed. New York: American Institute of Physics, 1990.

Psychology
American Psychological Association. *Publication Manual of the American Psychological Association.* 4th ed. Washington, DC: American Psychological Assn., 1994.

General
The Chicago Manual of Style. 14th ed. Chicago: U of Chicago P, 1993.

Gibaldi, Joseph. *MLA Handbook for Writers of Research Papers.* 4th ed. New York: Modern Language Assn. of America, 1995.

National Information Standards Organization. *Scientific and Technical Reports—Elements, Organization, and Design.* Bethesda, MD: 1995. ANSI/NISO Z39.18—1995.

Skillin, M. E., and Gay, R. M. *Words into Type.* 3rd ed. Englewood Cliffs, NJ: Prentice, 1974.

Turabian, K. L. *A Manual for Writers of Term Papers, Theses, and Dissertations.* 5th ed. Chicago: U of Chicago P, 1990.

United States Government Printing Office Style Manual. Washington, DC: U.S. GPO, 1984.

double negatives

A *double negative* is the use of an additional negative word to reinforce an expression that is already negative. It is an attempt to **emphasize** the negative, but the result is only awkward. Double negatives should never be used in writing.

> **CHANGE**   I *haven't* got *none.*
> **TO**   I have *none.*

Barely, hardly, and *scarcely* cause problems because writers sometimes do not recognize that these words are already negative.

> **CHANGE**   I *don't hardly* ever have time to read these days.
> **TO**   I *hardly* ever have time to read these days.

Not unfriendly, not without, and similar constructions are not double negatives because in such constructions two negatives are meant to suggest the gray area of meaning between negative and positive. Be careful, however, how you use these constructions; they are often confusing to the **reader** and should be used only if they serve a **purpose.**

EXAMPLES   He is *not unfriendly.* (meaning that he is neither hostile nor friendly)

It is *not without* regret that I offer my resignation. (implying mixed feelings rather than only regret)

The correlative **conjunctions** *neither* and *nor* may appear together in a **clause** without creating a double negative, so long as the writer does not attempt to use the word *not* in the same clause.

CHANGE   It was *not,* as a matter of fact, *neither* his duty *nor* his desire to fire the man.

TO   It was *neither,* as a matter of fact, his duty *nor* his desire to fire the man.

OR   It was not, as a matter of fact, *either* his duty *or* his desire to fire the man.

CHANGE   He did *not neither* care about *nor* notice the error.

TO   He *neither* cared about *nor* noticed the error.

Negative forms are full of traps that often entice inexperienced writers into errors of **logic,** as illustrated in the following example:

EXAMPLE   There is *nothing* in the book that has *not* already been published in some form, but some of it is, I believe, very little known.

In this sentence, "some of it," logically, can refer only to "*nothing* in the book that has *not* already been published." The sentence can be corrected in one of two ways. The pronoun *it* can be replaced by a specific noun.

EXAMPLE   There is nothing in the book that has not already been published in some form, but some of the *information* is, I believe, very little known.

Or the idea can be stated positively. (See also **positive writing.**)

EXAMPLE   Everything in the book has been published in some form, but some of it is, I believe, very little known.

drawings

A drawing can focus on details or relationships that a **photograph** cannot capture. A drawing can emphasize the significant part of a mechanism, or its function, and omit what is not significant. However

D

If You Do Fight the Fire, Remember the Word PASS

PULL

the pin: Some extinguishers require releasing a lock latch, pressing a puncture lever, or taking another first step.

AIM

low: Point the extinguisher nozzle (or its horn or hose) at the base of the fire.

SQUEEZE

the handle: This releases the extinguishing agent.

SWEEP

from side to side: Keep the extinguisher aimed at the base of the fire and sweep back and forth until it appears to be out. Watch the fire area. If fire breaks out again, repeat the process.

FIGURE 1   Drawing That Illustrates Instructions

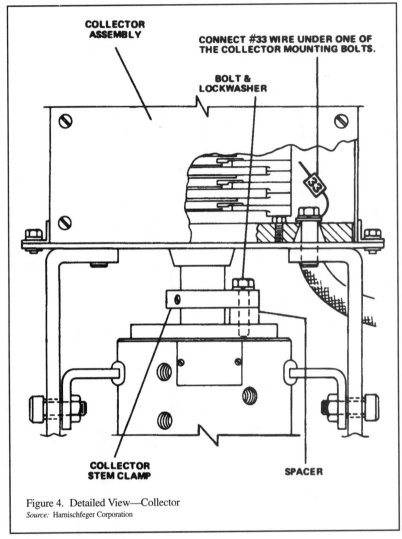

COLLECTOR ASSEMBLY

CONNECT #33 WIRE UNDER ONE OF THE COLLECTOR MOUNTING BOLTS.

BOLT & LOCKWASHER

COLLECTOR STEM CLAMP

SPACER

Figure 4. Detailed View—Collector
Source: Harnischfeger Corporation

FIGURE 2 Cutaway Drawing

if the actual appearance of an object is necessary to your **report** or document, include a photograph.

There are various types of drawings, each with unique advantages. The type of drawing used for an **illustration** should be determined by the specific purpose it is intended to serve. If your **reader** needs an impression of an object's general appearance or an overview of a series

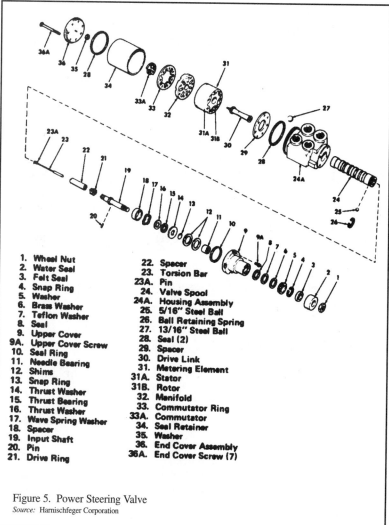

1. Wheel Nut
2. Water Seal
3. Felt Seal
4. Snap Ring
5. Washer
6. Brass Washer
7. Teflon Washer
8. Seal
9. Upper Cover
9A. Upper Cover Screw
10. Seal Ring
11. Needle Bearing
12. Shims
13. Snap Ring
14. Thrust Washer
15. Thrust Bearing
16. Thrust Washer
17. Wave Spring Washer
18. Spacer
19. Input Shaft
20. Pin
21. Drive Ring

22. Spacer
23. Torsion Bar
23A. Pin
24. Valve Spool
24A. Housing Assembly
25. 5/16" Steel Ball
26. Ball Retaining Spring
27. 13/16" Steel Ball
28. Seal (2)
29. Spacer
30. Drive Link
31. Metering Element
31A. Stator
31B. Rotor
32. Manifold
33. Commutator Ring
33A. Commutator
34. Seal Retainer
35. Washer
36. End Cover Assembly
36A. End Cover Screw (7)

Figure 5. Power Steering Valve
Source: Harnischfeger Corporation

FIGURE 3 Exploded-View Drawing

of steps or directions, a conventional drawing of the type illustrated by Figure 1 will suffice. To show the internal parts of a piece of equipment in such a way that their relationship to the overall equipment is clear, use a cutaway drawing as in Figure 2. To show the proper sequence in which parts fit together or to show the details of each individual part, use an exploded-view drawing such as Figure 3.

D

Tips for Creating and Using Drawings

Many organizations have their own **format** specifications for creating drawings. In the absence of such specifications, follow the general guidelines for creating and integrating illustrations with text in the **illustration** entry, in addition to the following guidelines specific to drawings:

1. Show equipment and other objects from the point of view of the person who will use them.
2. When illustrating a subsystem, show its relationship to the larger system of which it is a part.
3. Draw the different parts of an object in proportion to one another, unless you indicate that certain parts are enlarged.
4. When a sequence of drawings is used to illustrate a process, arrange them from left to right or from top to bottom.
5. Label parts in the drawing so that text references to them are clear and consistent.
6. Depending on the complexity of what is shown, place labels on the parts themselves, or give the parts letter or number symbols, with an accompanying key (see Figure 3).

due to/because of

Due to (meaning "caused by") is acceptable following a linking **verb.**

EXAMPLE   His short temper was *due to* work strain.

Due to is not acceptable, however, when it is used with a nonlinking verb to replace *because of.*

CHANGE   He went home *due to* illness.
TO   He went home *because of* illness.

E

each

When *each* is used as a subject, it takes a singular **verb** or **pronoun.**

EXAMPLES *Each* of the reports *is* to be submitted ten weeks after *it* is assigned.
Each worked as fast as *his* ability would permit.

When *each* occurs after a plural subject with which it is in apposition, it takes a plural verb or pronoun. (See also **agreement.**)

EXAMPLE The reports *each have* white embossed titles on *their* covers.

each and every

Although the phrase *each and every* is commonly used in speech in an attempt to emphasize a point, the phrase is redundant and should be eliminated from your writing. Replace it with *each* or *every.*

CHANGE *Each and every* part was accounted for.
TO *Each* part was accounted for.
OR *Every* part was accounted for.

economic/economical

Economic refers to the production, development, and management of material wealth. *Economical* simply means "not wasteful or extravagant."

EXAMPLES The strike at General Motors had an *economic* impact on the country.

Since gasoline will be in short supply, drivers should be as *economical* as possible with its use.

e.g./i.e.

The **abbreviation** *e.g.* stands for the Latin *exempli gratia,* meaning "for example"; *i.e.* stands for the Latin *id est,* meaning "that is." Since perfectly good English expressions exist for the same uses *(for example* and *that is),* there is no need to use a Latin expression or abbreviation except in notes and **illustrations** where you need to save space.

> CHANGE Some terms of the contract (*e.g.,* duration and job classification) were settled in the first two bargaining sessions.
>
> TO Some terms of the contract (*for example,* duration and job classification) were settled in the first two bargaining sessions.

> CHANGE We were a fairly heterogeneous group; *i.e.,* there were managers, foremen, and vice-presidents at the meeting.
>
> TO We were a fairly heterogeneous group; *that is,* there were managers, foremen, and vice-presidents at the meeting.

If you must use *i.e.* or *e.g.,* punctuate them as follows. If *i.e.* or *e.g.* connects two independent clauses, a **semicolon** should precede it and a **comma** should follow it. If *i.e.* or *e.g.* connects a **noun** and **appositive,** a comma should precede and follow it.

> EXAMPLES We were a fairly heterogeneous group, *i.e.,* managers, foremen, and vice-presidents.
>
> We were a fairly heterogeneous group; *i.e.,* we were managers, foremen, and vice-presidents.

electronic mail

Email* (electronic mail) refers to the transmission of messages from one person to another through a computer network. The messages are sent to and stored in a computer file called an electronic mail box until they are retrieved by the recipient. The recipient can display and read the message on the screen, print the message as a permanent record, forward the message to others, save it to a disk or hard drive, and send a reply to the original sender. Email is especially useful for employees within a large organization who must send **memorandums, reports, correspondence,** meeting notices, and even **questionnaires** to co-workers throughout the organization.

**Wired Style: Principles of English Usage in the Digital Age,* by the editors of *Wired* magazine, recommends spelling *email* with a lowercase *e* and without a hyphen.

This method of communication offers a variety of advantages. The message—whether sent across the hall or across the continent—reaches its audience within minutes at a cost well below overnight mail service. Because the information is transmitted in machine-readable form, the text can be printed, revised and sent back, or incorporated directly into another computer file without being retyped. This feature permits people working on **collaborative writing** projects to forward their drafts to other team members for comment and to receive and incorporate the comments electronically.

E

Writing Style

Although email messages are like conventional job-related writing in most respects, they tend to be less formal, more conversational—somewhere between a telephone conversation and a memorandum—because of the immediacy of the transmission and the fact that it usually occurs directly between two people. Construct your messages accordingly. Avoid long, complicated **sentences** that are dense with unnecessary technical **jargon** that can confuse the **reader** and convey the wrong message. Keep in mind, too, that although you want to keep sentences brief and words short, you should not use a telegraphic style that leaves important information unsaid or only partly said. Think carefully about the reader, the accuracy of your information, and the level of detail necessary to communicate your message clearly and concisely to that reader.

Review and Privacy Implications

Email communications require that the writer consider the inadvisability of sending unreviewed or unedited text and the privacy implications of what's being transmitted.

Most email messages written at a computer terminal will be sent directly to the recipient by the sender. This practice, although technically efficient, has pitfalls. Avoid the temptation to "dash off" a first draft and send it as is. Unless carefully reviewed, the information sent could contain errors of fact and of **grammar,** be ambiguous, contain unintended implications, or inadvertently omit crucial details.

Consider privacy, too. Because email messages are recorded on computer tape for retrieval and storage, they could be read or printed by someone other than the intended recipient. This may happen not only during transmission but subsequent to it. Consider all your messages in the light of these possibilities.

E

elegant variation

The attempt to avoid repeating a **noun** in a **paragraph** by substituting other words (often pretentious **synonyms**) is called *elegant variation*. Avoid this practice—repeat a word if it says what you mean, or use a suitable **pronoun**. (See also **affectation** and **long variants.**)

CHANGE   The use of modules in the assembly process has increased production. *Modular utilization* has also cut costs.

TO   The use of modules in the assembly process has increased production. The *use* of modules has also cut costs.

OR   The use of modules in the assembly process has increased production and also cut costs.

(See also **thesaurus.**)

ellipses

When you omit words in quoted material, use a series of three spaced periods—called *ellipsis dots*—to indicate the omission. However, such an omission should not detract from or alter the essential meaning of the **sentence.**

EXAMPLES   Technical material distributed for promotional use is sometimes charged for, particularly in high-volume distribution to schools, although prices for these publications are not uniformly based on the cost of developing them. (without omission)

Technical material distributed for promotional use is sometimes charged for . . . although prices for these publications are not uniformly based on the cost of developing them. (with omission)

When the omitted portion comes at the beginning of the sentence, begin the **quotation** with a lowercase letter.

EXAMPLES   "When the programmer has determined a system of runs, he must create a system flowchart to provide a picture of the data flow through the system." (without omission)

The letter states that the programmer ". . . must create a system flowchart to provide a picture of the data flow through the system." (with omission)

When the omission comes at the end of a sentence and you continue the **quotation** following the omission, use four periods to indicate both the final period to end the sentence and the omission.

EXAMPLES In all publications departments except ours, publications funds—once they are initially allocated by higher management—are controlled by publications personnel. Our company is the only one to have nonpublications people control funding within a budgeted period. In addition, all publications departments control printing funds as well. (without omission)

In all publications departments except ours, publications funds—once they are initially allocated by higher management—are controlled by publications personnel. . . . In addition, all publications departments control printing funds as well. (with omission)

Use a full line of **periods** across the page to indicate the omission of one or more **paragraphs.**

EXAMPLES A computer system operates with two types of programs: the software programs supplied by the manufacturer and the programs created by the user. The manufacturer's software consists of a number of programs that enable the system to perform complicated manipulations of data from the relatively simple instructions specified in the user's program.

Programmers must take a systematic approach to solving any data-processing problem. They must first clearly define the problem, and then they must define a system of runs that will solve the problem. For example, a given problem may require a validation and sort run to handle account numbers or employee numbers, a computation and update run to manipulate the data for the output reports, and a print run to produce the output reports.

When a system of runs has been determined, the programmer must create a system flowchart to provide a picture of the data flow through the system. (without omission)

A computer system operates with two types of programs: the software programs supplied by the manufacturer and the programs created by the user. The manufacturer's software consists of a number of programs that enable the system to perform complicated manipulations of data from the relatively simple instructions specified in the user's program.

. . . .

When a system of runs has been determined, the programmer must create a system flowchart to provide a picture of the data flow through the system. (with omission)

Do not use ellipsis dots for any purpose other than to indicate omission. Advertising copywriters, particularly those not skilled at achieving the effect they strive for with words, often use ellipsis dots to substitute for all marks of **punctuation**—and sometimes even use them where no punctuation is needed. This practice is a poor one to emulate.

eminent/imminent

E

Someone or something that is *eminent* is outstanding or distinguished.

> **EXAMPLE**   She is an *eminent* scientist.

If something is *imminent,* it is about to happen.

> **EXAMPLE**   He paced nervously, knowing that the committee's decision was *imminent.*

emphasis

Emphasis is the principle of stressing the most important ideas. The means of achieving emphasis are by:

- position within a **sentence, paragraph,** or document;
- the use of **repetition;**
- the selection of sentence type;
- variation in the length of sentences;
- the use of climactic order within a sentence;
- the use of the **dash;**
- the use of **intensifiers;**
- the use of mechanical devices, such as **italics** and **capital letters;** and
- the use of direct statement (such terms as *most important* and *foremost*).

Emphasis can be achieved by position because the first and last words of a sentence stand out in the **reader's** mind.

> **CHANGE**   Because they reflect geological history, moon craters are important to understanding the earth's history.
>
> **TO**   Moon craters are important to understanding the earth's history because they reflect geological history.

Notice that the revised version of the sentence emphasizes moon craters simply because the term is in the front part of the sentence and geological history because the term is at the end of the sentence. Similarly, the first and last sentences in a paragraph and the first and last paragraphs in a document tend to be the most emphatic to the reader.

Another way to achieve emphasis is to vary sentence length, following a very long sentence, or a series of long sentences, with a very short one that will stand out in the reader's mind.

EXAMPLE   We have already reviewed the problem the bookkeeping department has experienced during the past year. We could continue to examine the causes of our problems and point an accusing finger at all the culprits beyond our control, but in the end it all leads to one simple conclusion. *We must cut costs.*

Emphasis can be achieved by the **repetition** of key words and **phrases.**

E

EXAMPLE   Similarly, atoms *come and go* in a molecule, but the molecule *remains;* molecules *come and go* in a cell, but the cell *remains;* cells *come and go* in a body, but the body *remains;* persons *come and go* in an organization, but the organization *remains.*

Different emphasis can be achieved by the selection of a compound sentence, complex sentence, or simple sentence.

EXAMPLES   The report turned in by the police detective was carefully illustrated, and it covered five pages of single-spaced copy. (This compound sentence carries no special emphasis because it contains two coordinate independent clauses.)

The police detective's report, which was carefully illustrated, covered five pages of single-spaced copy. (This complex sentence emphasizes the size of the report.)

The carefully illustrated report turned in by the police detective covered five pages of single-spaced copy. (This simple sentence emphasizes that the report was carefully illustrated.)

Emphasis can be achieved by a climactic order of ideas or facts within a sentence.

EXAMPLE   Over subsequent weeks the industrial relations department worked diligently, management showed tact and patience, and the employees finally accepted the new policy.

Emphasis can be achieved by setting an item apart with a **dash.**

EXAMPLE   Here is where all the trouble begins—in the American confidence that technology is ultimately the medicine for all ills.

Emphasis can be achieved by the use of **intensifiers** (*most, very, really*), but this technique is so easily abused that it should be used only with caution.

EXAMPLE   The final proposal is *much* more persuasive than the first.

Although emphasis can be achieved by such mechanical devices as **italics** and **capital letters,** this technique is also easily abused and should be used with caution.

EXAMPLE   When we add the commonplace situations that allow these systems to function for hundreds of people *simultaneously* and on all five continents, the power of this new information medium is even more remarkable.

However, do not use all **capital letters** to try to show emphasis. Doing so can cause confusion because capital letters are used for so many other reasons.

The word *do* (or *does*) may be used for emphasis, but this is also easily overdone, so proceed with caution. Compare the following examples:

EXAMPLES   You believe weekly staff meetings are essential, don't you? (unemphatic)

You *do* believe weekly staff meetings are essential, don't you? (emphatic)

Direct address may also be used for emphasis in **correspondence.**

EXAMPLE   John, I believe we should rethink our plans.

(See also **subordination.**)

English, varieties of

There are two broad varieties of written English: standard and nonstandard. These varieties are determined through **usage** by those who write in the English language. Standard English (also called American edited English) is used to carry on the daily business of the nation. It is the language of business, industry, government, education, and the professions. Standard English is characterized by exacting standards of **punctuation** and capitalization, by accurate **spelling,** by exact **diction,** by an expressive vocabulary, and by knowledgeable usage choices. Nonstandard English, on the other hand, is the language of those not familiar with the standards of written English. This form of English rarely appears in printed material except when it is used for special effect by fiction writers. Nonstandard English is characterized by inexact or inconsistent punctuation, capitalization, spelling, diction, and usage choices. Both standard and nonstandard English find their way into everyday speech and writing in the forms of the following subcategories.

Colloquial

Colloquial English is spoken standard English or writing that attempts to re-create the flavor of this kind of speech by using words and expressions common to casual conversation. Colloquial English is appropriate to some kinds of writing (personal letters, notes, and the like) but not to most technical writing.

Vernacular

Vernacular English is the spoken form of the language, as opposed to its written form. It is the form used by the majority of those who speak the language. To write "in the vernacular" is to imitate this kind of language. Ordinarily, such writing is confined to fiction. Vernacular can also refer to the manner of expression common to a trade or profession (one might speak of "the legal vernacular"), although **jargon** more clearly expresses this meaning.

Dialect

Dialectal English is a social or regional variety of the language that is comprehensible to people of that social group or region but that may be incomprehensible to outsiders. Dialect, which is usually nonstandard English, involves distinct **word choices,** grammatical forms, and pronunciations. Technical writing, because it aims at a broad audience, should be free of dialect.

Localism

A localism is a word or phrase that is unique to a geographical region. For example, the words *poke, sack,* and *bag* all denote "a paper container," each term being peculiar to a different region of the United States. Such words should normally be avoided in writing because knowledge of their meanings is too narrowly restricted.

Slang

Slang refers to a manner of expressing common ideas in new, often humorous or exaggerated, ways. Slang often finds new use for familiar words ("He *crashed* early last night"—meaning that he went to bed early) or coins words ("He's a *kook*"—meaning that he is offbeat, unconventional). Slang expressions usually come and go very quickly. The fact that a fair number of words in the present standard English vocabulary were once considered slang indicates that when a slang word fills a legitimate need it is accepted into the language. *Skyscraper, bus,* and *date* (as in "to go on a date"), for example, were once considered slang expressions. Although slang may have a valid place in some writing, particularly in fiction, be careful not to let it creep into technical writing.

Barbarism

A barbarism is any obvious misuse of standard words, grammatical forms, or expressions. Examples include *ain't, irregardless, done got, drownded.*

E

English as a second language (ESL)

Learning to write well in a second language takes a great deal of effort and practice. The most effective way to improve your written command of English is to read widely beyond reports and professional articles: Read magazine and newspaper articles, novels, biographies, short stories, or any other writing that interests you.

This entry is a guide to some of the more common problems nonnative speakers experience when writing English. Other entries of particular interest to ESL readers are marked with an ESL icon: ◑. Use the references at the end of this entry for further study. For specific help when you are writing a memo or report, ask a native speaker.

Persistent problem areas for nonnative speakers include the following:

- count and noncount **nouns**
- articles
- gerunds and infinitives
- adjective **clauses**
- present perfect **verb** tense

Count and Noncount Nouns

Count nouns refer to things that can be counted: *tables, pencils, projects, specialists.* Noncount nouns identify things that cannot be counted: *electricity, water, oil, air, wood, loyalty, pride,* and *harmony.*

The distinction between whether something can or cannot be counted is important for nonnative speakers to master because it determines the form of the noun to use (singular or plural), the kind of article that precedes it (*a, an, the,* or no article), and the kind of limiting **adjective** it requires (*fewer* or *less, much* or *many,* and so on). This distinction can be confusing with such words as *electricity* or *oil.* Although we can count Watt hours of electricity or barrels of oil, counting becomes inappropriate when we use the words *electricity* and *oil* in a general sense, as in "*Oil* is a limited resource."

Articles

This discussion of articles applies only to common nouns (not to proper nouns, such as the names of people) because count and non-count nouns are always common nouns.

The general rule is that every count noun must be preceded by an article *(a, an,* or *the)*, a demonstrative **adjective** *(this, that, these)*, a possessive adjective *(my, her, their,* and so on), or some expression of quantity (such as *one, two, several, many, a few, a lot of, some,* and *no)*. The article, adjective, or expression of quantity appears either directly in front of the noun or in front of the whole noun **phrase.**

E

> **EXAMPLES** Mary read *a* book last week. (The article *a* appears directly in front of the noun *book.*)
>
> Mary read *a* long, boring book last week. (The article *a* precedes the noun phrase *long, boring book.*)
>
> *Those* books Mary read were long and boring. (The demonstrative adjective *those* appears directly in front of the noun *books.*)
>
> *Their* book was long and boring. (The possessive adjective *their* appears directly in front of the noun *book.*)
>
> *Some* books Mary read were long and boring. (The indefinite adjective *some* appears directly in front of the noun *books.*)

The articles *a* and *an* are used with nouns that refer to any one thing out of the whole class of those items.

> **EXAMPLE** Bill has *a* pen. (Bill could have *any* pen.)

The article *the* is used with nouns that refer to a specific item that both the reader and writer can identify.

> **EXAMPLE** Bill has *the* pen. (Bill has *a specific* pen and both the reader and the writer know which specific one it is.)

The only exception to this rule occurs when the writer is making a generalization. When making generalizations with count nouns, writers can either use *a* or *an* with a singular count noun, or they can use no article with a plural count noun. Consider the following generalization using an article.

> **EXAMPLE** *An* egg is a good source of protein. *(any egg, all eggs, eggs in general)*

However, the following generalization uses a plural noun with no article.

> **EXAMPLE** Eggs are good sources of protein. *(any egg, all eggs, eggs in general)*

When making generalizations with noncount nouns, do not use an article in front of the noncount noun.

>   **EXAMPLE**   Sugar is bad for your teeth.

E   *Gerunds and Infinitives*

Nonnative writers are often puzzled by which form of a **verbal** (a verb used as a noun) to use when it functions as the direct **object** of a verb. No consistent rule exists for distinguishing between the use of an infinitive or a gerund after a verb when it is used as an object. Sometimes a verb takes an infinitive as its object, sometimes it takes a gerund, and sometimes it takes either an infinitive or a gerund. At times, even the base form of the verb is used.

Using a Gerund as a Complement

>   **EXAMPLES**   He enjoys *working*.
>   She denied *saying* that.
>   Did Alice finish *reading* the report?

Using an Infinitive as a Complement

>   **EXAMPLES**   He wants *to attend* the meeting in Los Angeles.
>   The company expects *to sign* the contract soon.
>   He promised *to fulfill* his part of the contract.

Using a Gerund or an Infinitive as a Complement

>   **EXAMPLES**   It began *raining* soon after we arrived. (gerund)
>   It began *to rain* soon after we arrived. (infinitive)

Using the Basic Verb Form as a Complement

>   **EXAMPLES**   Let Maria *finish* the project by herself.
>   The president had the technician *reassigned* to another project.

To make these distinctions accurately, you must rely on what you hear native speakers use or on what you read.

Adjective Clauses

Because of the variety of ways adjective clauses are constructed in different languages, they can be particularly troublesome for nonnative speakers. You need to remember a few guidelines when using adjective clauses in order to form them correctly.

Place an adjective clause directly after the noun it modifies.

| INCORRECT | The tall man is a vice-president of the company *who is standing across the room.* |
| CORRECT | The tall man *who is standing across the room* is a vice-president of the company. |

The adjective clause *who is standing across the room* modifies *man,* not *company,* and thus comes directly after *man.*

E

Avoid using a relative pronoun with another pronoun in an adjective clause.

INCORRECT	The man *who he* sits at that desk is my boss.
CORRECT	The man *who* sits at that desk is my boss.
INCORRECT	The man *whom* we met *him* at the meeting is on the board of directors.
CORRECT	The man *whom* we met at the meeting is on the board of directors.

Present Perfect Verb Tense

As a general rule, use the present perfect tense when referring to events completed in the past, but at nonspecified times. When a specific time is mentioned, use the simple past. Notice the difference in the two sentences following.

| EXAMPLES | I *wrote* the letter yesterday. (simple past tense *wrote.* The time when the action took place is mentioned.) |
| | I *have written* the letter. (present perfect tense *have written.* No specific time is mentioned; it could have been yesterday, last week, or ten years ago.) |

Use the present perfect tense to describe actions that were repeated several or many times in the unspecified past.

| EXAMPLES | She *has written* that report three times. |
| | The president and his chief advisor *have met* many times over the past few months. |

Use the present perfect with a *since* or *for* phrase when describing actions that began in the past and continue up to the present.

| EXAMPLES | This company *has been* in business *for* ten years. |
| | This company *has been* in business *since* 1983. |

Additional Writing Problems

Other writing problems for nonnative speakers include subject-verb **agreement** and connectives between sentences and paragraphs (see transition). **Preposition** combinations as they appear in **idioms** or two-word verb structures also create problems for nonnative speakers.

equal/unique/perfect

E

Logically, *equal* (meaning "having the same quantity or value as another"), *unique* (meaning "one of a kind"), and *perfect* (meaning "a state of highest excellence") are **absolute words** and therefore should not be compared. However, colloquial usage of *more* and *most* as **modifiers** of *equal, unique,* and *perfect* is so common that an absolute prohibition against such use is impossible.

> **EXAMPLE**   Yours is a *more unique* (or *perfect*) coin than mine.

Some writers try to overcome the problem by using *more nearly equal (unique, perfect)*. When **clarity** and preciseness are critical, the use of comparative degrees with *equal, unique,* and *perfect* can be misleading. The best rule of thumb is to avoid using the comparative degrees with absolute terms.

> **CHANGE**   Ours is a *more equal* percentage split than theirs.
> **TO**   Our percentage split is 51–49; theirs is 54–46.

-ese

The **suffix** *-ese* is used to designate types of **jargon** or certain languages or literary **styles** (official*ese*, journal*ese*, Chin*ese*, Pentagon*ese*).

etc.

Etc. is an **abbreviation** for the Latin *et cetera*, meaning "and others" or "and so forth"; therefore, *etc.* should not be used with *and*.

> **CHANGE**   He brought pencils, pads, erasers, a calculator, *and etc.*
> **TO**   He brought pencils, pads, erasers, a calculator, *etc.*

Do not use *etc.* at the end of a list or series introduced by the **phrases** *such as* or *for example* because these phrases already indicate that there are other things of the same category that are not named.

> **CHANGE**   He brought camping items, *such as* backpacks, sleeping bags, tents, *etc.*, even though he didn't need them.
> **TO**   He brought backpacks, sleeping bags, tents, *etc.*, even though he didn't need them.
> **OR**   He brought camping items, *such as* backpacks, sleeping bags, and tents, even though he didn't need them.

In careful writing, *etc.* should be used only when there is logical progression (1, 2, 3, etc.) and when at least two items are named. It is often better to avoid *etc.* altogether, however, because there may be confusion as to the identity of the "class" of items listed.

> **EXAMPLE**   He brought backpacks, sleeping bags, tents, and other camping items.

E

euphemism

A *euphemism* is a word that is an inoffensive substitute for one that could be distasteful, offensive, or too blunt.

> **EXAMPLES**   *remains* for *corpse*
> *passed away* for *died*
> *marketing representative* for *salesperson*
> *previously owned* or *preowned* for *used*

Used judiciously, a euphemism might help you avoid embarrassing or offending someone. Overused, however, euphemisms can hide the facts of a situation (such as *incident* for *accident*) or be a form of **affectation.** (See also **word choice.**)

everybody/everyone

Both *everybody* and *everyone* are usually considered singular and so take singular **verbs** and **pronouns.**

> **EXAMPLES**   *Everybody is* happy with the new contract.
> *Everyone* here *eats* at 11:30 a.m.
> *Everybody* at the meeting made *his* or *her* proposal separately.
> *Everyone* went *his* or *her* separate way after the meeting.

But the meaning can be obviously plural.

> **EXAMPLE**   *Everyone* laughed at my sales slogan, and I really couldn't blame *them.*

When the use of singular verbs and pronouns would be offensive by implying sexual bias, it is better to use plural verbs and pronouns or to use the expression "his or her."

> **CHANGE**   *Everyone* went *his* separate way after the meeting.
> **TO**   *They* all went *their* separate ways after the meeting.
> **OR**   *Everyone* went *his or her* separate way after the meeting.

Although normally written as one word, *everyone* is written as two words when each individual in a group should be emphasized.

EXAMPLES   *Everyone* here comes and goes as he pleases.
           Every one of the team members contributed to this discovery.

E

exclamation marks

The purpose of the exclamation mark (!) is to indicate the expression of strong feeling. It can signal surprise, fear, indignation, or excitement. In technical writing, the exclamation mark is normally used only in cautions and warnings. It cannot make an argument more convincing, lend force to a weak statement, or call attention to an intended irony.

Uses

The most common use of an exclamation mark is after a word (**interjection**), **phrase, clause,** or sentence to indicate surprise. Interjections are words that express strong emotion.

EXAMPLES   Stop! Hurry! Wow!

An exclamation mark can also be used after a whole sentence, or even an element of a sentence.

EXAMPLES   The subject of this meeting—please note it well!—is our budget deficit.

           How exciting is Stravinsky's *The Rite of Spring!*

An exclamation mark is sometimes used after a title that is an exclamatory word, phrase, or sentence.

EXAMPLES   "Our Information Retrieval System Must Change!" is an article by Richard Moody.

           The Cancer with No Cure! is a book by Wilbur Moody.

When used with **quotation marks,** the exclamation mark goes outside, unless what is quoted is an exclamation.

EXAMPLE   The boss yelled, "Get in here!" Then Ben, according to Ray, "jumped like a kangaroo"!

executive summaries

The purpose of an *executive summary* is to consolidate the principal points of a **report** in one place. It must cover the information in the report in enough detail to reflect accurately its contents but concisely enough to permit an executive to digest the significance of the report without having to read it in full. They are called executive summaries because the intended audience is the busy executive who may have to make funding, personnel, or policy decisions based on findings or recommendations reported for a project.

E

The executive summary is a comprehensive restatement of the document's **purpose, scope,** methods, results, **conclusions,** findings, and recommendations. The executive summary condenses the entire work or explains how the results were obtained or why the recommendations were made. It simply states the results and recommendations, providing only enough information for a reader to decide whether to read the entire work.

Because they are comprehensive, executive summaries tend to be proportional in length to the larger work they summarize. The typical summary is 10 percent of the length of the report.

The executive summary is usually organized according to the sequence of chapters or sections of the report it summarizes. Note that the sample summary mirrors the structure of the report, as can be seen from its **table of contents,** shown in Figure 1.

The executive summary should be written so that it can be read independently of the report. It must not refer by number to figures, **tables,** or references contained elsewhere in the report. Executive summaries do occasionally contain a figure, table, or footnote, a practice appropriate as long as that information is integral to the summary. Because executive summaries are frequently read in place of the full report, all uncommon **symbols, abbreviations,** and **acronyms** must be spelled out. (For guidance on the location of executive summaries in **reports,** see **formal reports** and **abstracts.**)

The sample executive summary in Figure 2 is adapted from a thirty-page report that describes how electric utilities in a select group of foreign countries provide engineering expertise to control room operators for round-the-clock shift work at nuclear power plants.

Executive summaries differ from **abstracts** and **introductions.** An abstract is intended to be a snapshot of a longer work, most useful for **readers** deciding whether to read the work in full based on its subject matter and purpose. An executive summary, by contrast, may be the only section of a longer work read by many readers, so it must cover the information in more detail. Although abstracts also summarize and highlight the major points of a report or article, they do so in a highly condensed form, seldom running longer than 200 or 250 words. Abstracts may also vary in the type of information they pro-

TABLE OF CONTENTS

iii

FIGURE 1 Table of Contents

vide. *Descriptive abstracts,* for example, include information about the **purpose, scope,** and methodology of the longer work, whereas *informative abstracts* expand on the scope of descriptive abstracts to include any results, **conclusions,** and recommendations arrived at in the works being abstracted.

   Introductions differ from abstracts and executive summaries in that they provide readers with essential background information

EXECUTIVE SUMMARY

Introduction

Purpose
and scope

This report describes the experiences and practices of utilities in a selected group of foreign countries with providing engineering expertise on shift in nuclear power plants.[1] The report also discusses the extent to which engineering expertise is made available and the alternative models of providing such expertise. The implications of the foreign experience for plants in the United States is described, particularly with reference to the shift technical adviser position and to a proposed shift engineer position.

Method

The relevant information for this study came from the open literature, interviews with utility staff, and utility reports.

The major conclusions that emerge from this study include:

- The basis for the initial decision of whether to include a graduate engineer in the operations shift complement is unclear for most countries. However, the reasons for changes in a given system have been identified. Generally, changes have been introduced in response to specific problems (such as high turnover rates among crew members, or an accident seen as related to available shift engineering expertise).

Major
findings

- Two primary models have been used to provide engineering expertise on shift:

    1. the graduate engineer as line manager of shift operations (usually in a shift supervisor position)
    2. the graduate engineer in a nonsupervisory position on shift

- The comparison of these two models did not indicate that one inherently functions more effectively than the other. However, each alternative appears to affect the following specific areas differently:

    1. crew relationships and performance
    2. labor supply, recruitment, and retention
    3. system implementation problems

---

[1] For the purpose of this discussion, "engineering expertise on shift" refers to the use of university-degreed engineers during round-the-clock shift work.

1

FIGURE 2 Executive Summary

needed to understand better the detailed information in the longer work. They announce the purpose, scope of coverage, and methodology used. They also describe how the longer work is organized, but as with descriptive abstracts, introductions exclude findings, conclusions, and recommendations. As part of essential background information, introductions may also highlight the relationship of the work to a specific project or program and discuss any special circumstances leading to the study, especially when a foreword or preface is not included.

E   Findings

- Data were not available to analyze whether alternative approaches to providing engineering expertise on shift could be directly linked to plant safety in terms of accident frequency and severity. Indirect relationships to operational safety appear to be primarily through impacts on operational functioning (for example, turnover and crew relationships) that ultimately are likely to affect safe operations.

- The determination of which alterative model would be most appropriate for a given country needs to be made from a systems perspective.

- Decisions about engineering expertise on shift should not be made independently of staffing patterns and organizational design regarding the following issues:

  1. effects on incumbent reactor operations staff of changed career opportunities
  2. the availability of degreed engineers and reactor operators
  3. career paths for graduate engineers at nuclear power plants
  4. organizational relationships between the engineering and operations actions of nuclear power plants.

Engineering Expertise Alternatives

In the countries surveyed, essentially two approaches are employed to make engineering expertise available on shift: (1) a graduate engineer occupies a line management position, and (2) a specific engineering position was created to provide expertise to the operations staff. Both approaches are relevant to issues under consideration in the United States. In the United States, proposals have been made to require that all licensed reactor operators have engineering degrees. This proposed requirement would apply to reactor operators, senior reactor operators, and shift supervisors. Proposals have also been made to create a position for a degreed shift engineer in addition to the current requirement for a shift technical advisor.

Currently, Spain, Italy, and Japan do not, as a rule, use graduate engineers on shift. On the other hand, Canada, the Federal Republic of Germany, and the United Kingdom do have an engineering graduate on shift in operations line management positions or have made the decision to establish this policy. The use of a separate engineer position on shift is found in Switzerland, France, and Sweden.

2

FIGURE 2   Executive Summary *(continued)*

Additional Guidelines on Writing an Executive Summary

1. Write the executive summary after completing the report.
2. Avoid using technical terminology if your readers will include people not familiar with the topic.
3. Make the summary concise but not brusque. Be especially careful not to omit transitional words and **phrases** (such as *however, moreover, therefore, for example,* and *in summary*).
4. Finally, introduce no information not discussed in the report.

Conclusions and Recommendation

The foreign experiences described in this report suggest some key issues relevant to assessing the present requirement in the United States of a shift technical adviser and alternative proposals, including creating a shift engineer position and requiring baccalaureate engineering degrees for shift supervisors. Either model, operations line management or a separate engineering position, involves some trade-offs with regard to system advantages.

One important issue concerns whether the separation of engineers from routine operational functions enhances or detracts from their ability to problem-diagnose and make appropriate decisions in situations of operational failure. No sufficient evidence exists to answer this question; both options have adherents based on positive experience with a given system.

Using a degreed engineer as shift supervisor appears to have advantages for crew integration and the combination of technical expertise and operations experience in the major position of authority on shift. However, significant problems have been experienced with the retention of graduate engineers as shift supervisors. Additionally, if this requirement were imposed in the United States, it would limit the career prospects for incumbent reactor operators, which could have adverse consequences on crew performance.

Recommendation

Creating a separate nonsupervisory engineering position could have advantages in providing job functions and a career path that more fully utilize engineering expertise than the shift supervisor position does. In addition, such a position does not require a change in the established recruitment and career path open to existing reactor operators.

3

FIGURE 2   Executive Summary *(continued)*

explaining a process (see **process explanation**)

expletives

An *expletive* is a word that fills the position of another word, **phrase,** or **clause.** *It* and *there* are the usual expletives.

EXAMPLE   *It is* certain that he will go.

In this example, the expletive *it* occupies the position of subject in place of the real subject, *that he will go.* Although expletives are sometimes necessary to avoid **awkwardness,** they are commonly overused and most sentences can be better stated without them.

CHANGE   *There are* several reasons that I did it.
TO   I did it for several reasons.

CHANGE   *There were* many orders lost for unexplained reasons.
TO   Many orders were lost for unexplained reasons.
OR   We lost many orders for unexplained reasons.

In addition to its usage as a grammatical term, the word *expletive* means an exclamation or oath, especially one that is profane.

explicit/implicit

An *explicit* statement is one expressed directly, with precision and clarity. An *implicit* meaning may be found within a statement, even though it is not directly expressed.

EXAMPLES   His directions to the new plant were *explicit,* and we found it with no trouble.

Although he did not mention the nation's financial condition, the danger of an economic recession was *implicit* in the President's speech.

exposition

Exposition, or expository writing, informs the **reader** by presenting facts and ideas in direct and concise language that is usually not adorned with colorful or figurative words and **phrases.** Expository writing attempts to explain to the reader what its subject is, how it works, and how it relates to something else. Exposition is aimed at the reader's understanding, rather than at his or her imagination or emotions; it is a sharing of the writer's knowledge with the reader. The most important function of exposition is to give accurate and complete information to the reader and analyze it.

Because it is the most effective **form of discourse** for explaining difficult subjects, exposition is widely used in technical writing. To write exposition, you must have a thorough knowledge of your subject. As with all writing, how much of that knowledge you pass on to your reader should depend on the reader's needs and your **purpose.**

fact

Expressions containing the word *fact* ("due to the *fact* that," "except for the *fact* that," "as a matter of *fact*," or "because of the *fact* that") are often wordy substitutes for more accurate terms.

CHANGE   *Due to the fact that* the sales force has a high turnover rate, sales have declined.

TO   *Because* the sales force has a high turnover rate, sales have declined.

The word *fact* is, of course, valid when facts are what is meant.

EXAMPLE   Our research has brought out numerous *facts* to support your proposal.

Do not use the word *fact*, however, to refer to matters of judgment or opinion.

CHANGE   *It is a fact that* sales are poor in the Midwest because of insufficient market research.

TO   In my opinion, sales are poor in the Midwest because of insufficient market research.

(See also **logic.**)

facsimile (fax)

Facsimile transmission, commonly known as *fax*, uses telephone lines to transmit images of whole pages—text and graphics—from one fax machine to another at a different location. A fax page takes about 20 seconds to go from coast to coast and costs no more than a phone call to the same destination for the same amount of time. Transmitting text and graphics quickly to others in distant locations, much like the **Internet,** fax transmission facilitates **collaborative writing** and many other kinds of remote communications tasks.

Because it helps you communicate with other locations more quickly, a fax machine enables you to be more productive. In many offices, however, fax copies arrive in a central location and the faxed documents may be viewed by others. Keep that in mind and avoid faxing confidential or sensitive messages if they might be viewed by the wrong people.

F

feasibility reports

When the managers of an organization plan to undertake a new project—a move, the development of a new product, an expansion, or the purchase of new equipment—they try to determine the project's chances for success. A *feasibility report* is the study conducted to help them make this determination. This **report** presents evidence about the practicality of the proposed project: How much will it cost? Is sufficient manpower available? Are any legal or other special requirements necessary? Based on the evidence, the writer of the feasibility report recommends whether or not the project should be carried out. Management officials then consider the recommendation.

The most efficient way to begin work on a feasibility report is to state clearly and concisely the purpose of the study.

> **EXAMPLE**   The purpose of this study is to determine what type and how many new vans should be purchased to expand our present delivery fleet.

In a feasibility study, the **scope** should include the alternatives for accomplishing the purpose and the criteria by which each alternative will be examined.

A firm needing to upgrade its word-processing software, for example, might conduct a feasibility study to determine which software would best suit its requirements. The firm's requirements would establish the criteria by which each alternative is evaluated. The following example shows how the preliminary topic outline for such a study might be organized:

> I. Purpose: To determine which word-processing software would best serve our office needs.
> II. Alternatives: List of software and their features from various vendors.

III. Criteria:
    A. Current task requirements
        1. Memos and letters—100 per month
        2. Brief interoffice reports—8–10 per month
        3. One or two 30- to 50-page reports per month
        4. Numerous financial tables: need to link data between spreadsheet programs and text tables
        5. Occasional need to create and edit business graphics—bar and pie charts
    B. Compatibility with present hardware
        1. Need to upgrade hard-disk memory?
        2. Need to upgrade printers and purchase font cartridges?
    C. Costs
        1. Purchase of new software or upgrade of present software?
        2. Installation and transfer of existing working files
        3. Training of professional and secretarial staffs

In writing a feasibility report, you must first identify the alternatives and then evaluate each against your established criteria. After completing these analyses, summarize them in a conclusion. This summary of relative strengths and weaknesses usually points to one alternative as the best, or most feasible. Make your recommendation on the basis of this conclusion.

Although the order of elements may vary, every feasibility report should contain the following sections: (1) an **introduction,** (2) a body, (3) a **conclusion,** and (4) a recommendation.

Introduction

The introduction should state the purpose of the report, describe the problems that led to it, and include any pertinent background information. You may discuss the scope or extent of the report and any procedures or methods used in the analyses of alternatives.

Body

The body of the report should present a detailed evaluation of all alternatives under consideration. Evaluate each alternative according to your established criteria. Ordinarily, each evaluation would comprise a separate section of the report.

Conclusion

The conclusion should summarize the evaluation of each alternative, usually in the order in which they are discussed in the body of the report. In your conclusion you may also want to interpret the detailed evaluations.

Recommendation

This section must state the alternative that best meets the criteria.

Sample Feasibility Report

The sample feasibility report shown in Figure 1 opens with an introduction that states the **purpose** of the report, the problem that prompted the study, the alternatives that were examined, and the criteria used. The body presents a detailed discussion of each alternative, particularly in terms of the criteria stated in the introduction. The conclusion draws together and summarizes the details in the body of the study. The recommendation section suggests the course of action that the company should take.

Introduction

The purpose of this report is to determine which of two proposed computer processors would best enable the Jonesville Engineering and Manufacturing Branch to increase its data-processing capacity and thus to meet its expanding production requirements.

Problem

In October 19-- the Information Systems and Support Group at Jonesville put the MISSION System into operation. Since then the volume of processing transactions has increased fivefold. This increase has severely impaired system response time. Degraded performance is also apparent in the backlog of batch-processing transactions. During a recent check 70 real-time and approximately 2,000 secondary transactions were backlogged. In addition, the ABLE 98 Processor that runs MISSION is nine years old and frequently breaks down. Downtime caused by these repairs must be made up in overtime. In a recent 10-day period in January, processor downtime averaged 25 percent during working hours (7:30 a.m. to 6:00 p.m.). In February the system was down often enough that the entire plant production schedule was endangered.

Finally, because the ABLE 98 cannot keep up with the current workload, the following new systems, all essential to increased plant efficiency and productivity, cannot be implemented: shipping and billing, labor collection, master scheduling, and capacity planning.

Scope

Two alternative solutions to provide increased processing capacity have been investigated: (1) purchase of a new ABLE 98 Processor to supplement the first, and

---

FIGURE 1 Feasibility Report

(2) purchase of a Landmark I Processor to replace the current ABLE 98. The two alternatives will be evaluated primarily according to cost and, to a lesser extent, according to expanded capacity for future operations.

Purchasing a Second ABLE 98 Processor

This alternative would require additional annual maintenance costs, salary for an additional computer operator, increased energy costs, and a one-time construction cost for a new facility to house the processor.

Annual maintenance costs	$ 45,000
Annual salary for computer operator	28,000
Annual increased energy costs	7,500
Annual operating costs	$ 80,500
Facility cost (one-time)	$ 50,000
Total first-year cost	$130,500

These costs for the installation and operation of another ABLE 98 Processor are expected to produce the following anticipated savings in hardware reliability and system readiness.

Hardware Reliability

A second ABLE 98 would reduce current downtime periods from four to two per week. Downtime recovery averages 30 minutes and affects 40 users. Assuming that 50 percent of users require the system at a given time, we determined that the following reliability savings would result:

> 2 downtimes × 0.5 hours × 40 users × 50% × $12.00/
> hour overtime × 52 weeks = $12,480 (annual savings)

System Readiness

Currently, an average of one day of processing per week cannot be completed. This gap prevents online system readiness when users report to work and affects all users at least one hour per week. Improved productivity would yield these savings:

> 40 users × 1 hour/week × $9.00/hour average wage rate
> × 52 weeks = $18,720 (annual savings)

Summary of Savings

Hardware reliability	$12,480
System readiness	18,720
Total annual savings	$31,200

Costs and Savings for ABLE 98 Processor

Costs	
Annual	$ 80,500
One-time	50,000
First-year total	$130,500
	−50,000
Annual total	$80,500

FIGURE 1   Feasibility Report *(continued)*

F

Savings	
Hardware reliability	$ 12,480
System readiness	18,720
Total annual savings	$ 31,200
Annual Costs Less Savings	
Annual costs	$80,500
Annual savings	−31,200
Net additional annual operating cost	$49,300

ABLE 98 Capacity

By adding a second ABLE 98 processor, current capacity will be doubled. However, if new systems essential to increased plant productivity are added to the MISSION System, efficiency could be degraded to its present level in the next three to five years. This estimate is based on the assumption that the new systems will add a significant number of transactions per day immediately. These figures could increase tenfold in the next several years if current rates of expansion continue.

Purchasing a Landmark I Processor

This alternative will require additional annual maintenance costs, increased energy costs, and a one-time facility adaptation cost.

Annual maintenance costs	$ 75,000
Annual energy costs	9,000
Annual operating costs	$ 84,000
Cost of adapting existing facility	$ 24,500
Total first-year cost	$108,500

These costs for installation of the Landmark I Processor are expected to produce the following anticipated savings in hardware reliability, system readiness, and staffing for the Information Systems and Services Department.

Hardware Reliability

Annual savings will be the same as those for the ABLE 98 Processor: $12,480.

System Readiness

Annual savings will be the same as those for the ABLE Processor: $18,720.

Wages for the Information Systems and Services Department

New system efficiencies would permit the following wage reductions in the department:

One computer operator (wages and fringe benefits)	$28,000
One-shift overtime premium (at $200/week × 52 weeks)	10,400
Total annual savings	$38,400

Summary of Savings

Hardware reliability	$12,480
System readiness	18,720

FIGURE 1 Feasibility Report *(continued)*

Wages	38,400
Total annual savings	$69,600

Costs and Savings for Landmark I Processor

Costs

Annual	$84,000
One-time	24,500
First-year total	$108,500
	24,500
Annual total	$84,000

Savings

Hardware reliability	$ 12,480
System readiness	18,720
Wages	38,400
Total annual savings	$ 69,600

Annual Cost Less Savings

Annual costs	$ 84,000
Annual savings	−69,600
Net additional annual operating cost	$ 14,400

Landmark I Capacity

The Landmark I processor can process twice as many transactions per day as our current system. This increase in capacity over the present system would permit implementation of plans to add four new systems to MISSION.

Conclusion

A comparison of costs for both systems indicates that the Landmark I would cost $60,400 less in first-year costs

ABLE 98 Costs

Net additional operating	$49,300
One-time facility	50,000
First-year total	$99,300

Landmark I Costs

Net additional operating	$14,400
One-time facility	24,500
First-year total	$38,900

Installation of a second ABLE 98 Processor will permit the present information-processing systems to operate relatively smoothly and efficiently. It will not, however, provide the expanded processing capacity that the Landmark I Processor would for implementing new subsystems essential to improved production and record keeping.

Recommendation

The Landmark I Processor should be purchased because of the initial and long-term savings and because its expanded capacity will allow the addition of essential systems.

FIGURE 1 Feasibility Report *(continued)*

female

Female is usually restricted to scientific, legal, or medical contexts (a *female* patient or suspect). Keep in mind that this term sounds cold and impersonal. The terms *girl*, *woman*, and *lady* are acceptable substitutes in other contexts; however, be aware that these substitute words have connotations involving age, dignity, and social position. (See also **male.**)

few/a few

In certain contexts, *few* carries more negative overtones than the **phrase** *a few* does.

> **EXAMPLES** You have made *a few* good points in your report. (positive)
> You have made *few* good points in your report. (negative)

fewer/less

Fewer refers to items that can be counted (count nouns).

> **EXAMPLES** A good diet can mean *fewer* colds.
> *Fewer* members took the offer than we expected.

Less refers to mass quantities or amounts (mass nouns).

> **EXAMPLES** *Less* vitamin C in your diet may mean more, not fewer, colds.
> The crop yield decreased this year because we had *less* rain than necessary for an optimum yield.

figuratively/literally

These two words are often confused. *Literally* means "really" and should not be used in place of *figuratively*, which means "metaphorically." Do not say that someone "literally turned green with envy" unless that person actually changed color.

> **EXAMPLES** In the winner's circle the jockey was, *figuratively* speaking, ten feet tall.
> When he said, "Let's run it up the flagpole," he did not mean it *literally*.

Avoid the use of *literally* to reinforce the importance of something.

CHANGE   She was *literally* the best of the group.
TO   She was the best of the group.

figures of speech

A *figure of speech* is an imaginative **comparison,** either stated or implied, between two things that are basically unlike but have at least one thing in common. If a device is cone shaped with an opening at the top, for example, you might say that it looks like a volcano.

Technical people may find themselves using figures of speech to clarify the unfamiliar by relating a new and difficult concept to one with which the **reader** is familiar. In this respect, figures of speech help establish a common ground of understanding between the specialist and the nonspecialist. Technical people may also use figures of speech to help translate the abstract into the concrete; in the process of doing so, figures of speech also make writing more colorful and graphic.

Although figures of speech are not used extensively in technical writing, a particularly apt figure of speech may be just the right tool when you must explain or describe a complex concept. A figure of speech must be appropriate, however, to achieve the desired effect.

CHANGE   Without the fuel of tax incentives, our economic engine would operate less efficiently. (It would not operate at all without fuel.)
TO   Without the fuel of tax incentives, our economic engine would sputter and die. (This is not only apt, but it also states a rather dry fact in a colorful manner—always a desirable objective.)

A figure of speech must also be consistent to be effective.

CHANGE   We must get our research program back *on the track*, and we are counting on you to *carry the ball.* (inconsistent)
TO   We must get our research program back on the track, and we are counting on you to do it. (inconsistency removed)

A figure of speech should not, however, attract more attention to itself than to the point the writer is making.

EXAMPLE   The whine of the engine sounded like ten thousand cats having their tails pulled by ten thousand mischievous children.

Trite figures of speech, which are called **clichés,** defeat the purpose of a figure of speech—to be fresh, original, and vivid. A surprise that comes "like a bolt out of the blue" is not much of a surprise. It is better to use no figure of speech than to use a trite one.

Types of Figures of Speech

Analogy is a **comparison** between two objects or concepts to show ways in which they are similar. In effect, analogies say, "A is to B as C is to D."

EXAMPLE   Pollution is to the environment as cancer is to the body.

The resemblance between the concepts represented in an analogy must be close enough to illuminate the relationship the writer wants to establish. Analogies may be brief or extended, depending on the writer's purpose. Analogy often helps writers explain unfamiliar things by comparing them to things with which the reader is familiar.

Analogy can be a particularly useful tool to the technical person writing for an intelligent and educated but nontechnical audience, such as top management, because of its effectiveness in **defining terms** and explaining processes. Like all figurative language, analogy can provide a shortcut means of communication if it is used with care and restraint.

The following example explains a computer search technique for finding information in a data file by comparing it to the method used by most people to find a word in a **dictionary.**

> The search technique used in this kind of file processing is similar to the search technique used to look up a word in a dictionary. To locate a specific word, you scan the key words located at the top of each dictionary page (these key words identify the first and last words on the page) until you find the key words that confine your word to a specific page. Assume that all the key words that reference the last word on each page of a dictionary were placed in a file, along with their corresponding page numbers, and that all the words in the dictionary were placed in another file. To locate any word, you would simply scan the first file until you found a key word greater in alphabetical sequence than the desired word and go to the place in the second file indicated by the key word to find the desired word. This search technique is called *indexed sequential processing.*

Metaphor is a figure of speech that points out similarities between two things by referring to one of them as being the other. Metaphor states that the thing being described is the thing to which it is being compared.

EXAMPLE   The building site was a *beehive*, with iron workers still at work on the tenth floor, plumbers installing fixtures on the floors just beneath them, electricians busy on the middle floors, and carpenters putting the finishing touches on the first floor.

The use of metaphor often helps clarify complex theories or objects. For example, the life-sustaining tube connecting space-walking astronauts to their oxygen supply in the spacecraft is called an *umbilical cord.* The person who first drew this comparison knew that such a life-support tube was new and unfamiliar, so to make its function immediately apparent he used a metaphor that compared it to something with which people were already familiar. A similar purpose was served when the term *window* was used to describe the abstract concept of the point at which a spacecraft reenters the earth's atmosphere. Here the strange and faraway was made familiar and near by the metaphorical use of a term familiar to everyone. Both of these expressions have gained widespread use because they helped clarify new and unfamiliar concepts.

A word of caution: Some writers sometimes mix metaphors, which usually results in an illogical statement.

> **EXAMPLE**   Billingham's proposal *backfired,* and now she is *in hot water.*

This example makes an incongruous statement because a backfiring automobile does not land someone in hot water.

A *simile* is a direct comparison of two essentially unlike things, making the comparison with the word *like* or *as.* Similes state that A is like B.

> **EXAMPLE**   Constructing the frame is *like* piecing together the borders of a jigsaw puzzle.

Like metaphors, similes can help illuminate difficult or obscure ideas.

> **EXAMPLE**   The odds against having your plant or business destroyed by fire are at least as great as the odds against hitting the jackpot on a Las Vegas slot machine. Because of the long odds, top management may consider sophisticated fire protection and alarm systems more costly than the risk merits and refuse to approve the expenditure.
>
> And yet, plants are destroyed by fire, just as gamblers sometimes hit the jackpot on a one-armed bandit. Losing big if fire strikes the plant or winning big on a slot machine both depend on the chance alignment of three events or factors. . . .

Hyperbole is exaggeration to achieve an effect or **emphasis.** It is a device used to emphasize or intensify a point by going beyond the literal facts. It distorts to heighten effect.

> **EXAMPLES**   They *murdered* us at the negotiating session.
> The hail was *like boulders.*

Used cautiously, hyperbole can magnify an idea without distorting it; therefore, it plays a large role in advertising.

    EXAMPLE   A box of Clenso detergent can clean a mountain of clothes.

Everyday speech abounds in hyperbole.

    EXAMPLES   I'm starved. (meaning "hungry")
                 I'm dead. (meaning "tired")

Because technical writing needs to be as accurate and precise as possible, always avoid hyperbole.

Antithesis is a statement in which two contrasting ideas are set off against each other in a balanced syntactical structure.

    EXAMPLES   Man proposes but God disposes.
                 Art is long but life is short.

Litotes are understatements, for emphasis or effect, achieved by denying the opposite of the point you are making.

    EXAMPLES   Einstein was no dummy.
                 Seventy dollars is not a small price for a book.

Metonymy is a figure of speech that uses one aspect of a thing to represent it, such as *the red, white, and blue* for the American flag, *the blue* for the sky, and *wheels* for an automobile. This device is common in everyday speech because it gives our expressions a colorful twist.

    EXAMPLE   *The hard hat* area of the labor force was especially hurt by the current economic recession.

Personification is a figure of speech that attributes human characteristics to nonhuman things or abstract ideas. One characteristically speaks of the *birth* of a planet and the *stubbornness* of an engine that will not start.

    EXAMPLE   Early tribes of human beings attributed scientific discoveries to gods. To them, fire was not *a child of man's brain* but a gift from Prometheus.

fine

When used in expressions such as "I feel *fine*" or "a *fine* surf," *fine* is colloquial. The colloquial use of *fine*, like that of *nice*, is too vague for technical writing. In writing, the word *fine* should retain the sense of "refined," "delicate," or "pure."

    EXAMPLES   A *fine* film of oil covered the surface of the water.

There was a *fine* distinction between the two possible meanings of the disputed passage.

Fine crystal is currently made in Austria.

first/firstly

Firstly—like *secondly, thirdly . . . lastly*—is an unnecessary attempt to add the *-ly* form to an adverb. *First* is an **adverb** in its own right, and sounds much less stiff than *firstly*.

F

> CHANGE *Firstly,* we should ask for an estimate.
> TO *First,* we should ask for an estimate.

flammable/inflammable/nonflammable

Both *flammable* and *inflammable* mean "capable of being set on fire." Since the *-in* **prefix** usually causes the word following to take its opposite meaning *(incapable, incompetent), flammable* is preferable to *inflammable* because it avoids possible misunderstanding.

> EXAMPLE The cargo of gasoline is *flammable.*

Nonflammable is the opposite, meaning "not capable of being set on fire."

> EXAMPLE The asbestos suit was *nonflammable.*

flowcharts

A *flowchart* is a diagram of a process that involves stages, with the sequence of stages shown from beginning to end. The flowchart presents an overview of the process that allows the **reader** to grasp the essential steps of the process quickly and easily. The process being illustrated could range from the steps involved in assembling a bicycle to the stages by which bauxite ore is refined into aluminum ingots for fabrication.

Flowcharts can take several forms to represent the steps in a process. They can consist of labeled blocks (Figure 1), pictorial representations

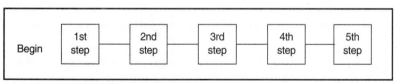

FIGURE 1 Example of a Block Flowchart

F

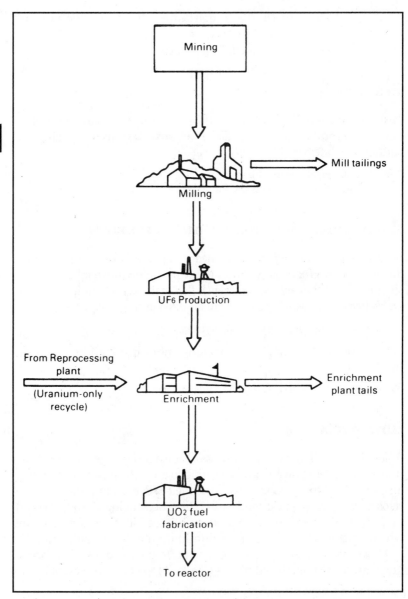

FIGURE 2   Pictorial Flowchart of Light-Water Reactor Uranium Fuel Cycle

(Figure 2), or standardized symbols (Figure 3); the items in any flow-chart are always connected according to the sequence in which the steps occur. The normal direction of flow in a chart is left to right or

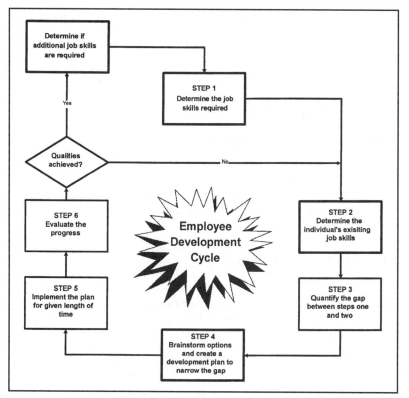

FIGURE 3   Flowchart of an Employee Development Cycle

top to bottom. When the flow is otherwise, be sure to indicate it with arrows.

Flowcharts that document computer programs and other information-processing procedures use standardized **symbols.** The standards are set forth in *Information Processing—Documentation Symbols and Conventions for Data, Program and System Flowcharts, Program Network Charts and System Resources Charts,* published by the International Organization for Standardization, publication ISO 5807–1985 (E).

When creating a flowchart, follow the general guidelines for creating and integrating **illustrations** with text in the illustration entry, in addition to the following guidelines specific to flowcharts:

1. Label each step in the process, or identify it with a conventional symbol. Steps can also be represented pictorially or by labeled blocks.
2. With flowcharts made up of labeled blocks and standardized symbols, use arrows to show the direction of flow only if the flow is oppo-

site to the normal direction. With pictorial representation, use arrows to show the direction of all flow.

3. Include a key if the flowchart contains symbols your readers may not understand.

forceful/forcible

F

Although *forceful* and *forcible* both are **adjectives** meaning "characterized by or full of force," *forceful* is usually limited to persuasive ability and *forcible* to physical force.

> **EXAMPLES**   John made a *forceful* presentation at the committee meeting. The thief made a *forcible* entry into my apartment.

foreign words in English

The English language has a long history of borrowing words from other languages. Most of these borrowings occurred so long ago that we seldom recognize the borrowed terms (also called *loan words*) as being of foreign origin.

> **EXAMPLES**   whiskey (Gaelic), animal (Latin), church (Greek)

Guidelines for Use of Foreign Words in English

Foreign expressions should be used only if they serve a real need. The overuse of foreign words in an attempt to impress your **reader** or to be elegant is **affectation.** Your goal of effective communication can be accomplished only if your reader understands what you write; choose foreign expressions, therefore, only when they make an idea clearer.

Words not fully assimilated into the English language are set in **italics** if printed (underlined in typed manuscript).

> **EXAMPLES**   *sine qua non, coup de grâce, in res, in camera*

Words and abbreviations that have been fully assimilated need not be italicized. When in doubt, consult a current dictionary.

> **EXAMPLES**   cliché, etiquette, vis-à-vis, de facto, résumé, etc., i.e., e.g.

As foreign words become current in English, their plural forms give way to English plurals.

> **EXAMPLES**   *formulae* becomes *formulas*
> *agenda* becomes *agendas*

In addition, accent marks tend to be dropped from words (especially from French words) the longer they are used in English. But because not all foreign words shed their accent marks with the passage of time, consult a dictionary whenever you are in doubt. (See also **e.g./i.e.** and **etc.**)

foreword/forward

Although the pronunciation is the same, the spellings and meanings of these two words are quite different. The word *foreword* is a **noun** meaning "introductory statement at the beginning of a book or other work."

> EXAMPLE   The department chairperson was asked to write a *foreword* for the professor's book.

The word *forward* is an **adjective** or **adverb** meaning "at or toward the front."

> EXAMPLES   Move the lever to the *forward* position on the panel. (adjective)
> Turn the dial until the needle begins to move *forward*. (adverb)

foreword/preface (see **formal reports**)

formal reports

Formal reports are the written accounts of major projects. Projects that are likely to produce formal reports include research into new developments in a field, explorations of the advisability of launching a new product or an expanded service, or an end-of-year review of developments within an organization. The scope and complexity of the project will determine how long and how complex the report should be. Most formal reports—certainly those that are long and complex—require a carefully planned structure that offers the **readers** an easy-to-recognize guide to the material in the reports. Such aids as a **table of contents,** an **abstract,** and an **executive summary** make the information in the report more accessible. Making a formal topic outline that lists the major facts and ideas in the report and indicates their relationship to one another should help you write a well-organized report.

Most formal reports are divided into three major parts—front matter, body, and back matter—each of which contains a number of elements. Just how many elements are needed for a particular report depends on the subject, the length of the report, and the kinds of material covered. In fact, the number and arrangement of the elements in a formal report may vary.

Order of Elements in a Formal Report

Many companies and governmental and other institutions have a preferred style for formal reports and furnish guidelines that staff members must follow. If your employer has prepared a set of style guidelines, follow it; if not, use the format recommended in this entry. The following list includes most of the elements a formal report might contain:

Front Matter
> Title page
> Abstract
> Table of contents
> List of figures
> List of tables
> Foreword
> Preface
> List of abbreviations and symbols

Body
> Executive summary
> Introduction
> Text (including headings)
> Conclusions
> Recommendations
> References

Back Matter
> Bibliography
> Appendixes
> Glossary
> Index

Front Matter

The front matter, which includes all the elements that precede the body of the report, serves several purposes: it gives the reader a general idea of the author's purpose in writing the report; it indicates whether the report contains the kind of information the reader is looking for; and it lists where in the report the reader can find specific chapters, **headings, illustrations,** and **tables.** Not all formal reports require every element of front matter. A title page and **table of contents** are usually mandatory, but whether an **abstract,** a preface, and **lists** of figures, **tables, abbreviations,** and **symbols** are included will depend on the **scope** of the report and its intended audience. The front matter pages are numbered with lowercase roman numerals. Throughout the report, page numbers should be centered near the bottom of each page.

Title Page. The formats of title pages for formal reports vary, but the page should include the following information:

F

1. The full **title** of the report. The title should indicate the topic and announce the **scope** and **purpose** of the report. Titles often provide the only basis on which readers can decide whether to read a report. Titles that are too vague or too long not only hinder readers but also can prevent efficient filing and information retrieval by librarians and other information specialists. Follow these guidelines when creating the title:

   - Do not use "Report on . . .," "Technical Report on . . .," or "XYZ Corporation Report on . . .," in the title, since the fact that the information appears in a report will be self-evident to your reader.
   - Do not use **abbreviations** in the title. Use **acronyms** only when the report is intended for an audience familiar enough with the topic that the acronym will not confuse them.
   - Do not include the period covered by a report in the title; include that information in a subtitle:

     EFFECTS OF PROPOSED HIGHWAY
     CONSTRUCTION ON PROPERTY VALUES
     Annual Report, 19—

2. The name of the writer, principal investigator, or compiler. Frequently, contributors simply list their names and almost never list their academic degrees. Sometimes they identify themselves by their job title in the organization (Jane R. Doe, Cost Analyst; Jack T. Doe, Head, Research and Development). Sometimes contributors identify themselves by their tasks in contributing to the report (Jane R. Doe, Compiler; Jack T. Doe, Principal Investigator).

3. The date or dates of the report. For one-time reports, list the date when the report is to be distributed. For periodic reports, which may be issued monthly or quarterly, list in a subtitle the time period that the present report covers. Elsewhere on the title page, list the date when the report is to be distributed.

4. The name of the organization for which the writer works.

5. The name of the organization to which the report is being submitted, if the work is being done for a customer or client.

These categories are standard on most title pages, but some organizations may require additional information.

The title page, although unnumbered, is considered page i (small Roman numeral 1). The back of the title page, which is blank and unnumbered, is considered page ii, and the abstract then falls on page iii so that it appears on a right-hand (i.e., odd-numbered) page. For reports with printing on both sides of each sheet of paper, right-hand pages are always odd-numbered and left-hand pages are always even-numbered. (Note the pagination in this book.) New sections and chapters of reports typically begin on a new right-hand page. Reports with printing on only one side of each sheet can be numbered consecutively regardless of where new sections begin.

Abstract. An abstract, which normally follows the title page and is page iii, is a condensed version of a longer piece of writing that summarizes and highlights the major points, enabling the prospective reader to decide whether to read the entire report. Usually 200 to 250 words or less, abstracts must make sense independently of the works they summarize. (For a complete discussion of abstract writing supported with samples, see **abstracts.**)

Table of Contents. A table of contents lists all the major **headings** or sections of the report in their order of appearance, along with their page numbers. It lists all front matter and back matter except the title page and the table of contents itself. The table of contents begins on a new page, numbered page iv, because it follows the title page (page i) and the abstract (page iii). By convention, the table of contents and its page number are not listed in the table of contents page.

Along with the abstract, a table of contents permits the reader to assess the value of a report. It also aids a reader who may want to look at only certain sections of the report. For this reason, the wording of chapter and section titles in the table of contents should be identical to those in the text. (For a sample contents page, see **table of contents.**) The table of contents need not list all headings in the report. Generally no more than three levels are listed. However, when any heading of a given level is listed, all headings of that level must also be listed in the contents.

Sometimes the table of contents is followed by lists of the figures and tables contained in the report. These lists should always be presented separately, and a page number should be given for each item.

List of Figures. Figures include all **illustrations—drawings, photographs, maps,** charts, and **graphs**—contained in the report. When a report contains more than five figures, they should be listed, along with their page numbers, in a separate section immediately following the table of contents. The section should be entitled "List of Figures" and should begin on a new page. Figure numbers, titles, and page numbers should be identical to those in the text. In most reports figures are numbered consecutively with Arabic numbers.

List of Tables. When a report contains more than five tables, they should be listed, along with their titles and page numbers, in a separate section entitled "List of Tables," immediately following the list of figures (if there is one). Tables are numbered consecutively with Arabic numbers throughout most reports.

Foreword. A foreword is an optional introductory statement written by someone other than the author. The writer of the foreword is usually an authority in the field, whose name and affiliation (as well as the date the statement was written) appear at the end of the foreword. The foreword provides background information about the study's significance or places it in the context of other works written in the field. If the work was done under contract to a sponsoring organization, the manager of the project at the sponsoring organization may indicate how the work pertains to specific programs or goals of that organization. The foreword always precedes the preface when a work has both.

Preface. The preface is an optional introductory statement, usually written by the author, that announces the purpose, background, and scope of the report. Sometimes the preface specifies the audience for whom the report is intended; it may also highlight the relationship of the report to a given project or program. A preface may contain acknowledgments of help received during the course of the project or in the preparation of the report. Finally, it may cite permission obtained for the use of copyrighted works. If the report does not require a preface, place this type of information, if it is essential, in the introduction.

The preface follows the table of contents (and the lists of figures or tables if they are included) and it begins on a separate page entitled "Preface."

List of Abbreviations and Symbols. When the abbreviations and symbols used in the report are numerous and there is a chance that the reader will not be able to interpret them, the front matter should include a list of all symbols and abbreviations (including **acronyms**) and what they stand for in the report. Such a list, which follows the preface, is particularly helpful for reports on specialized subjects whose audience will include nonspecialists. Such a list, appropriately titled, begins on a new page.

Body (Text)

F

The body is that portion of the report in which the author introduces the subject, describes in detail the methods and procedures used in the study, demonstrates how results were obtained, and draws conclusions on which any recommendations are based.

The first page of the body is numbered 1. This and all subsequent pages in the report are usually numbered in Arabic numbers, unlike the front matter, which is numbered in lowercase Roman numerals.

Executive Summary. The body of the report begins with an executive summary that provides a more complete overview of a report than either a descriptive or an informative abstract does. The summary states the purpose and nature of the investigation and gives major findings, conclusions, and—if any are to be made—recommendations. It also provides an account of the procedures used to conduct the study. Although more complete than an abstract, the summary should not contain a detailed description of the work on which the findings, conclusions, and recommendations were based. The size of an executive summary is proportional to the size of the report; a rule of thumb is that the length of the executive summary should be approximately 10 percent of the length of the report.

Some summaries are written to follow the organization of the report. Others highlight the findings, conclusions, and recommendations by summarizing them first before going on to discuss procedures or methodology.

A summary enables people who may not have time to read a lengthy report to scan its primary points and then decide whether they need to read the entire report. Because they are often intended for busy executives, summaries are sometimes called *executive summaries*. Addressed primarily to this audience, summaries often appear as the first section of the body of a report. Like abstracts, summaries should not contain tables, illustrations, or bibliographic citations. The summary is considered part of the body of the report. (For a more detailed discussion of summaries and a sample summary, see **executive summaries.**)

Introduction. The purpose of an introduction is to give the readers any general information they must have in order to understand the detailed information in the rest of the report. An introduction sets the stage for the report, providing a frame of reference for the details contained in the body of the report. In writing the introduction, you need to state the subject, the purpose, the scope, and the way you plan to develop the topic. You may also describe how the report will be organized. (For a detailed discussion of introductions, with samples, see **introductions.**)

Text. Generally the longest section of the report, the text (or body) presents the details of how the topic was investigated, how the problem was solved, how the best choice from among alternatives was selected, or whatever else the report covers. This information is often clarified and further developed by the use of illustrations and tables and may be supported by references to other studies.

Most formal reports have no single best organization. How the text is organized will depend on the topic and on how you have investigated it (survey, analysis, feasibility study, laboratory experiment, or whatever). The text is ordinarily divided into several major sections, comparable to the chapters in a book. These sections are then subdivided to reflect the logical divisions in your main sections. (For a discussion of the writing process, see "Five Steps to Successful Writing" at the beginning of this book.)

Conclusions. The conclusion section pulls together the results of your study in one place. Here you tell the reader what you have learned and its implications. You show how the results follow from the study objectives and method. You should also point out any unexpected results. (For a fuller discussion of this section of a report, see **conclusions.**)

Recommendations. Recommendations, which are sometimes combined with the conclusions section, state what course of action should be taken based on the results of the study. Of the three possible locations for a new warehouse, which is the best? Which make of delivery van should be purchased to replace the existing fleet? Which contractor has submitted the best proposal for remodeling the company cafeteria? Which type of microcomputer equipment should the office buy? Should the equipment be leased rather than purchased? The recommendations section says, in effect, "I think we should purchase this, or do that, or hire them."

The emphasis is on the verb *should.* Recommendations advise the reader on the best course of action based on the researcher's findings. Generally someone up the organizational ladder, or a customer or client, makes the final decision about whether to accept the recommendations.

References. If in your report, you refer to material in or quote directly from a published work or other research source, you must provide a list of references in a separate section called *references.* If your employer has a preferred reference style, follow it; otherwise, use the guidelines provided in **documenting sources.**

F

   For a relatively short report, the references should go at the end of the report. For a report with a number of sections or chapters, the reference section should fall at the end of each major section or chapter. In either case, every reference section should be labeled as such and should start on a new page. If a particular reference appears in more than one section or chapter, it should be repeated in full in each appropriate reference section.

Back Matter

The back matter of a formal report contains supplemental information, such as where to find additional information about the topic (**bibliography**) and how the information in the report can be easily located (**index**), clarified (**glossary**), and explained in more detail (**appendix**).

Bibliography. A bibliography is a list, usually in alphabetical order, of all sources that were consulted in researching the report but that are not cited in the text. A bibliography may not be necessary, however, when the reference listing contains a complete list of sources. When a report has both, the bibliography follows the reference section. This section is often entitled "Works Cited," especially when the sources include nonprint information like computer codes, microfiche, or video material.

   Like the other elements in the front and back matter, the bibliography starts on a new page and is labeled by name. (For detailed information about creating a bibliography, see **bibliography.**)

Appendixes. An appendix contains information that clarifies or supplements the text. This type of information is placed at the back of the report because it is too detailed or voluminous to appear in the text without impeding the orderly presentation of ideas. Material typically placed in an appendix includes long charts and supplementary **graphs** or **tables,** copies of **questionnaires** and other material used in gathering information, texts of interviews, pertinent **correspondence,** and explanations too long for explanatory footnotes but helpful to the reader who is seeking further assistance or clarification.

A report may have one or more appendixes; generally, each appendix contains one type of material. For example, a report with two appendixes may have a questionnaire in one and a detailed computer printout tabulating the questionnaire's results in the other.

Place the first appendix on a new page directly after the bibliography. Additional appendixes also begin on a new page. Identify each with a title and a heading. Appendixes are ordinarily labeled Appendix A, Appendix B, and so on. If your report has only one appendix, simply label it Appendix, followed by its title. To call it Appendix A would imply that an Appendix B will follow. (For a detailed discussion of appendixes, see **appendix/appendixes/appendices.**)

Glossary. A glossary is a list of selected terms on a particular subject that are defined and explained. Include a glossary only if your report contains many words and expressions that will be unfamiliar to your intended readers. Arrange the terms alphabetically, with each entry beginning on a new line. The definitions then follow each term, dictionary style. The glossary, labeled as such, appears directly after the appendix and begins on a new page.

Even though the report may contain a glossary, you should also define all unfamiliar terms or terms having a special meaning in your study when they are first mentioned in the text. (See **defining terms.**)

Index. An index is an alphabetical list of all the major topics discussed in the report. It cites the pages where each topic can be found and thus allows readers to find information on topics quickly and easily. The index is always the final section of the report. The index to this volume can be found at the back of the book.

An index is an optional finding device in a report because a detailed table of contents usually gives readers adequate information about topics covered. Indexes are most useful for reference works like handbooks and manuals. (For guidance on creating an index, see **indexing.**)

format

Format has at least two distinct, but related, meanings: it can refer (1) to the sequence in which information is presented in a publication, and (2) to the physical arrangement of information on the page and the general physical appearance of the finished publication. The first meaning applies to the standard arrangement or layout of information in many of the following types of job-related writing:

- **formal reports**
- **memorandums**
- **questionnaires**
- **proposals**
- **correspondence**
- **trip reports**
- **résumés**
- **test reports**
- **laboratory reports**

F

These types of writing are characterized by format conventions that govern where each section will be placed. In **formal reports,** for example, the **table of contents** precedes the preface but follows the title page and **abstract.** Likewise, although variations exist, letters typically are written according to the following standard pattern:

- letterhead
- date
- inside address
- salutation
- body
- complimentary close
- signature
- typed name and title
- additional information (initials, enclosure, and "carbon copy" notation)

These conventions also pertain to the use of heads, **illustrations,** and **indentation.** (For detailed information about numbered and un-numbered heading systems in reports and manuals, see **headings.**)

In recent years, a body of information has been published that expands the concept of format to encompass all typographic and graphic principles used to depict information in a document. (See **design and layout.**)

former/latter

Former and *latter* should be used to refer to only two items in a sentence or paragraph.

EXAMPLE   The president and his trusted aide emerged from the conference, the *former* looking nervous and the *latter* looking downright glum.

Because these terms make the **reader** look back to previous material to identify the reference, they impede reading and are best avoided.

forms of discourse

There are four *forms of discourse:* exposition, description, persuasion, and narration. **Exposition** is the straightforward presentation of facts and ideas with the **purpose** of informing the **readers; description** is an attempt to re-create an object or situation with words so that the reader can visualize it mentally; **persuasion** attempts to convince the reader that the writer's point of view is the correct or desirable one; and **narration** is the presentation of a series of events in chronological order. These types of writing rarely exist in pure form; rather, they usually appear in combination.

formula/formulae/formulas

The plural form of *formula* is either *formulae* (a Latin derivative) or *formulas. Formulas* is more common, however, since it is the more natural English plural. (See also **foreign words in English.**)

> **EXAMPLE** You may present the underlying theory in your introduction, but save proofs or *formulas* for the body of your report.

fortuitous/fortunate

When an event is *fortuitous,* it happens by chance or accident and without plan. Such an event may be lucky, unlucky, or neutral.

> **EXAMPLE** My encounter with the general manager in Denver was entirely *fortuitous;* I had no idea he was there.

When an event is *fortunate,* it happens by good fortune or happens favorably.

> **EXAMPLE** Our chance meeting had a *fortunate* outcome.

fragment (see sentence fragments)

functional shift

Many words shift easily from one **part of speech** to another, depending on how they are used. When they do, the process is called a *functional shift,* or a shift in function.

EXAMPLES It takes ten minutes to *walk* from the sales office to the accounting department. (**verb**)

The long *walk* from the sales office to the accounting department reduces efficiency. (**noun**)

Let us *run* the new data through the computer. (verb)
May I see the last computer *run*? (noun)

I talk to the Chicago office on the *telephone* every day. (noun)
Sometimes I have to call from a *telephone* booth. (**adjective**)
He will *telephone* the home office from London. (verb)

After we discuss the project, we will begin work. (**conjunction**)
After lengthy discussions, we began work. (**preposition**)
The partners worked well together forever *after*. (**adverb**)

Do not arbitrarily shift the function of a word in an attempt to shorten a **phrase** or expression.

EXAMPLES In medical jargon, an *attending physician* becomes an "attending." (a shift from an adjective to a noun)

In computer jargon, *to pass a signal through a computer logic gate* becomes "to gate." (a shift from a noun to a verb)

In nuclear energy jargon, a *reactor containment building* becomes a "containment." (a shift from an adjective to a noun)

garbled sentences

A *garbled sentence* is one that is so tangled with structural and grammatical problems that it cannot be repaired by simply replacing words or rewriting **phrases.** The following nearly unreadable sentence appeared in an actual **memorandum:**

> EXAMPLE My job objectives are accomplished by my having a diversified background which enables me to operate effectively and efficiently, consisting of a degree in mechanical engineering, along with twelve years of experience, including three years in Staff Engineering-Packaging sets a foundation for a strong background in areas of analyzing problems and assessing economical and reasonable solutions.

A garbled sentence often results from an attempt to squeeze too many ideas into one sentence. Do not try to patch such a sentence; rather, analyze the ideas it contains, list them in a logical sequence, and then construct one or more entirely new sentences. An analysis of the previous example yields the following five ideas:

> EXAMPLES My job requires that I analyze problems and find economical and workable solutions to them.
>
> My diversified background helps me accomplish my job.
>
> I have a mechanical engineering degree.
>
> I have twelve years of job experience.
>
> Three of these years have been in Staff Engineering-Packaging.

Using these ideas, the writer might have described his job as follows:

> EXAMPLE My job requires that I analyze problems and find economical and workable solutions to them. Both my training and my expe-

235

rience help me achieve this goal. Specifically, I have a mechanical engineering degree and twelve years of job experience, three of which have been in the Staff Engineering-Packaging Department.

Notice that the revised job description contains three sentences with logical development and clear **transitions.** Such analysis and rewriting is useful for any sentence or **paragraph** that you cannot repair with minor editing. (See also **clarity** and **sentence construction.**)

G

gender

In **grammar,** *gender* is a term for the way words are formed to designate sex. The English language provides for recognition of three genders: masculine, feminine, and neuter. The gender of most words can be identified only by the choice of the appropriate **pronoun** *(he, she, it).* Only these pronouns and a select few **nouns** *(waiter/waitress)* reflect gender.

Gender is important to writers because they must be sure that nouns and pronouns within a grammatical construction agree in gender. A pronoun, for example, must agree with its noun antecedent in gender. We must refer to a woman as *she* or *her,* not as *it;* to a man as *he* or *him,* not as *it;* to a barn as *it,* not *he* or *she.*

An antecedent that includes both sexes, such as *everyone* and *student,* has traditionally taken a masculine pronoun. Since this usage might be offensive by implying sexual bias, however, it is better to use plural nouns and pronouns.

CHANGE   Every *employee* should be aware of *his* insurance benefits.
TO   *All* employees should be aware of *their* insurance benefits.

It is best (though not always possible) to avoid the awkward "his or her," "he or she" type of construction. (See also **he/she.**)

general-to-specific method of development

In a *general-to-specific method of development,* you begin with a general statement and then provide facts or examples to develop and support that statement. For example, if you begin a report with the general statement "Companies that diversify are more successful than those that do not," the remainder of the report would offer examples and statistics that prove to your **reader** that companies that diversify are, in fact, more successful than companies that do not.

A document organized in a general-to-specific sequence discusses only one point—the point made in the opening general statement. All other information in the memo or report supports the general statement, as shown in Figure 1.

General
statement

Specific
supporting
information

Locating Additional Flat Panel Display Suppliers

Based on information presented at the Supply Committee meeting on April 14, we recommend that the company locate additional suppliers of flat panel displays. Several related events make such an action necessary.

Our current supplier, ABC Electronics, is reducing its output. Specifically, we can expect a reduction of between 800 and 1,000 units per month for the remainder of this fiscal year. The number of units should stabilize at 15,000 units per month thereafter.

Domestic demand for our laptop computers continues to grow. Demand during the current fiscal year is up 25,000 units over the last fiscal year. Sales Department projections for the next five years show that demand should peak next year at 50,000 units and then remain at that figure for at least the following four years.

Finally, our overseas expansion into England and Germany will require additional shipments of 5,000 units per quarter to each country for the remainder of this fiscal year. Sales Department projections put laptop sales for each country at double this rate, or 20,000 units in a fiscal year, for the next five years.

FIGURE 1   General-to-Specific Method of Development

gerunds (see verbals)

glossary

A *glossary* is a selected list of terms defined and explained for a particular field of knowledge. An alphabetical glossary, such as is often found at the end of a textbook or a **formal report,** can be helpful for quick reference.

If you are writing a report that will go to people who are not familiar with many of your technical terms, you may want to include a glossary. Inclusion of a glossary does not relieve you of the responsibility of defining in the text any terms you are certain your reader will not know, however. If you do, keep the entries concise and be sure they are so clear that any reader can understand the definitions. (See also **defining terms.**)

EXAMPLES *Amplitude Modulations:* Varying the amplitude of a carrier current with an audio-frequency signal.

Carbon Microphone: A microphone that uses carbon granules as a means of varying resistance with sound waves.

Overmodulation: Distortion created when the amplitude of the modulator current is greater than the amplitude of the carrier current.

G gobbledygook

Gobbledygook is writing that suffers from an overdose of traits guaranteed to make it stuffy, pretentious, and wordy. These traits include the overuse of big and mostly **abstract words, affectation,** inappropriate **jargon, long variants,** stale expressions, **euphemisms,** jammed **modifiers, vogue words,** and deadwood. Gobbledygook is writing that attempts to sound official (officialese), legal (legalese), or scientific; it tries to make a "natural elevation of the geosphere's outer crust" out of a molehill. Consider the following statement from an auto repair release form:

CHANGE I hereby authorize the above repair work to be done along with the necessary material and hereby grant you and/or your employees permission to operate the car or truck herein described on streets, highways, or elsewhere for the purpose of testing and/or inspection. An express mechanic's lien is hereby acknowledged on above car or truck to secure the amount of repairs thereto.

TO You have my permission to do the repair work listed on this work order and to use the necessary material. You may drive my vehicle to test its performance. I understand that you will keep my vehicle until I have paid for all repairs.

(See also **conciseness/wordiness,** and **word choice.**)

good/well

The confusion about the use of *good* and *well* can be cleared up by remembering that *good* is an **adjective** and *well* is an **adverb.**

EXAMPLES John presented a *good* plan.
The plan was presented *well.*

However, *well* can also be used as an adjective to describe someone's health.

EXAMPLES   She is not a *well* woman.
           Jane is looking *well.*

(See also **bad/badly.**)

government proposals (see **proposals**)

grammar

Grammar is the systematic description of the way words work together to form a coherent language; in this sense, it is an explanation of the structure of a language. However, grammar is popularly taken to mean the set of "rules" that governs how a language ought to be spoken and written; in this sense, it refers to the **usage** conventions of a language.

These two meanings of grammar—how the language functions and how it ought to function—are easily confused. To clarify the distinction, consider the expression *ain't.* Unless used purposely for its colloquial flavor, *ain't* is unacceptable to careful speakers and writers because a convention of usage prohibits its use. Yet taken strictly as a **part of speech,** the term functions perfectly well as a **verb;** whether it appears in a declarative sentence ("I ain't going.") or an interrogative sentence ("Ain't I going?"), it conforms to the normal pattern for all verbs in the English language. Although we may not approve of its use in a sentence, we cannot argue that it is ungrammatical.

To achieve **clarity,** writers need to know both grammar (as a description of the way words work together) and the conventions of usage. Knowing the conventions of usage helps writers select the appropriate over the inappropriate word or expression. A knowledge of grammar helps them diagnose and correct problems arising from how words and **phrases** function in relation to one another. For example, knowing that certain words and phrases function to modify other words and phrases gives the writer a basis for correcting those **modifiers** that are not doing their job. Understanding **dangling modifiers** helps the writer avoid or correct a construction that obscures the intended meaning. In short, an understanding of grammar and its special terminology is valuable for writers chiefly because it enables them to recognize problems and thus communicate clearly and precisely.

graphs

A *graph* presents numerical data in visual form. This method has several advantages over presenting data in **tables** or within the text. Trends, movements, distributions, and cycles are more readily apparent in graphs than they are in tables. By providing a means for ready comparisons, a graph often shows a significance in the data not otherwise immediately apparent. Be aware, however, that although graphs present statistics in a more interesting and comprehensible form than tables do, they are less accurate. For this reason, they are often accompanied by tables giving the exact data. The main types of graphs are line graphs, bar graphs, pie graphs, and picture graphs. When creating graphs, follow the general guidelines for identifying graphs and integrating them with text in the **illustration** entry, in addition to following the guidelines throughout this entry specific to the types of graphs described.

Line Graphs

The line graph shows the relationship between two sets of numbers by means of points plotted in relation to two axes drawn at right angles. The points, once plotted, are connected to form a continuous line. In this way, what was merely a set of dots having abstract mathematical significance becomes graphic, and the relationship between the two sets of figures can easily be seen.

The line graph's vertical axis usually represents amounts, and its horizontal axis usually represents increments of time, as shown in Figure 1.

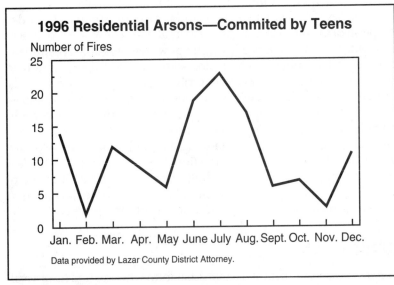

FIGURE 1 Single-Line Graph

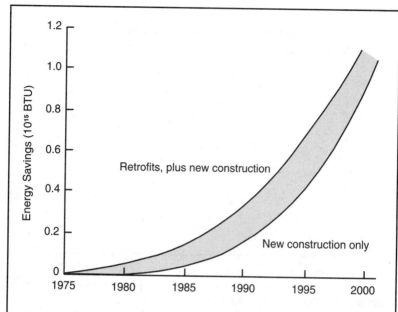

Figure 14. Projected energy savings for residences heating by solar power
Source: Electric Power Research Institute

FIGURE 2 Double-Line Graph with Difference Shaded

Line graphs with more than one line are common because they allow for comparisons between two sets of statistics for the same period of time. In creating such graphs, be certain to identify each line with a label or a legend, as shown in Figure 2. The difference between the two lines can be emphasized by shading the space between them.

Tips on Preparing Line Graphs

1. Indicate the zero point of the graph (the point where the two axes meet). If the range of data shown makes it inconvenient to begin at zero, insert a break in the scale, as in Figure 3.
2. Graduate the vertical axis in equal portions from the least amount at the bottom to the greatest amount at the top. Ordinarily, the caption for this scale is placed at the upper left.
3. Graduate the horizontal axis in equal units from left to right. If a caption is necessary, center it directly beneath the scale.
4. Graduate the vertical and horizontal scales so that they give an accurate visual impression of the data, since the angle at which the curved

G

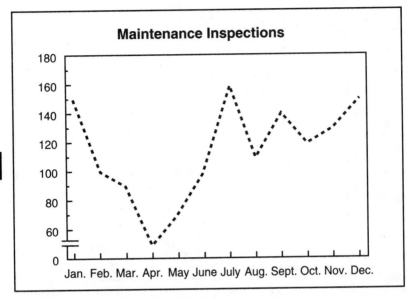

FIGURE 3 Line Graph with Vertical Axis Broken

line rises and falls is determined by the scales of the two axes. The curve can be kept free of distortion if the scales maintain a constant ratio with each other. See Figures 4 and 5.

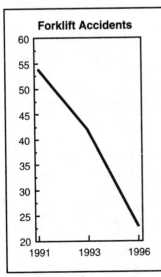

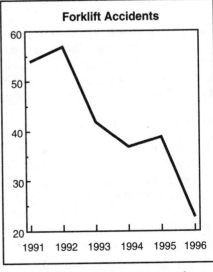

FIGURE 4 Distorted Expression of Data

FIGURE 5 Distortion-Free Expression of Data

5. Hold grid lines to a minimum so that curved lines stand out. Since precise values are usually shown in a table of data accompanying a graph, detailed grid lines are unnecessary.
6. Include a key (which lists and explains symbols) when necessary. At times a label will do just as well.

Bar Graphs

Bar graphs consist of horizontal or vertical bars of equal width but scaled in length to represent some quantity. They are commonly used to show (1) quantities of the same item at different times, (2) quantities of different items for the same time period, or (3) quantities of the different parts of an item that make up the whole.

Figure 6 is an example of a bar graph showing varying quantities of the same item (ratio of sales closings to sales calls) at the same time. Here each bar represents a different quantity of the same item.

Some bar graphs show the quantities of different items for the same period of time. See Figure 7. (A bar graph with vertical bars is also called a column graph.)

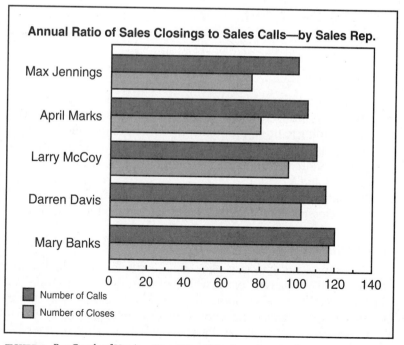

FIGURE 6   Bar Graph of Varying Quantities of the Same Item at the Same Time

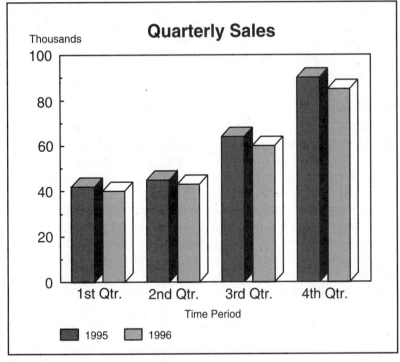

FIGURE 7   Bar Graph of Performance for Two Separate Years

Bar graphs can also show the different portions of an item that make up the whole. Here the bar is equivalent to 100 percent. It is then divided according to the appropriate proportions of the item sampled. This type of graph can be constructed vertically or horizontally and can indicate more than one whole when comparisons are necessary. See Figures 8 and 9.

If the bar is not labeled, each portion must be marked clearly by shading or crosshatching. Include a key that identifies the subdivisions.

Pie Graphs

A pie graph presents data as wedge-shaped sections of a circle. The circle must equal 100 percent, or the whole, of some quantity (a tax dollar, a bus fare, the hours of a working day), with the wedges representing the various ways in which the whole is divided. In Figure 10, for example, the circle stands for a city tax dollar, and it is divided into units equivalent to the percentage of the tax dollar spent on various city services.

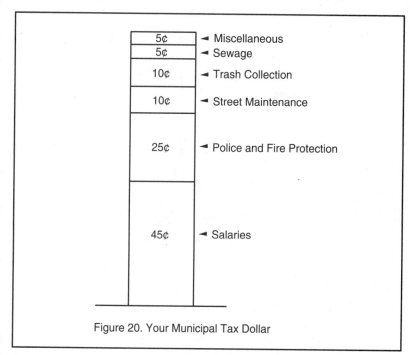

Figure 20. Your Municipal Tax Dollar

FIGURE 8 Bar Graph of Quantities of Different Parts Making Up a Whole

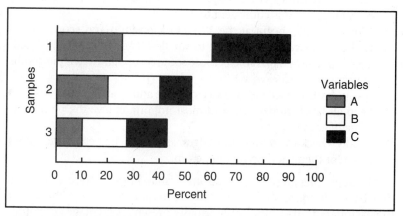

FIGURE 9 Bar Graph Showing Variables in Three Samples

Pie graphs provide a quicker way of presenting the same information that can be shown in a table; in fact, a table often accompanies a pie graph with a more detailed breakdown of the same information.

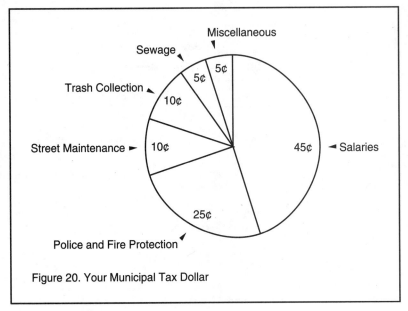

Figure 20. Your Municipal Tax Dollar

FIGURE 10 Pie Graph

Tips on Preparing Pie Graphs

1. The complete 360° circle is equivalent to 100 percent; therefore, each percentage point is equivalent to 3.6°.
2. When possible, begin at the 12 o'clock position and sequence the wedges clockwise, from largest to smallest. (Adherence to this guidance is not always possible because some computer-drawn pie graphs appear counterclockwise.)
3. If you shade the wedges, do so clockwise and from light to dark.
4. Keep all labels horizontal, and most important, give the percentage value of each wedge.
5. Finally, check to see that all wedges, as well as the percentage values given for them, add up to 100 percent.

Although pie graphs have strong visual impact, they also have drawbacks. If more than five or six items are presented, the graph looks cluttered. Also, since they usually present percentages of something, they must often be accompanied by a table listing precise statistics. Further, unless percentages are shown on the sections, the reader cannot compare the values of the sections as accurately as with a bar graph. (The terms that identify each segment of the graph are referred to as callouts.)

Picture Graphs

Picture graphs are modified bar graphs that use picture symbols of the item presented. Each symbol corresponds to a specified quantity of the item and should be self-explanatory. If each symbol represents more than a single unit, label it appropriately. See Figure 11. Note that precise figures are included, since the graph can present only approximate figures.

Picture graphs often are not as accurate as other graphs, but they do provide visual interest. They can be particularly effective in presenting data to **readers** who may not be well-acquainted with the information. **Oral presentations** often benefit from the use of visual aids that incorporate picture graphs.

G

Focus on Ease of Interpretation

With the widespread use of computer software that can produce dozens of graph styles to display the same data, focus on helping readers interpret the data rather than on dazzling them with unnecessarily complex visuals. To do so, select graphs that accurately present the

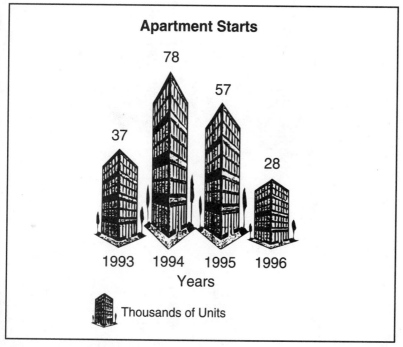

FIGURE 11   Picture Graph

data in the simplest form available so that readers can grasp their significance with the least amount of effort.

Consider the differences between Figures 12 and 13, which depict four variables for a three-month period.

Figures 12 and 13 represent "floating" data. Despite the visually interesting design in Figure 12, the reader would soon be defeated in trying to interpret the data. The stripes are too difficult to match with the key showing the dollar ranges and the expenditures cannot be easily correlated with the month in which they were spent. In fact, February data for courier and guard force services are completely obscured. Figure 13 at least separates the data and provides grid lines. However, the reader still cannot interpret the funds spent for February and March, except for mail services. Moreover, the courier services data completely obscure the guard force data for January. Using either of these graphs to depict such expenditures for twelve months would greatly compound these problems.

Figures 14 and 15 also display the data in three dimensions to poor effect. Guard force expenditures are obscured in both figures. Figure 15

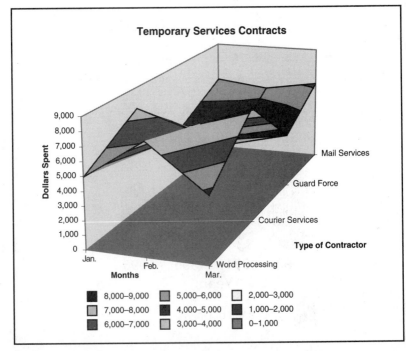

FIGURE 12 Three-Dimensional Surface Graph

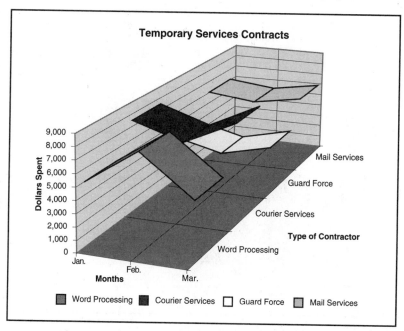

FIGURE 13   Three-Dimensional Line Graph

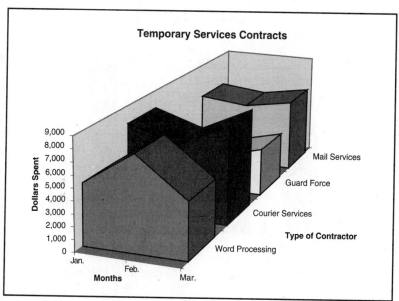

FIGURE 14   Three-Dimensional Area Graph

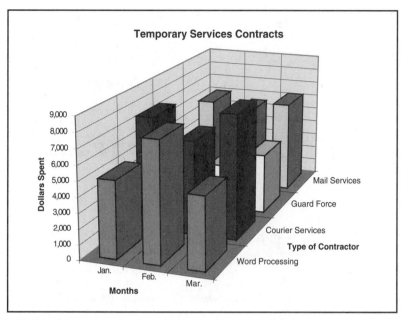

FIGURE 15 Three-Dimensional Column Graph 1

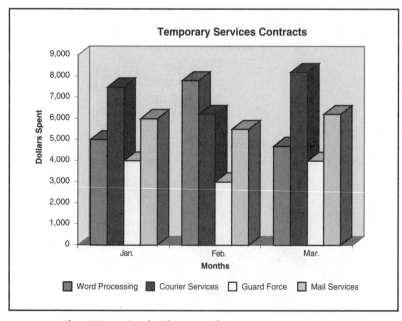

FIGURE 16 Three-Dimensional Column Graph 2

also makes the trends of expenditures over three months nearly impossible to see at a glance.

In Figures 16 and 17, the data begin to emerge with more clarity. However, the three-dimensional perspective in Figure 16 is still confusing for at least two reasons. At first glance, we are tricked into interpreting the spaces between the clusters of columns as relevant. More confusing still, are we to interpret the front or back of the column tops as the correct data points? Figure 17 avoids these ambiguities by presenting the data in two dimensions. Even so, the *x* axis contains four sets of figures for each month, which permits a comparison of expenditures for a given month but does not easily permit an impresson of one set of service expenditures for three months. Again, these problems are compounded on graphs showing data over longer periods.

Figure 18, a line graph representation of the data, is the clearest rendering yet because it shows readers trends in contract expenditures at a glance. Each expenditure has its own line with an identifying sym-

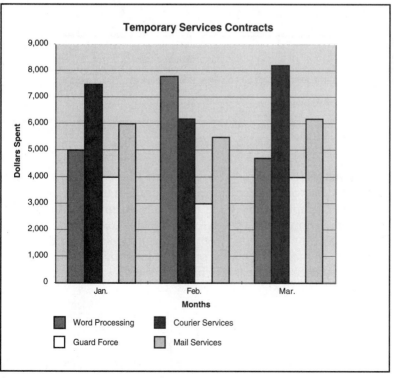

FIGURE 17 Sheet Graph

G

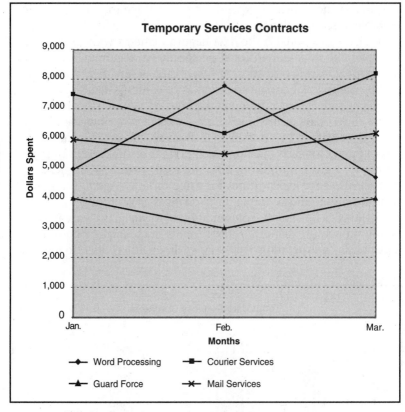

Temporary Services Contracts

FIGURE 18   Line Graph

bol that makes its pattern easy to interpret. When necessary, you can provide a table with precise figures to support the trends shown.

As these samples show, there is virtually an inverse relationship between how impressive a graph looks and how easy it is to interpret. On balance, simple is better. Keeping this principle in mind, review your computer-generated graphs on screen in several styles before you make a final decision about which to use.

half

Half a, half an, and *a half* all are correct idiomatic uses of the word *half*.

> **EXAMPLES** Wait *half a* minute before stirring the solution.
> Allow *half an* hour for the compound to solidify.
> It takes *a half second* for the signal to reach your receiver.

A half a (or *a half an*) is colloquial, however, and should be avoided in writing.

> **CHANGE** The subroutine is executed in *a half a* minute.
> **TO** The subroutine is executed in *half a* minute.

he/she

Since there is no singular personal **pronoun** in English that refers to both sexes, the word *he* was traditionally used when the sex of the antecedent is unknown.

> **EXAMPLE** Whoever is appointed [either a man or woman] will find *his* task difficult.

The use of a masculine pronoun to refer to both sexes can be offensive, however, and it is better to rewrite the sentence in the plural or avoid use of a pronoun altogether.

> **EXAMPLES** *Employees* should take advantage of *their* insurance benefits.
> *Whoever* is appointed will find *the* task difficult.

You could also use the **phrases** *he or she* and *his or her.*

> **EXAMPLE** Whoever is appointed will find *his or her* task difficult.

Unfortunately, *he or she* and *his or her* are clumsy when used repeatedly and *s/he* is awkward; the best advice is to reword the **sentence** to use a plural pronoun. Be sure also to change the noun to which the pronoun refers to its plural form. (See also **gender** and **Ms./Miss/Mrs.**)

> **CHANGE**   The engineer cannot do *his or her* job until *he or she* understands the concept.
>
> **TO**   Engineeers cannot do *their* jobs until *they* understand the concept.

headers and footers

A *header* in a **report, technical manual,** or **specification** is identifying information carried at the top of each page. The header normally contains the topic (or **topic** and subtopic) dealt with in that section of the report, manual, or specification (specifications also include the identification **numbers** of sections and **paragraphs**).

A *footer* is identifying information carried at the bottom of each page. The footer generally contains the date of the document, the page number, and sometimes the manual name and section title.

Although the types of information included in headers and footers vary greatly from one organization to the next, the example in Figure 1 is fairly typical (see p. 255).

headings

Headings (also called *heads*) are **titles** or subtitles within the body of a long piece of writing that serve as guideposts for the **reader.** They divide the material into manageable segments, call attention to the main topics, and signal changes of topic. If a **formal report, proposal,** or other document you are writing is long or complicated, you may need several levels of heads to indicate major divisions, subdivisions, and even smaller units of those. In extremely technical material as many as five levels of heads may be appropriate, but as a general rule it is rarely necessary (and usually confusing) to use more than three.

The best advice on the effective use of headings is to be sure to prepare a good topic outline before writing your draft and then use the major topic listings of the topic outline (without Roman numerals and capital letters) as in the headings of your draft. In a short document, you probably will use only the major divisions of your topic outline as heads, but in a longer document, you may use both the

Header

Header

| 4 | **Establishing your password** |

☐ note: If you entered any fields incorrectly, the advisory message doesn't appear. Instead, a message appears indicating the nature of the error. Correct the password as indicated by the message. You may also ask your supervisor for assistance.

3. Enter the same password you entered in step 1.

4. Press POST. The Establish Password Screen appears, showing your employee record. The message area of the screen advises that you must change your password.

```
                  ESTABLISH PASSWORD SCREEN  Updated . . . By . . .

ACTION CODE: C   A = add,  D = delete,  C = change,  I = inquire
PASSWORD: _____
PASSWORD: _____
DATE PASSWORD
   LAST CHANGED: _____ Zero if no password or N/A.

STATUS CODE: I   Blank = operative,  I = inoperative


SECURITY EXCEPTION COUNTERS:
   # Timeouts    # Password exceptions in 1 session
   # Trans       # Password exceptions today
                 # Transactions attempted with insufficient authority

ADV    YOU MUST CHANGE YOUR PASSWORD
```

5. Enter information in the following fields. If you don't completely fill a field, press TAB to move to the next field.

ACTION CODE   Change the **I** (inquire) to a **C** (change).

PASSWORD   In each Password field, enter your new password.

               ☐ note: For security reasons, the password does not appear as you type it.

FIGURE 1 Header and Footer

major and minor divisions. Be careful, however, not to use too many headings; four or five on a page would certainly be too many under most circumstances.

On page 256, the example (Figure 1), greatly shortened, is intended only to illustrate the use of the heads—in this case, three levels (beyond the title of the entire report).

As you can infer from this example, the heads you use grow naturally from the divisions and subdivisions prepared in **outlining.**

H

Interim Report of the Committee to Investigate New Factory Locations

The committee initially considered thirty possible locations for the proposed new factory. Of these, twenty were eliminated almost immediately for one reason or another (unfavorable tax structure, remoteness from rail service, inadequate labor supply, etc.). Of the remaining ten locations, the committee selected for intensive study the three that seemed most promising: Chicago, Minneapolis, and Salt Lake City. These three cities we have now visited, and our observations on each of them follow.

1st level head

CHICAGO

Of the three cities, Chicago presently seems to the committee to offer the greatest advantages, although we wish to examine these more carefully before making a final recommendation.

2nd level head

Location

Though not at the geographical center of the United States, Chicago is centrally located in an area that contains more than three-quarters of the U.S. population. It is within easy reach of our corporate headquarters in New York. And it is close to several of our most important suppliers of components and raw materials—those, for example, in Columbus, Detroit, and St. Louis.

2nd level head

Transportation

Rail Transportation. Chicago is served by the following major railroads. . . .
Sea Transportation. Except during the winter months when the Great Lakes are frozen, Chicago is an international seaport. . . .

3rd level heads

Air Transportation. Chicago has two major airports (O'Hare and Midway) and is contemplating building a third. Both domestic and international air cargo service are available. . . .
Transportation by Truck. Virtually all of the major U.S. carriers have terminals in Chicago. . . .

FIGURE 1 The Use of Heads

The decimal numbering system uses a combination of numbers and decimal points to subordinate levels of heads in a report. The following outline shows the correspondence between different levels of headings and the decimal numbers used:

1.   FIRST-LEVEL HEAD
1.1   Second-Level Head
1.2   Second-Level Head
1.2.1   Third-Level Head
1.2.2   Third-Level Head

1.2.2.1 Fourth-Level Head
1.2.2.2 Fourth-Level Head
1.3   Second-Level Head
1.3.1   Third-Level Head
1.3.2   Third-Level Head
2.   FIRST-LEVEL HEAD

Although the second-, third-, and fourth-level heads are indented in an outline or **table of contents,** as headings they are flush with the left margin in the body of the report when the decimal numbering system is used. Every head is typed on a separate line, with an extra line of space above and below the head.

There is no one correct format for headings. Sometimes a company settles on a standard **format,** which everyone within the company is then expected to follow. Or a customer for whom a **report** or **proposal** is being prepared may specify a particular format. In the absence of such guidelines, the system used in the example should serve you well. Note the following format characteristics: (1) The first-level head is in all **capital letters,** typed flush to the left margin on a line by itself and separated by a line space from the material it introduces. (2) The second-level head is in capital and lowercase letters, also typed flush to the left margin on a line by itself and also separated by a line space from the material it introduces. (3) The third-level head is in capital and lowercase letters, but it is "run in" right on the same line with the first sentence of the material it introduces. Therefore, it is followed by a period to set it apart from what follows, and it is underlined or italicized so that it will stand out clearly on the page.

The following are some important points to keep in mind about heads:

1. They should signal a shift to a new topic (or, if they are lower level heads, a new "subtopic" within the larger topic).
2. Within the unit they subdivide, all heads at one level should be consistent in their relationship to the topic of the larger unit.
3. The fact that one unit at a particular level is subdivided by lower-level heads does not mean that every unit at that particular level must also include lower-level heads.
4. All heads at any one level should be parallel with one another in structure (see **parallel structure**). For instance, most heads are **nouns** or noun **phrases.**
5. Too many heads, or too many levels of heads, can be as bad as too few. Keep in mind the needs of your reader.

6. If your document needs a **table of contents,** create it from your heads and subheads. Be sure that the wording in your table of contents is the same as the wording in your text.

7. The head does not substitute for the discussion; the text should read as if the head were not there.

(See also **design and layout.**)

healthful/healthy

If something is *healthful,* it promotes good health. If something is *healthy,* it has good health. Carrots are *healthful;* people can be *healthy.*

> EXAMPLE   We provide *healthful* lunches in the cafeteria because we want our employees to be *healthy.*

helping verbs (see **verbs**)

herein/herewith

The words *herein* and *herewith* are a type of **gobbledygook** that should be avoided.

> CHANGE   I have *herewith* enclosed a copy of the contract.
> TO   I have enclosed a copy of the contract.

hyperbole (see **figures of speech**)

hypertext

Hypertext is a computer-based system that allows readers to move from one document to another (navigate) and find online information in a sequence determined by the reader. World Wide Web pages on the **Internet** often contain hypertext indicators—words, graphics, or icons—that when clicked, yield additional electronic pages.

A hypertext system includes three main components: texts, nodes, and links. Texts, the informational contents of the system (words, graphics, sound, and video), are collected into self-contained units called *nodes.* The nodes are connected in a variety of ways by electronic *links.* In a way, hypertext is similar to the structure of the book you're reading. Each entry in this handbook is a discrete unit (node) that contains text or graphical information. In many cases, the entries include a cross-reference (link) in the form of a boldfaced word or a "see also" statement that serves as a pointer to related information. That's the basic structure of a hypertext document, although more advanced systems include animation segments, audio clips, video clips, and real-time conferencing capabilities.

Hypertext systems can be categorized as either *modest* (not modifiable) or *robust* (modifiable), depending on the degree to which users are encouraged to make modifications.

In modest versions of hypertext, readers navigate information by following existing paths created by hypertext designers. This read-only approach enables readers to retrieve the specific amount of information they need by navigating the predetermined paths. Readers can recover and retrace what they have read, as well as view the ways they explored ideas. In on-the-job writing situations, applications common to such modest approaches include tutorials, guided tours, help systems, procedures manuals, and reports.

In robust versions of hypertext, readers can add to, delete from, and edit the text within nodes, as well as construct, eliminate, or reconstruct the links between them. Using such modifiable approaches to hypertext, a user can become both writer and reader. In on-the-job writing situations, applications common to robust hypertext approaches include real-time conferencing sessions and **collaborative writing** activities.

Some applications combine degrees of both modest and robust hypertext systems by allowing users to modify some portions of the hypertext while restricting their ability to modify others. The online help systems used by software programs that allow users to create and link their own text to existing help screens are examples of such a hybrid approach.

hyphens

Although the *hyphen* functions primarily as a **spelling** device, it also functions to link and to separate words; in addition, it occasionally replaces the **preposition** *to* (0-100 for 0 *to* 100). The most common use of the hyphen, however, is to join **compound words.**

EXAMPLES able-bodied, self-contained, carry-all, brother-in-law

A hyphen is used to form compound **numbers** from twenty-one to ninety-nine and fractions when they are written out.

EXAMPLES   *twenty-one, one forty-second*

When in doubt about whether or where to hyphenate a word, check your **dictionary.**

Hyphens Used with Modifiers

Two-word and three-word unit **modifiers** that express a single thought are hyphenated when they precede a **noun** (an *out-of-date* car, a *clear-cut* decision). If each of the words can modify the noun without the aid of the other modifying word or words, however, do not use a hyphen (a *new digital* computer—no hyphen). Also, if the first word is an **adverb** ending in *-ly*, do not use a hyphen (a *hardly* used computer, a *badly* needed micrometer).

The presence or absence of a hyphen can alter the meaning of a sentence.

EXAMPLE   We need a biological waste management system.
COULD MEAN   We need a biological-waste management system.
OR   We need a biological waste-management system.

A modifying **phrase** is not hyphenated when it follows the noun it modifies.

EXAMPLE   Our office equipment is *out of date.*

A hyphen is always used as part of a letter or number modifier.

EXAMPLES   5-cent, 9-inch, A-frame, H-bomb

In a series of unit modifiers that all have the same term following the hyphen, the term following the hyphen need not be repeated throughout the series; for greater smoothness and brevity, use the term only at the end of the series.

CHANGE   The third-floor, fourth-floor, and fifth-floor rooms have recently been painted.
TO   The third-, fourth-, and fifth-floor rooms have recently been painted.

Hyphens Used with Prefixes and Suffixes

A hyphen is used with a **prefix** when the root word is a proper **noun.**

EXAMPLES   pre-Sputnik, anti-Stalinist, post-Newtonian

A hyphen may optionally be used when the prefix ends and the root word begins with the same vowel. When the repeated vowel is *i*, a hyphen is almost always used.

EXAMPLES re-elect, re-enter, anti-inflationary

A hyphen is used when *ex-* means "former."

EXAMPLES ex-partners, ex-wife

A hyphen may be used to emphasize a prefix.

EXAMPLE He was anti-everything.

The **suffix** *-elect* is hyphenated.

EXAMPLES president-elect, commissioner-elect

Other Uses of the Hyphen

H

To avoid confusion, some words and modifiers should always be hyphenated. *Re-cover* does not mean the same thing as *recover,* for example; the same is true of *re-sent* and *resent, re-form* and *reform, re-sign* and *resign.*

Hyphens should be used between letters showing how a word is spelled.

EXAMPLE In his letter he misspelled *believed* b-e-l-e-i-v-e-d.

Hyphens identify prefixes, suffixes, or written syllables.

EXAMPLE *Re-, -ism,* and *ex-* are word parts that cause spelling problems.

A hyphen can stand for *to* or *through* between letters and numbers.

EXAMPLES pp. 44-46
the Detroit-Toledo Expressway
A-L and M-Z

I

idea

The word *idea* means the "mental representation of an object of thought."

> **EXAMPLE**   He didn't accept my *idea* about how to reorganize the branch.

Use *impression* or *intention* to convey more precisely what you mean to say.

> **CHANGE**   We started out with the *idea* of returning early.
> **TO**   We started out with the *intention* of returning early.

> **CHANGE**   She left me with the *idea* that he was in favor of the proposal.
> **TO**   She left me with the *impression* that he was in favor of the proposal.

idioms

Idioms are groups of words that have a special meaning apart from their literal meaning. Someone who "runs for office" in the United States, for example, need not be a track star. The same person would "stand for office" in England. In both nations, the individual is seeking public office; only the idioms of the two countries differ. This difference indicates why idioms give foreign writers and speakers trouble (see also **English as a second language**).

> **EXAMPLES**   The judge *threw the book* at the convicted arsonist.
>
> The company's legal advisers debated whether to *stand fast* or to *press forward* with the lawsuit.

The foreigner must memorize such expressions; they cannot logically be understood. The native writer has little trouble understanding idioms and need not attempt to avoid them in writing, provided the

reader is equally at home with them. Idioms are often helpful short-cuts, in fact, that can make writing more vigorous and natural.

If there is any chance that your writing might be translated into another language or read in other English-speaking countries, eliminate obvious idiomatic expressions that might puzzle readers.

Idioms are the reason that certain **prepositions** follow certain **verbs, nouns,** and **adjectives.** Since there is no sure system to explain such **usages,** the best advice is to check in a **dictionary.** The following are some common idiomatic expressions that give writers trouble:

absolve from (responsibility)
absolve of (crimes)
accordance with
according to
accountable for (actions)
accountable to (a person)
accused by (a person)
accused of (a deed)
acquaint with

adapt for (a purpose)
adapt to (a situation)
adapt from (change)
adhere to
adverse to
affinity between, with
agree on (terms)
agree to (a plan)
agree with (a person)

angry with (a person)
angry at, about (a thing)
approve of
apply for (a position)
apply to (contact)
argue for, against (a policy)
argue with (a person)
arrive in (a city, country)
arrive at (a specific location, conclusion)
based on, upon
blame for (an action)
blame on (a person)

compare to (things that are similar but not the same kind)
compare with (things of the same kind to determine similarities and differences)
comply with
concur in (consensus)
concur with (a person)
conform to
consist of
convenient for (a purpose)
convenient to (a place)
correspond to, with (a thing)
correspond with (a person)

deal with
depend on
deprive of
devoid of
differ about, over (an issue)
differ from (a thing)
differ with (a person)
differ on (amounts, terms)
different from

disagree on (an issue, plan)
disagree with (a person)
disappointed in
disapprove of
disdain for
divide between, among
divide into (parts)
engage in

exclude from
expect from (things)
expect of (people)
expert in

identical with, to
impatient for (something)
impatient with (someone)
imply that
impose on
improve on
inconsistent with
independent of

infer from
inferior to
necessary for (an action)
necessary to (a state of being)
occupied by (things, people)
occupied with (actions)
opposite of (qualities)
opposite to (positions)
part from (a person)

part with (a thing)
proceed with (a project)
proceed to (begin)
proficient in

profit by (things)
profit from (actions)
prohibit from
qualify as (a person)
qualify by (experience, actions)
qualify for (a position, award)

rely on
responsibility for, of
reward for (an action)
reward with (a gift)
similar to
surrounded by (people)
surrounded with (things)
talk to (a group)
talk with (a person)
wait at (a place)
wait for (a person, event)

(See also **English as a second language** and **international correspondence.**)

illegal/illicit

If something is *illegal*, it is prohibited by law. If something is *illicit*, it is prohibited by law or custom. *Illicit* behavior may or may not be *illegal*, but it does violate custom or moral codes and therefore usually has a clandestine or immoral **connotation.**

> **EXAMPLES**   Their *illicit* behavior caused a scandal, and the district attorney charged that *illegal* acts were committed.
>
> Explicit sexual material may be *illicit* without being *illegal.*

illustrations

The objective of using an *illustration* is to help your **reader** absorb the facts and ideas you are presenting. When used well, an illustration can convey an idea that words alone could never really make clear. Notice how the description in Figure 1 depends on its illustration. The passage describes the function of a type of safety control on machinery that prevents a worker's hands and arms from coming in contact with dangerous moving parts.

Illustrations should never be used as ornaments, however; they should always be functional working parts of your writing. Be careful not to overillustrate; use an illustration only when you are sure that it makes a direct contribution to your reader's understanding of your subject. When creating illustrations, always consider your objective and your reader. You would use different illustrations of an X-ray

I

Two-Hand Control

The two-hand control requires constant, concurrent pressure by the operator to activate the machine. This kind of control requires a part-revolution clutch, brake, and a brake monitor if used on a power press as shown in Figure 46. With this type of device the operator's hands are required to be at a safe location (on control buttons) and at a safe distance from the danger area while the machine completes it closing cycle.

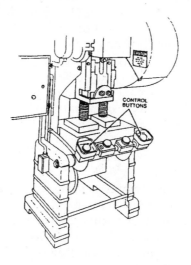

Figure 46. Two-hand control buttons on part-revolution clutch power press
Source: U.S. Dept. of Labor, Occupational Safety and Health Administration

FIGURE 1 Drawing Illustrating Safety Equipment

machine for a high school science class than you would for a group of X-ray service technicians. Many of the attributes of good writing—simplicity, **clarity, conciseness,** directness—are equally important in creating and using illustrations.

The most common types of illustrations are **photographs, graphs, tables, drawings, flowcharts, organizational charts, schematic diagrams,** and **maps.** Your material will normally suggest one of these types when an illustration is needed.

Creating Illustrations and Integrating Them with Text

Each type of illustration has its unique strengths and weaknesses. The guidelines presented here apply to most visual material you use to supplement or clarify the information in your text. (The term *illustrations* in this entry refers to both graphics and tables.) The following tips will help you create and present your illustrations to good effect.

1. Be sure to make clear in the text why the illustration is included. The amount of discussion that each illustration requires will vary with its importance to the text. An illustration showing an important feature or system may be the central focus of an entire discussion. The complexity of the illustration will also affect the discussion, as will the background your readers bring to the information. Nonexperts require lengthier explanations than experts do, as a rule.
2. Keep the illustration as brief and simple as possible by including only information necessary to the discussion in the text.
3. Try to present only one type of information in each illustration.
4. Keep terminology consistent. Do not refer to something as a "proportion" in the text and as a "percentage" in the illustration.
5. Specify the proportions used or include a scale of relative distances, when appropriate.
6. Position the lettering of any explanatory text on the illustration horizontally for ease of reading, if possible.
7. Give each illustration a concise title that clearly describes its contents.
8. Assign a figure or table number, particularly if your document contains five or more illustrations. The figure or table number precedes the title: *Figure 1 Widget Production for Fiscal 19___.* Note that graphics illustrations (photographs, drawings, maps, etc.) are generically referred to as "figures," while tables are referred to simply as "tables."
9. Refer to the illustration in your text by its figure or table number.
10. If the illustration is central to the discussion, illuminating or strongly reinforcing it, place it as close as possible to the text that discusses it. However, no illustration should precede its first text mention. Its

appearance without an introduction in the text will confuse readers. But, if the illustration is lengthy, detailed, and peripheral to the discussion, place it in an appendix or elsewhere at the back of the text. However, even material in an appendix should be referred to in the text. Otherwise, the illustration is superfluous and need not be included. (See **appendix/appendixes/appendices.**)

11. Allow adequate white space on the page around and within the illustration. (See **design and layout.**)

12. In documents with more than five illustrations, list them by title, together with figure and page numbers, or table and page numbers, following the report's **table of contents.** The figures so listed should be titled "List of Figures." The tables so listed should be titled "List of Tables."

13. If you wish to use an illustration from a copyrighted publication, first obtain a written release to do so from the **copyright** holder. Acknowledge such borrowings in a source or credit line below the caption for a figure and below any footnotes at the bottom of a table. Illustrated materials in publications of the federal government are not copyrighted. You need not obtain written permission to reproduce them, although you should acknowledge their source in a credit line. (Such a source line appears in Figure 1.)

imply/infer

If you *imply* something, you hint or suggest it. If you *infer* something, you reach a conclusion on the basis of evidence.

> EXAMPLES   His memo *implied* that the project would be delayed.
>
> The general manager *inferred* from the memo that the project would be delayed.

in/into

In means "inside of"; *into* implies movement from the outside to the inside.

> EXAMPLE   The equipment was *in* the test chamber, so she sent her assistant *into* the chamber to get it.

in order to

The **phrase** *in order to* is sometimes essential to the meaning of a sentence.

EXAMPLE   If the vertical scale of a graph line would not normally show the zero point, use a horizontal break in the graph in *order to* include the zero point.

The phrase *in order to* also helps control the pace of a sentence, even when it is not essential to the meaning of the sentence.

EXAMPLE   The committee must know the estimated costs *in order to* evaluate the feasibility of the project.

Most often, however, *in order to* is just a meaningless filler phrase that is dropped into a sentence without thought.

CHANGE   *In order to* start the engine, open the choke and throttle and then press the starter.
TO   *To* start the engine, open the choke and throttle and then press the starter.

Search for these thoughtless uses of *in order to* in your writing and eliminate them.

in terms of

When used to indicate a shift from one kind of language or terminology to another, the **phrase** *in terms of* can be useful.

EXAMPLE   *In terms of* gross sales, the year has been relatively successful; however, *in terms of* net income, it has been discouraging.

When simply dropped into a sentence because it comes easily to mind, however, *in terms of* is meaningless affectation.

CHANGE   She was thinking *in terms of* subcontracting much of the work.
TO   She was thinking *about* subcontracting much of the work.
OR   She was thinking *of* subcontracting much of the work.

inasmuch as/insofar as

Inasmuch as, meaning "because," is a weak connective for stating causal relationships.

CHANGE   *Inasmuch as* the heavy spring rains delayed construction, the office building will not be completed on schedule.
TO   *Because* the heavy spring rains delayed construction, the office building will not be completed on schedule.

Insofar as, meaning "to the extent that," should be reserved for that explicit **usage** rather than used to mean *since.*

EXAMPLE   *Insofar as* the report deals with causes of highway accidents, it will be useful in preparing future plans for highway construction.

CHANGE   *Insofar as* you are here, we will review the material.
TO   *Since* you are here, we will review the material.

increasing-order-of-importance method of development

When you want the most important of several ideas to be freshest in your **reader's** mind at the end of your writing, organize your infor-

MEMORANDUM

To: William D. Vane, Vice President for Operations
From: Harry Matthews, Personnel Department   ⟶M
Date: December 3, 19—
Subject: Recruiting Qualified Electronics Technicians

To keep our company staffed with qualified electronics technicians, we will have to redirect our recruiting program.

Least important source

In the past dozen years, we have relied heavily on the recruitment of skilled veterans. An end to the military buildup, as well as attractive reenlistment bonuses for skilled technicians now in uniform, has all but eliminated veterans as a source of trained employees. Our attempts to reach this group through ads in service newspapers and daily newspaper ads has not been successful recently. I think that the want ads should continue, although each passing month brings fewer and fewer veterans as job applicants.

Our in-house apprentice program has not provided the needed personnel either. High-school enrollments in the area are continually dropping. Each year, fewer high-school graduates, our one source of trainees for the apprentice program, enter the shop. Even vigorous Career Day recruiting has yielded disappointing results. The number of students interested in the apprentice program has declined proportionately as school enrollment goes down.

Most important source

The local and regional technical schools produce the greatest number of qualified electronics technicians. These graduates tend to be highly motivated, in part because many have obtained their education at their own expense. The training and experience they have received in the tech program, moreover, are first-rate. Competition for these graduates is keen, but I recommend that we increase our recruiting efforts and hire a larger share of this group than we have been doing.

I would like to meet with you soon to discuss the details of a more dynamic recruiting program in the technical schools. I am certain that with the right recruitment campaign we can find the skilled personnel essential to our expanding role in electronics products and service.

mo

FIGURE 1   Increasing-Order-of-Importance Method of Development

mation by *increasing order of importance.* This sequence begins with the least important point or fact, then moves to the next least important, and builds finally to the most important point at the end. Increasing order of importance is often used when the writer is leading up to an unpleasant or unexpected conclusion.

Writing organized by increasing order of importance has the disadvantage of beginning weakly, with the least important information, which can cause your reader to become impatient before reaching your main point. But when you build a case in which the ideas lead, point by point, to an important **conclusion,** increasing order of importance is an effective method of **organization. Reports** on production or personnel goals are often arranged by this method, as are **oral presentations.**

In the example given in Figure 1, the writer begins with the least productive source of applicants and builds up to the most productive source.

incredible/incredulous

Something that is unbelievable or nearly unbelievable is *incredible.*

EXAMPLE   The precision of the new instrument is truly *incredible.*

Incredulous means that a person is skeptical or unbelieving about something or some event.

EXAMPLE   He was *incredulous* at the reports of the instrument's precision.

indefinite adjectives (see adjectives)

indefinite pronouns (see pronouns)

indentation

Type that is *indented* is set in from the margin. The most common use of indentation is at the beginning of a **paragraph,** where the first line is usually indented five spaces in a typed manuscript and one inch in a longhand manuscript.

EXAMPLE   A hydraulic crane is not like a lattice boom friction crane in one very important way. In most cases, the safe lifting capacity of a lattice boom crane is based on the weight needed to tip the machine. Therefore, operators of friction machines sometimes depend on signs that the machine might tip to warn them of impending danger.

This is a very dangerous thing to do with a hydraulic crane. Hydraulic crane ratings are based on the strength of the material of the boom (and other components). Therefore, the hydraulic crane operator who waits for signs of tipping to warn him of an overloaded condition will often bend the boom or cause severe damage to his machine before any signs of tipping occur.

Another use of indentation is in **outlining,** in which each subordinate entry is indented under its major entry.

EXAMPLE   II. Types of Outlines
  A. The Sentence Outline
    1. A sentence outline provides order and establishes the relationship of topics to one another to a greater degree than a topic outline.

Finally, a block **quotation** may be indented within a manuscript instead of being enclosed in **quotation marks;** it must, in this case, be single-spaced. Set off the indented material from the text by indenting five spaces from the left margin and five spaces from the right margin and by double-spacing above and below the passage.

EXAMPLE   *The Operator's Manual* states:

> This is a very dangerous thing to do with a hydraulic crane. Hydraulic crane ratings are based on the strength of the material of the boom (and other components). . . .
>
> A hydraulic crane is not like a lattice boom friction crane in one very important way. In most cases, the safe lifting capacity of the lattice boom crane is based on the weight needed to tip the machine. Therefore, operators of friction machines sometimes depend on signs that the machine might tip to warn them of impending danger.

This warning was given because many operators were accustomed to . . .

independent clauses (see clauses)

indexing

An *index* is an alphabetical list of all the major topics discussed in a written work. It cites the pages where each topic can be found and thus allows **readers** to find information on particular topics quickly and easily. The index falls at the very end of the work.

The key to compiling a useful index is selectivity. Instead of listing every possible reference to a topic, select references to passages where the topic is discussed fully or where a significant point is made about it. For actual index entries choose those words or **phrases** that best represent a **topic.** For example, the key terms in a reference to the development of legislation about environmental impact statements would probably be *legislation* and *environmental impact statement,* not *development.* Key terms are those that a reader would most likely look for in an index. In selecting terms for index entries, use chapter or section titles only if they include such key words. For index entries on **tables** and **illustrations,** use the key words in their **titles.**

Compiling an Index

Do not attempt to compile an index until the final manuscript is completed, because terminology and page numbers will not be accurate before then. The best way to compile your list of topics is to read through your written work from the beginning; each time a key term appears in a significant context, list the term and its page number through your keyboard (or 3×5 card if you are doing it manually). An index entry can consist solely of a main entry and its page number.

> **EXAMPLE**   Aquatic monitoring programs, 42

An index entry can also include a main entry, subentries, and even sub-subentries, as in Figure 1. A subentry indicates pages where a specific subcategory or subdivision of a main topic can be found. Indexing application programs are available for indexing, or you can use

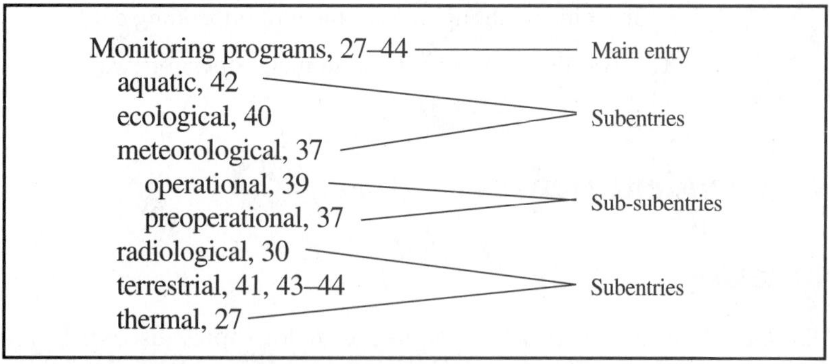

FIGURE 1   Index Entry with Main Entry, Subentries, and Sub-Subentries

your word processing program to create your index. The "search and find," "search and replace," and "cut" and "paste" features are especially useful. (When compiling index entries on cards, enter a subentry on its own separate card, but include the main entry as well.)

When you have completed this process for the entire work, sort the main entries alphabetically, and then sort subentries and sub-subentries alphabetically beneath the proper main entry.

Wording Index Entries

The first word of an index entry should be the principal word because the reader will look for topics alphabetically by their main words. Selecting the right word to list first is of course easier for some topics than for others. For instance, *tips on repairing electrical wire* would not be a suitable index entry because a reader looking for information on electrical wire would not look under the word *tips*. Whether you select *electrical wire; wire, electrical;* or *repairing electrical wire* depends on the **purpose** and **scope** of your report. Ordinarily, an entry with two key words, like *electrical wire*, should be indexed under each word *(electrical wire* and *wire, electrical)*. An index entry should be written as a **noun** or a noun **phrase** rather than as an **adjective** alone.

CHANGE   Electrical
          grounding, 21
          insulation, 20
          repairing, 22
          size, 21
          wire, 20-22

TO   Electrical wire, 20-22
          grounding, 21
          insulation, 20
          repairing, 22
          size, 21

Cross-Referencing

Cross-references in an index guide the reader to other pertinent topics in the text. A reader looking up *technical writing*, for example, might find a cross-reference to *memorandum*. Cross-references do not include page numbers; they merely direct the reader to another index entry, which of course gives pages. There are two kinds of cross references: *see* references and *see also* references.

See references are most commonly used with topics that can be identified by several different terms. Listing the topic page numbers by only one of the terms, the indexer then lists the other terms throughout the index as *see* references.

EXAMPLE   Economic costs. *See* Benefit-cost analyses

See references also direct readers to index entries where a topic is listed as a subentry.

EXAMPLE   L-shaped fittings. *See* Elbows, L-shaped fittings

See also references indicate other entries that include additional information on a topic.

EXAMPLE   Ecological programs, 40-49
              See also Monitoring programs

Styling the Index

Capitalize the first word of a main entry and all other terms normally capitalized in the text. Do not capitalize the first words in subentries and sub-subentries unless they appear that way in the text. Italicize (underline) the cross-reference terms *see* and *see also*.

Place each subentry in the index on a separate line, indented from its main entry. Indent sub-subentries from the preceding subentry. These indentations allow readers to scan a column quickly for pertinent subentries or sub-subentries.

Separate entries from page numbers with **commas.** Type the index in a double-column format, as is done in the index to this book.

indiscreet/indiscrete

Indiscreet means "lacking in prudence or sound judgment."

EXAMPLE   His public discussion of the proposed merger was *indiscreet*.

Indiscrete means "not divided or divisible into parts."

EXAMPLE   The separate units, once combined, become *indiscrete*.

(See also **discreet/discrete.**)

individual

Avoid using *individual* as a **noun** if *person* is more appropriate.

CHANGE   Several *individuals* on the panel did not vote.
    TO   Several *persons* on the panel did not vote.
    OR   Several *people* on the panel did not vote.

Individual is most appropriate when used as an **adjective** to distinguish a single person from a group.

> EXAMPLE   The *individual* employee's obligations to the firm are detailed in the booklet that describes company policies.

(See also **persons/people.**)

infinitive phrases (see **phrases**)

infinitives (see **verbals**)

ingenious/ingenuous

Ingenious means "marked by cleverness and originality," and *ingenuous* means "straightforward" or "characterized by innocence and simplicity."

> EXAMPLES   Wilson's *ingenious* plan, which streamlined production in Department L, was the beginning of his rise within the company.
>
> I believe that the *ingenuous* co-op students bring some freshness into the company.

inquiry letters and responses

An *inquiry letter* may be simple, requesting a free brochure, or complex, asking a financial consultant to define the specific requirements for floating a multimillion-dollar bond issue.

There are two broad categories of inquiry letters. One kind provides a benefit (or potential benefit) to the reader: you may ask, for instance, for information about a product that a company has recently advertised. The second kind of inquiry letter primarily benefits the writer; an example is a request to a public utility for information on the energy-related project you are developing. This kind of letter requires the use of persuasion and special consideration of your **reader's** needs.

Writing an Inquiry Letter

Your **purpose** in writing the letter will probably be to obtain, within a reasonable period of time, answers to specific questions. You will be more likely to receive a prompt, helpful reply if you make it easy for the reader to respond by following these guidelines:

1. Keep your questions concise but specific and clear.
2. Phrase your questions so that the reader will know immediately what type of information you are seeking, why you are seeking it, and how you will use it.

P.O. Box 113
University of Dayton
Dayton, OH 45409
March 11, 19—

Ms. Jane Metcalf
Engineering Services
Miami Valley Power Company
P.O. Box 1444
Miamitown, OH 45733

Dear Ms. Metcalf:

I am an architectural student at the University of Dayton and I need your help.

Could you please send me some information on heating systems for a computer-controlled, energy-efficient, middle-priced house that our systems design class at the University of Dayton is designing. The house, which contains 2,000 square feet of living space (17,6000 cubic feet), meets all the requirements stipulated in your brochure "Insulating for Efficiency."

We need the following information:

1. The proper size heat pump to use in this climate for such a home;

2. The wattage of the supplemental electrical furnace that would be required for this climate; and

3. The estimated power consumption, and current rates, of these units for one year.

We will be happy to send you a copy of our preliminary design report. Thank you very much.

Sincerely yours,

Kathryn J. Parsons

Kathryn J. Parsons

FIGURE 1 Inquiry Letter

3. If possible, present your questions in a numbered list.
4. Keep the number of questions to a minimum, to save the reader's time.
5. Offer some inducement for the reader to respond, such as promising to share the results of your research.
6. Promise to keep responses confidential, especially if you are asking the recipient to complete a questionnaire.

MIAMI VALLEY POWER COMPANY

P.O. BOX 1444
MIAMITOWN, OH 45733
(513) 264-4800
(513) 264-9780 (FAX)
email: mico@aol.com

March 15, 19—

Ms. Kathryn J. Parsons
P.O. Box 113
University of Dayton
Dayton, OH 45409

Dear Ms. Parsons:

Thank you for inquiring about the heating system we would recommend for use in homes designed according to the specifications outlined in our brochure "Insulating for Efficiency."

Since I cannot answer your specific questions, I have forwarded your letter to Mr. Michael Stott, Engineering Assistant in our development group. He should be able to answer the questions you have raised.

Sincerely,

Jane E. Metcalf

Jane E. Metcalf
Director of Public Information

JEM/mk
cc: Michael Stott

FIGURE 2   Letter Indicating That Inquiry Has Been Forwarded

At the end of the letter, thank the reader for taking the time and trouble to respond, and do not forget to include the address to which the material is to be sent. Your chances of getting a reply will improve if you enclose a stamped, self-addressed return envelope, especially if you are writing to someone who is self-employed. Figure 1 (page 276) is a typical inquiry letter.

MIAMI VALLEY POWER COMPANY
P.O. BOX 1444
MIAMITOWN, OH 45733
(513) 264-4800
(513) 264-9780 (FAX)
email: mico@aol.com

March 24, 19—

Ms. Kathryn J. Parsons
P.O. Box 113
University of Dayton
Dayton, OH 45409

Dear Ms. Parsons:

Jane Metcalf has forwarded your letter of March 11 about the house that your systems design class is designing. I can estimate the insulation requirements of a typical home of 17,600 cubic feet, as follows:

1. We would generally recommend, for such a home, a heat pump capable of delivering 40,000 BTUs. Our model AL-42 (17 kilowatts) meets this requirement.
2. With the efficiency of the AL-42, you would not need a supplemental electrical furnace.
3. Depending on usage, the AL-42 unit averages between 1,000 and 1,500 kilowatt hours from December through March. To determine the current rate for such usage, check with the Dayton Power and Light Company.

I can give you an answer that would apply *specifically* to your house only with information about its particular design (such as number of stories, windows, and entrances). If you would send me more details, I would be happy to provide more precise figures—your project sounds interesting.

Sincerely,

Michael Stott

Michael Stott
Engineering Assistant

MS/mo

FIGURE 3   Response to an Inquiry

Response to an Inquiry Letter

When you receive an inquiry letter, first read it quickly to determine whether you are the right person in your organization to answer it—that is, whether you are the one who has both the information and authority to respond. If you are in a position to answer, do so as promptly as you can, and be sure to answer every question the writer has asked. How long your responses should be and how much technical language you should use depend, of course, on the nature of the question and on what information the writer has provided about himself or herself. Even if the writer has asked a question to which the answer is obvious or a question that seems inappropriate, answer it as completely as you can, and do so courteously. You may point out that the reader has omitted or misunderstood something, but be tactful.

If you have received a letter that you feel you cannot answer, find out who can. Then forward the letter to that person. This person should state in the first paragraph of his or her letter that although the letter was addressed to you, it is being answered by someone else in the firm because he or she has the information needed to respond to the inquiry. Figure 2 (page 277) indicates to the inquirer that her letter has been forwarded, and Figure 3 shows a typical response to an inquiry. (See also **correspondence.**)

inside/inside of

In the **phrase** *inside of*, the word *of* is redundant and should be omitted.

> **CHANGE**   The switch is just *inside of* the door.
> **TO**   The switch is just *inside* the door.

Using *inside of* to mean "in less time than" is colloquial and should be avoided in writing.

> **CHANGE**   They were finished *inside of* an hour.
> **TO**   They were finished *in less than* an hour.

insoluble/unsolvable

Although *insoluble* and *unsolvable* are sometimes used interchangeably to mean "incapable of being solved," careful writers distinguish between them. *Insoluble* means "incapable of being dissolved." *Unsolvable* means "incapable of being solved."

EXAMPLES   The plastic is *insoluble* in most household solvents.
Until yesterday, the production problem seemed *unsolvable.*

instructions

When you explain how to perform a specific task—however simple or complex it may be—you are giving *instructions*. Written instructions that are based on clear thinking and careful planning should enable a **reader** to carry out the task successfully.

To write instructions that are accurate and easily understandable, you first must thoroughly understand the task you are describing. Otherwise, your instructions could prove confusing or even dangerous. If you are unfamiliar with the task you are describing, you should watch someone who is familiar with the task go through each step. As you watch, ask questions about any step that is not clear to you. Such direct observation should help you to write instructions that are exact, complete, and easy to follow.

Keep in mind your reader's level of knowledge and experience. Is he or she at all skilled in the kind of task for which you are writing instructions? If you know that your reader has a good background in the topic, you might feel free to use fairly specialized words. If, on the other hand, he or she has little or no knowledge of the subject, you would more appropriately use simple, everyday language—and avoid specialized terms as much as possible.

If the task requires any special tools or materials, say so at the beginning of the instructions. List all essential equipment at the beginning, in a section labeled "Tools Required" or "Materials Required."

The clearest, simplest instructions are written as commands, in the imperative **mood.**

CHANGE   The operator should raise the access lid. (indicative)
TO   Raise the access lid. (imperative)

Phrase instructions concisely but not as if they were telegrams. You can make sentences shorter by leaving out articles *(a, an, the)*, some **pronouns** *(you, this, these)*, and some **verbs,** but sometimes such sentences sound very **telegraphic** and are difficult to understand. The first version of the following instruction for cleaning a computer punch card assembly, for example, is confusing and difficult to understand.

CHANGE   Pass plunger through channel for debris.
TO   Pass *the* plunger through *the* channel *to clear away any* debris.

The meaning of the **phrase** *for debris* needed to be made clearer. The revised instruction is easily and quickly understood.

One good way to make instructions easy to follow is to divide them into short, simple steps. Be sure to order the steps in the proper sequence. Steps can be organized in one of two ways: with numbers or with words. You can label each step with a sequential number.

> EXAMPLE   1. Connect each black cable wire to a brass terminal. . . .
> 2. Attach one 4-inch green jumper wire to the back. . . .
> 3. Connect both jumper wires to the bare cable wires. . . .

Or you can use words that indicate time or sequence.

> EXAMPLE   *First,* determine what the customer's problem is with the computer. *Next,* observe the system in operation. *At that time,* question the operator until you are sure that the problem has been explained completely.

Plan ahead for your reader. If the instructions in step 2 will affect a process in step 9, say so in step 2. Otherwise, your reader may reach step 9 before discovering that an important piece of information—that should have been given in advance—is missing.

Sometimes your instructions have to make clear that two operations must be performed at the same time. Either state this fact in an **introduction** to the specific instructions or include both operations in the appropriate step.

> CHANGE   4. Hold the CONTROL key down.
> 5. Press the BELL key before releasing the CONTROL key.
> TO   4. While holding the CONTROL key down, press the BELL key.

In any operation certain steps must be performed with more preciseness than others. Alert your reader to those steps that require especially precise timing or measurement. Also warn of potentially hazardous steps or materials. If the instructions call for materials that are flammable or that give off noxious fumes, let the reader know before he or she reaches the step for which the material is needed.

If your instructions involve a great many steps, break them into stages, each with a separate head so that each stage begins again with step 1. Using heads as dividers is especially important if your reader is likely to be performing the operation as he or she reads the instructions.

Clear and well-planned **illustrations** can make even complex instructions easy to understand. Illustrations often simplify instructions by reducing the number of words necessary to explain a process or procedure. Appropriate **drawings** and diagrams will enable your reader to identify parts and the relationships between parts more easily than will long explanations. They will also free you, the writer, to focus on the steps making up the instructions rather than on the descriptions of parts. Not all instructions require illustrations, of course. Whether or not illustrations will be useful depends on your reader's needs as well as on the nature of the project. Instructions for inexperienced readers need to be more heavily illustrated than do those for experienced readers.

INSTRUCTIONS

Distribute the inoculum over the surface of the agar in the following manner:

(1) Beginning at one edge of the saucer, thin the inoculum by streaking back and forth over the same area several times, sweeping across the agar suface until approximately one quarter of the surface has been covered.
(2) Sterilize the loop in an open flame.
(3) Streak at right angles to the originally inoculated area, carrying the inoculum out from the streaked areas onto the sterile surface with only the first stroke of the wire. Cover half of the remaining sterile agar surface.
(4) Sterilize the loop.
(5) Repeat as described in step (3), covering the remaining sterile agar surface.

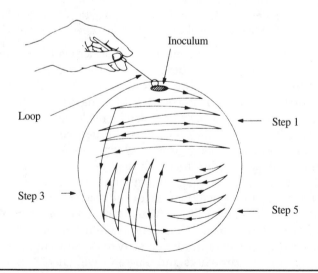

FIGURE 1 Illustrated Instructions

To test the accuracy and **clarity** of your instructions, ask someone who is not familiar with the operation to follow your written directions. A novice will quickly spot missing steps or point out passages that should be worded more clearly.

The instructions in Figure 1 guide the reader through the steps of "streaking" a saucer-sized disk of material (called *agar*) used to grow bacteria colonies. The object is to thin out the original specimen (the *inoculum*) so that the bacteria will grow in small, isolated colonies. The streaking process makes certain that part of the saucer is inoculated heavily, whereas its remaining portions are inoculated progressively more lightly. The streaking is done by hand with a thin wire, looped at one end for holding a small sample of the inoculum. (See also **process explanation.**)

insure/ensure/assure

Insure, ensure, and *assure* all mean "make secure or certain." *Assure* refers to persons, and it alone has the **connotation** of setting a person's mind at rest. *Ensure* and *insure* also mean "make secure from harm." Only *insure* is widely used in the sense of guaranteeing the value of life or property.

> **EXAMPLES**   I *assure* you that the equipment will be available.
> We need all the data to *ensure* the success of the project.
> We should *insure* the contents of the building.

intensifiers

Intensifiers are **adverbs** that emphasize degree, such as *very, quite, rather, such,* and *too.* Although they serve a legitimate and necessary function, they can also seduce the unwary writer who is not on guard against overusing them. Too many intensifiers weaken your writing. When revising your draft, either eliminate intensifiers that do not make a definite contribution, or replace them with specific details.

> **CHANGE**   The team was *quite* happy to receive the *very* good news that it had been awarded a *rather* substantial monetary prize for its design.
> **TO**   The team was happy to receive the good news that it had been awarded a $5,000 prize for its design.

The difference is not that the first example is wrong and the second one right but that the intensifiers in the first example add nothing to the sentence and are therefore superfluous.

Some words (such as *unique, perfect, impossible, final, permanent, infinite,* and *complete*) do not logically permit intensification because they do not permit degrees of comparison. Although **usage** often ignores this logical restriction, the writer should be aware that to ignore it is, strictly speaking, to defy the basic meanings of these words.

    CHANGE   It was *quite* impossible for the part to fit into its designated position.

        TO   It was impossible for the part to fit into its designated position.

interface

An *interface* is the "surface providing a common boundary between two bodies or areas." The bodies or areas may be physical (the *interface* of piston and cylinder) or conceptual (the *interface* of mathematics and statistics). Do not use *interface* as a substitute for the **verbs** *cooperate, interact,* or even *work.*

    CHANGE   The Water Resources Department will *interface* with the Department of Marine Biology on the proposed project.

        TO   The Water Resources Department will *work* with the Department of Marine Biology on the proposed project.

interjections

An *interjection* is a word or **phrase** of exclamation that is used independently to express emotion or surprise or to summon attention. *Hey! Ouch!* and *Wow!* are strong interjections. *Oh, well,* and *indeed* are mild ones. An interjection functions much as do *yes* or *no,* in that it has no grammatical connection with the rest of the sentence in which it appears. When an interjection expresses a sudden or strong emotion, punctuate it with an **exclamation mark.**

    EXAMPLE   His only reaction was a resounding *Wow!*

Because they get their main expressive force from sound, interjections are more common in speech than in writing. They are rarely appropriate to technical writing.

internal proposals (see **proposals**)

international correspondence

Many businesses rely on overseas markets and suppliers, employ workers and managers from different countries, and maintain plants and offices abroad. Such companies need to communicate effectively with readers from diverse cultural and linguistic backgrounds.

Just as American **correspondence** has changed over time, ideas about appropriate style vary from culture to culture. Writers must be alert to the special needs and expectations of readers from different cultural and linguistic backgrounds.

Style and Usage in International Correspondence

Because English is not your reader's native language, it is especially important to write clear, complete sentences. Unusual syntax, jammed **modifiers,** and rambling sentences will hinder a nonnative user of English (see **sentence construction** and **sentence faults**).

Take special care to avoid American **idioms, affectation,** unusual **figures of speech,** allusions, and localism, any of which could easily confuse your **reader.** Humor, particularly sarcasm, and **euphemisms** should also be avoided because they are easily misunderstood outside their cultural context.

Elegant variation, especially the use of **long variants** or pretentious synonyms, will also impede the reader's understanding. Ask yourself whether the words you choose might be found in the abbreviated English-language dictionary that your reader would likely have close at hand.

Proofread your letters carefully; a misspelled word will be particularly troublesome for someone reading in a second language (especially if that reader turns to a dictionary for help and cannot find the word because it is misspelled). Finally, read your writing aloud to identify overly long sentences and to eliminate any possible **ambiguity.**

Although it is important to choose your words carefully and to write clear, complete sentences, avoid using an overly simplified style. A reader who has studied English as a second language might be insulted by a condescending tone and childish language. Also, avoid writing in short, choppy sentences and in a **telegraphic style.**

Translation

Because English is increasingly being taught and used as an international business language, most international correspondence will not require translation. If you do employ a translator, however, be sure that the translator understands the **objective** of the document you are preparing. It is also prudent to let your reader know (in the letter itself or in a postscript) that a translator helped write the document. For first-time contacts, consider sending a translation along with the English version of the letter.

Dates, Time, and Measurement

Different countries use a variety of formats to represent dates, time, and other kinds of measurement. To represent **dates,** Europeans typically write the day before the month and year. In England, for example, 1/11/97 means 1 November 1997; in the United States, it means January 11, 1997. The strictly numerical form for dates, therefore, should never be used in international correspondence. Writing out the name of the month makes the entire date immediately clear to any reader, whichever format you use. Time poses similar problems: particularly in this age of instantaneous communication and high-speed transportation, you may need to specify time zones or to refer to international standards, such as Greenwich Mean Time, to ensure clarity.

Your use of other international standards, such as the metric system, will also help your reader. For up-to-date information about accepted conventions for numbers and symbols in chemical, electrical, data processing, radiation, pharmaceutical, and other fields, consult guides and manuals specific to the subject matter.

Culture, Tradition, and Writing Style

In American **correspondence,** traditional salutations like *Dear* and cordial closings like *Yours truly* have, through custom and long use, acquired meanings quite distinct from their dictionary definitions. Understanding the unspoken meanings of these forms and using them "naturally" is a significant mark of being a part of American business and technical culture. Be aware that such customary expressions and formal amenities vary from culture to culture; when you read foreign correspondence, be alert to these differences.

Japanese business and technical writers, for example, often use traditional openings that reflect on the season, compliment the reader's success, and offer hopes for the reader's continued prosperity. These traditional openings may strike some American readers as being overly elaborate or "literary." Where Americans value directness and employ a forthright **tone** in their **correspondence,** Japanese readers might find the writing blunt and tactless. Remember that our ideas about technical writing have evolved in a particular educational, social, economic, and cultural context. Be sensitive to the expectations of readers who judge effective and appropriate communication from the vantage of a different cultural tradition.

Decision-making styles also vary from culture to culture and from country to country. Consequently, the correspondence, **reports,** and **memorandums** that support decision-making processes will vary as well. Indeed, readers from different backgrounds may interpret the objective of a particular letter or memo in dramatically different ways. The different decision-making strategies, and the stylistic preferences they inspire, reflect larger cultural differences. Whereas an American business writer might consider one brief letter sufficient to communicate a need or transmit a request, for example, a writer in another cultural setting may expect an exchange of three or four longer letters in order to pave the way for action.

Internet

The *Internet* is a worldwide computer network (composed of smaller public and private networks) that allows people around the world to communicate, find and share information, and offer commercial services online. The structure of the Internet (which is supported by dedicated lines that link computers, local-area networks, and wide-area networks) is often compared to an interstate highway system: to get from one location to another, information travels at high speeds along physical paths that interconnect geographically distant sites. Business applications for the Internet include marketing, sales, and collaboration between people in different locations.

A wide range of resources is available on the Internet, including electronic mail **(email),** discussion groups, chat environments, the World Wide Web (WWW, or Web), Gopher, File Transfer Protocol (FTP), and telnet. There are also search engines for most of these resources that make finding information relatively easy (a search engine is a software program that can locate documents on the Internet based on a search of key words that these documents contain).

Discussion Groups

You can use the Internet to join discussion groups, which are electronic communities that exchange messages on a particular topic (such as business ethics or fly fishing). Discussion groups exist for almost every subject imaginable. The Internet offers two primary types of discussion groups: LISTSERV and USENET. If you join a LISTSERV discussion group, you're added to a mailing list. When a message is posted to the group it is automatically distributed to each person on the list. This message arrives in your mailbox along with your other email. If you join a USENET discussion group, you go to the messages (they don't come to you). Instead of using your email program to read messages, you use a newsreader—a software program that allows you to access and manage discussion group messages.

Chat Environments

In addition to discussion groups, in which users read messages at their leisure, chat environments exist for real-time conversations. One popular application is Internet Relay Chat (IRC), topic-oriented channels on the Internet used for public and private talk. Other more elaborate environments are Multi-User Dungeons (MUDs) and Multi-User Dungeons Object Oriented (MOOs), text-based virtual environments often based on themes or topics. In these environments, users assume online personas, chat with other people, navigate "rooms" and other settings, and, in more advanced cases, design their own virtual worlds.

World Wide Web (WWW, or Web)

The Web is a powerful Internet resource that supports applications not only for email, discussion groups, and chat environments, but also for searching, retrieving, and providing information. In fact, the Web is an umbrella application. Operating from within the Web, users can accomplish a variety of Internet tasks once supported by discrete programs. The Web combines multimedia capabilities and **hypertext** links, making available worldwide pages of connected information that may contain written texts, graphics, animation segments, audio clips, and video clips. Each page on the Web has a Uniform Resource Locator (URL), a unique address such as the following:

EXAMPLE   http://www.jones.com/retail/hats.html

In this address, *http* stands for HyperText Transfer Protocol, the method used on the Internet for accessing hypertext documents. The rest of the information identifies the host computer supporting the page *(www.jones.com)*, the directory in which the page is stored *(retail)*, and the file name associated with the page *(hats.html)*. Fortunately, you don't have to remember URLs for most Web pages: these addresses are often automated as hypertext links. Web pages are developed using Hypertext Markup Language (HTML), tags of standard ASCII characters identifying data types and logical relations. Web pages are accessible to anyone with an Internet connection and a client program called a "browser."

Gopher

As with Web pages, Gopher sites allow you to search for and retrieve information on the Internet. But in Gopher sites, information is organized using hierarchical menu structures, as opposed to the often non-hierarchical hypertext structures of the Web. When you use Gopher, you don't need to know addresses: just navigate up and down menus, select an item, and the client software "tunnels" through the Internet and finds it for you. Gopher is useful because it finds different kinds of Internet resources while keeping that complicated process transparent. In addition to Gopher client software, you can use a Web browser to access sites. The address for a Gopher site available on the Web begins with *gopher://*.

File Transfer Protocol (FTP)

FTP is a resource for moving files around the Internet. Using an FTP client program, you can download software applications and other files to your personal computer. Although some FTP sites require passwords, most allow for anonymous FTP connections. When using FTP, you log into a host computer, find the files you want, and transfer those files to your own computer. Because filenames can be cryptic, INDEX and README files often exist to describe file contents. Once files are downloaded, you might in certain cases need to convert them into a readable format. In addition to FTP client software, you can use a Web browser to access sites. The address for an FTP site available on the Web begins with *ftp://*. Some Gopher sites provide access to popular FTP sites.

Telnet

Telnet allows you to connect to other computers on the Internet and use those computers and their software applications as if you were physically there. If you are away from home and want to check your email, you can use a telnet client to log into the computer supporting your account and gain access to your regular mail program. Or if you are away from work and want to edit your Web pages, you can use telnet to log into your account, open the text editor you use for HTML coding, make the edits in your HTML file, and use the operating system to update permission settings. You can also use telnet to connect to public sites such as libraries and databases. These more limited access sites may not require a password, although like anonymous FTP sites some require a user name or other login language that the opening screen usually provides. In addition to telnet client software, you can use a Web browser to access sites. The address for a telnet site available on the Web begins with *telnet://*.

Searching and Using the Internet

Using the Internet for communicating, finding and sharing information, and offering commercial services online is made much easier with the robust search engines often provided with many of these resources. On the Web, for instance, there are dozens of engines that find key words and phrases, support complex search strategies, and provide extensive categories and indexes. You can even delimit a search by specifying which resources you want to examine: for example, only USENET discussion groups. Veronica and Jughead are search engines available in Gopher systems, and Archie is useful for finding FTP sites. Wide Area Information Server (WAIS) is an engine that searches selected databases primarily associated with science and technological topics, although the scope of this engine is expanding. For both novice and advanced users, it would be difficult to spend too much time learning about search engines and search strategies for the Internet. So much information is available that without them the Internet is much less useful.

The Internet, however, is not solely a collection of computer hardware and software. It also consists of groups of people learning and working online, some cooperating and some competing. Because the Internet is a social space, conventions exist for both discussion and design. For example, in exchanging email messages it's common to quote back the particular part of a message to which you're responding to provide context. And in creating Web pages for a site, it's common to include a home page, an index of what's available and how it's organized, with an automated email link to the Web master (the person responsible for maintaining the site and answering questions). If you're a new Internet user, notice the conventions practiced by users of the resources that you want to use before getting started.

In addition to conventions, there are many complicated legal and ethical issues to consider: censorship versus free speech, surveillance versus privacy, and so forth. Even the relatively simple task of creating a modest Web page suggests some difficult issues; for example, the information provided on a Web page is available not only worldwide for viewing but also for downloading. Once Web-based materials are downloaded they can be reproduced or manipulated quite easily because of their digital form. **Copyright** tensions therefore exist: authors and publishers want to protect the royalties associated with published materials, while teachers and researchers (who are also often authors) seem intent on protecting and expanding fair-use policies. As these few examples suggest, the Internet has a social as well as a technological dimension.

interviewing for information

The **scope** of coverage that you have established for your writing project may require more information (or more current information) than is available in written form (in the library, in corporate specifications, in trade journals, and so on). When you reach this point, consider a personal interview with an expert in your subject.

A discussion of interviewing can be divided into three parts: (1) determining the proper person to interview, (2) preparing for the interview, and (3) conducting the interview.

Determining the Proper Person to Interview

Many times your subject, or your **purpose** in writing about the subject, logically points to the proper person to interview. If you were writing about the use of a freeway onramp metering device on Highway 103, for example, you would want to interview the Director of the Traffic Engineering Department; if you were writing about the design of the device, you would want to interview the manufacturer's design engineer. You may interview a professor who specializes in your subject, in addition to getting leads from him or her as to others you might interview. Other sources available to help you determine the appropriate person to interview are (1) the city directory in the library, (2) professional societies, (3) the yellow pages in the local telephone directory, or (4) a local firm that is involved with all or some aspects of your subject.

Preparing for the Interview

After determining the name of the person you want to interview, you must request the interview. You can do this either by telephone or by letter, although a letter may be too slow to allow you to meet your deadline.

Learn as much as possible about the person you are going to interview and about the company or agency for which he or she works. When you make contact with your interviewee, whether by letter or by telephone, explain (1) who you are, (2) why you are contacting him or her, (3) why you chose him or her for the interview, (4) the subject of the interview, (5) that you would like to arrange an interview at his or her convenience, and (6) that you will allow him or her to review your draft.

Prepare a list of specific questions to ask your interviewee. The natural temptation for the untrained writer is to ask general questions rather than specific ones. "What are you doing about air pollution?" for example, may be too general to elicit specific and useful information. "Residents in the east end of town complain about the black smoke that pours from your east-end plant; are you doing anything to relieve the problem?" is a more specific question. Analyze your questions to be certain that they are specific and to the point.

Conducting the Interview

When you arrive for the interview, be prepared to guide the discussion. The following list of points should help you to do so:

1. Be pleasant, but purposeful. You are there to get information, so don't be timid about asking leading questions on the subject.
2. Use the list of questions you have prepared, starting with the less controversial aspects of the topic to get the conversation started and then going on to the more controversial aspects.
3. Let your interviewee do the talking. Don't try to impress him or her with your knowledge of the subject on which he or she is likely an expert.
4. Be objective. Don't offer your opinions on the subject. You are there to get information, not to debate.
5. Some answers prompt additional questions; ask them. If you do not ask these questions as they arise, you may find later that you have forgotten to ask them at all.
6. When the interviewee gets off the subject, be ready with a specific and direct question to guide him or her back onto the track.
7. Take only memory-jogging notes that will help you recall the conversation later. Do not ask your interviewee to slow down so you can take detailed notes. To do so would not only be an undue imposition on the interviewee's time but might also disturb or destroy his or her train of thought.
8. The use of a tape recorder is often unwise, since it often puts people on edge and requires tedious transcription after the interview. On the other hand, a tape recorder does allow you to listen more intently instead of taking notes. If you do use one, do not let it lure you into relaxing so that you neglect to ask crucial questions.

Immediately after leaving the interview, use your memory-jogging notes to help you mentally review the interview and record your detailed notes. Do not postpone this step. No matter how good your memory is, you will forget some important points if you do not do this at once.

interviewing for a job

A job interview may last for thirty minutes, or it may take several hours; it may be conducted by one person or by several, either at one time or in a series of interviews. Since it is impossible to know exactly what to expect, it is important to be as well prepared as possible.

Before the Interview

Learn as much as you can about your potential employer before the interview. What kind of business is it? Is the company locally owned? Is it a nonprofit organization? If the job is public employment, at what level of government is it? Does the business provide a service, and, if so, what kind? How large is the firm? Is the owner self-employed? Is it a subsidiary of a larger operation? Is it expanding? Where will you fit in? This kind of information can be obtained from current employees, from company publications, or from back issues of the local newspaper's business section (available at the public library). You should try to learn about the company's size, sales volume, products and new products, credit rating, and subsidiary companies from its annual reports, and from other business reference sources such as *Moody's Industrials, Dunn and Bradstreet, Standard and Poor's,* and *Thomas' Register.* Such preparation will help you to speak knowledgeably about the firm or industry as well as to ask intelligent questions of the interviewer.

It is a good idea to try to anticipate the questions an interviewer might ask and to prepare your answers in advance. The following is a list of some typical questions posed in job interviews:

What are your short-term and long-term occupational goals?
What are your major strengths and weaknesses?
Do you work better alone or with others?
Why do you want to work for this company?
How do you spend your free time?

During the Interview

Promptness is very important. Be sure to arrive for an interview at the appointed time. It is usually a good idea, in fact, to arrive ahead of time, since you may be asked to fill out an application before meeting the interviewer. Take along an extra copy of your **résumé.**

Remember that the interview will actually begin before you are seated. What you wear and how you act will be closely observed. The way you dress matters: It is usually best to dress conservatively and to be well-groomed. Remain standing until you are offered a seat. Then sit up straight—good posture suggests self-assurance—and look at the interviewer, trying to appear relaxed and confident. It is natural to be nervous during an interview, but be careful to remain alert. Listen carefully and make an effort to remember especially important information. (See **listening.**) Do not attempt to take extensive notes during an interview, although it is acceptable to jot down a few facts and figures.

When answering questions, don't ramble or stray from the subject. Take a minute to think before you answer difficult questions; not only will the time help you collect your thoughts, but it will also make you appear careful in your answer. Say only what you must in order to answer each question and then stop; however, avoid giving just yes and no answers, which usually don't permit the interviewer to learn enough about you. Some interviewers allow a silence to fall just to see how you will react. The burden of conducting the interview is the interviewer's, not yours—and he or she may interpret it as a sign of insecurity if you rush in to fill a void in the conversation. But if such a silence would make you uncomfortable, be ready to ask an intelligent question about the company.

Highlight your qualifications for the job, but admit obvious limitations as well. Remember also that the job, the company, and the location must be right for you. Ask about such factors as opportunity for advancement, fringe benefits (but don't create the impression that your primary interest is security), educational opportunity and assistance, and community recreational and cultural activities (if the job would require you to relocate).

If the interviewer overlooks important points, bring them up. But if possible, let the interviewer mention salary first. If you are forced to bring up the subject, ask it as a straightforward question. Knowing the prevailing salaries in your field will make you better prepared to discuss salary. It is usually unwise to bargain, especially if you are a recent graduate. Many companies have inflexible starting salaries for beginners.

Interviewers look for self-confidence and an understanding of the field in which the candidate is applying. Less is expected of a beginner, but even a newcomer must show some knowledge of the field. One way to impress your interviewer is to ask questions about the company that are related to your line of work. Interviewers respond favorably to people who can communicate easily and present themselves well. Jobs today require interactions of all kinds: person-to-person, department-to-department, division-to-division.

At the conclusion of the interview, thank your interviewer for his or her time. Indicate that you are interested in the job (if true), and try tactfully to get an idea of when you can expect to hear from the company.

After the Interview

When you leave, jot down pertinent information you learned during the interview. (This information will be especially helpful in compar-

2647 Sitwell Road
Charlotte, NC 28210
March 17, 19—

Mr. F. E. Vallone
Personnel Manager
Calcutex Industries, Inc.
3275 Commercial Park Drive
Raleigh, NC 27609

Dear Mr. Vallone:

Thank you for the information and pleasant interview we had last Wednesday. Please extend my thanks to Mr. Wilson of the Servocontrol Group as well. I came away from our meeting most favorably impressed with Calcutex Industries.

I find the position to be an attractive one and feel confident that my qualifications would enable me to perform the duties to everyone's advantage.

I look forward to hearing from you soon.

Sincerely yours,

Philip Ming

Philip Ming

FIGURE 1 Follow-up Letter to Job Interview

ing job offers.) A day or two later, send the interviewer a brief note of thanks, saying that you find the job attractive (if true) and feel you can fill it well. Figure 1 is typical. (See also **job search, résumés, listening, application letters,** and **acceptance letters.**)

introductions

The purpose of an *introduction* is to give your readers enough general information about your **topic** to enable them to understand the detailed information in the body of the document. If you don't need to set the stage this way, turn to the entry on **openings** for a variety of interesting ways to begin without writing an introduction. If you are writing a **newsletter article,** a **proposal,** a **memorandum,** or something similar, you will probably find one of the various types of openings more appropriate than an introduction. If you are preparing a **formal report,** a **laboratory report,** a **technical manual,** a **specification,** a **trade journal article,** a conference paper, or some similar writing project, however, you will need to present your **readers** with some general information before getting into the body of your writing.

You may find that you need to write one kind of introduction for a **report** or trade journal article and a different kind for a technical manual or specification.

Writing Introductions for Reports and Trade Journal Articles

In writing an introduction for a report or trade journal article, you need to state the subject, the purpose of your report or article, its scope, and the way you plan to develop the topic.

Stating the Subject. In addition to stating the subject, you may need to define it if it is one with which some of your readers may be unfamiliar. (See **definition method of development.**)

Stating the Purpose. The statement of purpose should make your readers aware of your goal as they read your supporting statements and examples. It should also tell them why you are writing about the subject: whether your material provides a new perspective or clarifies an existing perspective.

Stating the Scope. Stating the **scope** of your article or report tells your reader how much or how little detail to expect. Does your article or report present a broad survey of your topic, or does it concentrate on one facet of the topic?

Stating the Development of the Subject. In a larger work, it may be helpful to your reader if you state how you plan to develop the subject. Providing such information makes it easier for your readers to anticipate how the subject will be presented (see **methods of development**), and it gives them a basis for evaluating how you arrived at your conclusions or recommendations.

The following introduction is from a trade journal article:

Subject and background

Toxic waste sites are viewed by the public as one of the most serious environmental hazards according to opinion polls (*New York Times* 1991). The Environmental Protection Agency's (EPA) current process for ranking sites uses the "best engineering judgment and does not consider socioeconomic factors" (Mitre Corporation 1981); the cost to physically clean the site is estimated, but the social costs from the existence of the site (the social benefits from the cleaning) are not calculated. The decline in house values has been well documented (for example, Kohlhase 1991; Michaels and Smith 1990), but the role of the EPA's actions in mitigating the costs has not been thoroughly examined. The EPA could potentially speed the market's adjustment process by disseminating information about the toxicity of the site and how effective cleaning is expected to be, thereby reducing adjustment costs. Social costs may also be eliminated by cleaning the site.

Purpose

Method of development

One way to evaluate the social benefits obtained by removing a toxic site is to look at the impact of the site on adjacent house values over the period of adjustment. In theory, house values are determined by the discounted value of future rents that could be earned by that unit. If the existence of a toxic site causes rents to fall because the unit is less desirable, house prices close to the site will decrease. If cleaning the site completely removes all concern about locating near the site, the prices of neighboring houses should increase. By comparing the house values before and after the cleaning, the cost captured by house values can be measured.

Scope and method of development

This study uses a unique data set, consisting of information on the sales price, sales date, house and neighborhood characteristics, and distance from toxic sites, for over 2,000 houses in the Boston metropolitan area from 1975–92. The area under study (Woburn, Massachusetts) has two Superfund sites. Both were announced as sites by the EPA in the early 1980's, and the cleaning process has begun. Using regression techniques, the effect on house values of the EPA announcement of a discovered toxic site and of subsequent cleanup efforts can be measured.

K.A. Kiel, "Measuring the Impact of the Discovery and Cleaning of Identified Hazardous Waste Sites on House Values." *Land Economics* 71 (4)

Writing Introductions for Technical Manuals and Specifications

In writing an introduction for a technical manual or set of specifications, identify the topic and its primary purpose or function in the first sentence or two. Be specific, but do not get into details. Your introduction sets the stage for the entire document, and it is important that your reader get from the introduction a broad frame of reference. Only later, when your reader has some understanding of the overall topic and can appreciate its details in proper perspective, should you introduce technical details. The following is an example:

> The System Constructor is a program that can be used to create operating systems for a specific range of microcomputer systems. The constructor selects requested operating software modules from an existing file of software modules and combines those modules with a previously compiled application program to create a functional operating system designed for a specific hardware configuration. The constructor selects the requested software modules, generates the necessary linkage between the modules, and generates the necessary control tables for the system according to parameters specified at run time.

How technical your introduction should be depends on your readers: What is their technical background? What kind of information are they seeking in the manual or specification? A computer programmer, for example, has different interests in an application program or utility routine than does an operator—and the topic should be introduced accordingly. The assumptions you can make without providing explanations and the terminology you can use without providing definitions differ greatly from one reader to another.

You may encounter a dilemma that is common in technical writing: Although you can't explain topic A until you have explained topic B, you can't explain topic B before explaining topic A. The solution is to explain both topics in broad, general terms in the introduction. Then, when you need to write a detailed explanation of topic A, you will be able to do so because your reader will know just enough about both topics to be able to understand your detailed explanation.

EXAMPLE   The NEAT/3 programming language, which treats all peripheral units as file storage units, allows your program to perform data input or output operations depending on the specific unit. Peripheral units from which your program can only input data are referred to as *source units*. Units to which your program can only output data are referred to as *destination units* or as combination *source-destination units*.

Thus, when you need to explain any one of the three units in detail, your readers will have at least a general knowledge of the topic.

Finally, consider writing the introduction last. Many writers find it helpful to write the introduction last because they feel that only then do they have a full enough perspective on the writing to introduce it adequately.

investigative reports

When a person investigates a particular topic, he or she might write up the results in an *investigative report*. For example, the report could sum-

---

MEMORANDUM

To: Noreen Rinaldo, Training Manager
From: Charles Lapinski, Senior Instructor *CL*
Date: February 14, 19—

Subject: Addison Corporation's Basic English Program

Purpose

    As you requested, I have investigated Addison's program to determine whether we might also adopt such a program. The purpose of the Addison Basic English course is to teach foreign mechanics who do not speak or read English to understand repair manuals written in a special 800-word vocabulary called "Basic English," and thus eliminate the need for Addison to translate its manuals into a number of different languages. The Basic English Program does not attempt to teach the mechanics to be fluent in English but, rather, to recognize the 800 basic words that appear in the repair manuals.

    The course does not train mechanics. Students must know, in their own language, what a word like *torque* means; the course simply teaches them the English term for it. As prerequisites for the course, students must have a basic knowledge of their trade, must be able to identify a part in an illustrated parts book, must have served as a mechanic on Addison products for at least one year, and must be able to read and write in their own language.

Scope

    Students are given the specially prepared instruction manual, an illustrated book of parts and their English names, and a pocket reference containing all 800 words of the Basic English vocabulary plus the English names of parts (students can write the corresponding word in their language beside the English words and then use the pocket reference as a bilingual dictionary). The course consists of thirty two-hour lessons, each lesson introducing approximately 27 words. No effort is made to teach pronunciation; the course teaches only recognition of the 800 words, which include 450 nouns, 70 verbs, 180 adjectives and adverbs, and 100 articles, prepositions, conjunctions, and pronouns.

    The 800-word vocabulary enables the writers of the manuals to provide mechanics with any information that might be required, because the area of communication is strictly limited to maintenance, inspection, troubleshooting, safety, and the operation of Addison equipment. All nonessential words (such as *apple, father, mountain,* and so on) have been eliminated, as have most synonyms (for example *under* appears, but *beneath* does not).

Conclusions

I see three possible ways in which we might be able to use some or all of the elements of the Basic English Program: (1) in the preparation of all our student manuals, (2) in the preparation of student manuals for the international students in our service school, or (3) as Addison uses the program.

    I think it would be unnecessary to use the Basic English methods in the preparation of student manuals for *all* our students. Most of our students are English-speaking people to whom an unrestricted vocabulary presents no problem.

**Recommen-
dations**

    In conjunction with the preparation of student manuals for international students, the program might have more appeal. Students would take the Basic English course either before coming to this country to attend school or after arriving but before beginning their technical training.

    As for our initiating a Basic English Program similar to Addison's, we could create our own version of the Basic English vocabulary and write our service manuals in it. Since our product lines are much broader than Addison's, however, we would need to create illustrated parts books for each of the different product lines.

FIGURE 1   Investigative Report

marize the findings of an opinion survey, of a product evaluation, or of a marketing study. It could review the published work on a particular topic or compare the different procedures used to perform the same operation. An investigative report gives a precise analysis of its topic and then offers the writer's conclusions and recommendations.

Open such a **report** with a statement of its **purpose.** In the body of the report, first define the **scope** of your investigation. If the report is on a survey of opinions, for example, you might need to indicate the number of people and the geographical areas surveyed as well as the income categories, the occupations, the age groups, and possibly even the racial groups of those surveyed. Include any information that is pertinent in defining the extent of the investigation. Then report your findings and, if necessary, discuss their significance. End your report with your conclusions and, if appropriate, any recommendations.

Notice that the sample investigative report shown in Figure 1 opens by stating its purpose and ends with its recommendations.

irregardless/regardless

Irregardless is nonstandard English because it expresses a double negative. The word *regardless* is already negative, meaning, "unmindful." Always use *regardless.*

> CHANGE   *Irregardless* of the difficulties, we must increase the strength of
> the outer casing.
>
> TO   *Regardless* of the difficulties, we must increase the strength of
> the outer casing.

it

The **pronoun** *it* has a number of uses. First, it can refer to a preceding **noun** that names an object or idea or to a baby or animal whose sex is unknown or unimportant to the point. *It* should have a clear antecedent.

> EXAMPLE   Darwinism made an impact on nineteenth-century American
> thought. *It* even influenced economics.

It can also serve as an **expletive.**

> EXAMPLE   *It* is necessary to sand the hull before you paint it.

Avoid overusing expletives, however.

> CHANGE   *It* is seldom that we go.
> TO   We seldom go.

(See also **its/it's** and **pronoun reference.**)

italics

Italics is a style of type used to denote **emphasis** and to distinguish foreign expressions, book titles, and certain other elements. *This sentence is printed in italics.* You may need to italicize words that require special emphasis in a sentence.

> EXAMPLE   Contrary to projections, sales have *not* improved since we started the new procedure.

Do not overuse italics for emphasis, however.

> CHANGE   This will hurt *you* more than *me.*
> TO   This will hurt you more than me!

Titles

Italicize the titles of books, periodicals, newspapers, movies, and paintings.

> EXAMPLES   The book *Statistical Methods* was published in 1991.
> *The Cincinnati Enquirer* is one of our oldest newspapers.
> *Chemical Engineering* is a monthly journal.

Abbreviations of such titles are italicized if their spelled-out forms would be italicized.

> EXAMPLE   *The Journal of QA Technology* is an informative publication.

Titles of chapters or articles within publications and titles of reports are placed in **quotation marks,** not italicized.

> EXAMPLE   "Clarity, the Technical Writer's Tightrope" was an article in *Technical Communications.*

Titles of holy books and legislative documents are not italicized.

> EXAMPLE   The Bible and the Magna Carta changed the history of Western civilization.

Titles of long poems and musical works are italicized, but titles of short poems and musical works and songs are enclosed in quotation marks.

> EXAMPLES   Milton's *Paradise Lost* (long poem)
> Handel's *Messiah* (long musical work)
> T.S. Eliot's "The Love Song of J. Alfred Prufrock" (short poem)
> Leonard Cohen's "Suzanne" (song)

Words, Letters, and Figures

Words, letters, and figures discussed as such are italicized.

EXAMPLES   The word *inflammable* is often misinterpreted.
The *s* and the *6* keys on my keyboard tend to stick.

Foreign Word

Foreign words that have not been assimilated into the English language are italicized.

EXAMPLES   *sine qua non, coup de grace, in res, in camera*

Foreign words that have been fully assimilated into the language, however, need not be italicized.

EXAMPLES   cliché, etiquette, vis-à-vis, de facto, siesta

When in doubt about whether or not to italicize a foreign word, consult a current **dictionary.** (See also **foreign words in English.**)

Subheads

Subheads in a report are sometimes italicized—or underlined in traditional typescript. (See also **headings.**)

EXAMPLE   There was no publications department as such, and the writing groups were duplicated at each plant or location. Wellington, for example, had such a large number of publications groups that their publication efforts can only be described as disorganized. Their duplication of effort must have been enormous.

Training Writers
We are certainly leading the way in developing first-line managers (or writing supervisors) who are not only technically competent but can also train the writers under their direction and be responsible for writing quality as well.

its/it's

Be careful never to confuse these two words: *Its* is a possessive **pronoun,** whereas *it's* is a **contraction** of *it is.*

EXAMPLE   *It's* important that the factory meet *its* quota.

Although nouns normally form the possessive by the addition of an **apostrophe** and an *s*, the contraction of *it is (it's)* has already used that device; therefore, the possessive form of the pronoun *it* is formed by adding only the *s.*

J

jammed modifiers (see **modifiers**)

jargon

Jargon is a highly specialized technical slang that is unique to an occupational group. If all your **readers** are members of a particular occupational group, jargon may provide a time-saving and efficient means of communicating with them. For example, finding and correcting the errors in a computer program is referred to by programmers as *debugging*. If you have any doubt that your entire reading audience is a part of this group, however, avoid using jargon.

When jargon becomes so specialized that it applies only to one company or a subgroup of an occupation, it is referred to as *shop talk*. Obviously, shop talk is appropriate only for those familiar with its special vocabulary and should be reserved for speech or informal **memorandums.** (See **affectation** and **gobbledygook.**)

job descriptions

Most large companies and many small ones specify, in a formal *job description,* the duties of and requirements for many of the jobs in the firm. Job descriptions fill several important functions: They provide information on which equitable salary scales can be based; they help management determine whether all responsibilities within a company are adequately covered; and they let both prospective and current employees know exactly what is expected of them. Together, all of the job descriptions in a firm present a picture of the organization's structure.

Sometimes plant or office supervisors are given the task of writing the job descriptions of the employees assigned to them. In many organizations, though, an employee may draft his or her own job description, which the immediate superior then checks and approves.

A Format for Writing Job Descriptions

Although job description formats vary from organization to organization, they commonly contain the following sections:

Accountability. This section identifies, by title only, the person to whom the employee reports.

Scope of responsibilities. This section provides an overview of the primary and secondary functions of the job and states, if applicable, who reports to the employee.

Specific duties. This section gives a detailed account of the specific duties of the job, as concisely as possible.

Personal requirements. This section lists the education, training, experience, and licensing required or desired for the job.

Tips for Writing Job Descriptions

If you have been asked to prepare a job description for your position, the following guidelines should help:

1. Before attempting to write your job description, make a list of all the different tasks you do for a week or a month. Otherwise, you will almost certainly leave out some of your duties.
2. Focus on content. Remember that you are describing your job, not yourself.
3. **List** your duties in decreasing order of importance. Knowing how your various duties rank in importance makes it easier to set valid job qualifications.
4. Begin each statement of a duty with a **verb,** and be specific. Write "Answer and route incoming telephone calls" rather than "Handle telephone calls."
5. Review existing job descriptions that have been successful.

The job description shown in Figure 1, which is a typical one, never mentions the person holding the job described; it focuses, instead, on the job and on the qualifications any person must possess to fill the position. (See page 305.)

Manager, Technical Publications
Acme Electrical Corporation

Accountability

Reports directly to the Vice President, Customer Service.

Scope of Responsibilities

The Manager of Technical Publications is expected to plan, coordinate, and supervise the design and development of technical publications and documentation required in the support of the sale, installation, and maintenance of Acme products. The manager is responsible for the administration and morale of the staff. The supervisor for instruction manuals and the supervisor for parts manuals report directly to the manager.

Specific Duties

Direct an organization presently composed of twenty people (including two supervisors), over 75 percent of whom are writing professionals and graphics specialists.

Screen, select, and hire qualified applicants for the department.

Prepare a formal program designed to orient writing trainees to the production of reproducible copy and graphic arts.

Evaluate the performance of departmental members and determine salary adjustments for all personnel in the department.

Plan documentation to support new and existing products.

Determine the need for subcontracted publications and act as a purchasing agent when they are needed.

Offer editorial advice to supervisors.

Develop and manage an annual budget for the Technical Publications Department.

Cooperate with the Engineering, Parts, and Service Departments to provide the necessary repair and spare parts manuals upon the introduction of new equipment.

Serve as a liaison between technical specialists, the publications staff, and field engineers.

Recommend new and appropriate uses for the department within the company.

Keep up with new techniques in printing processes, typesetting, computerized text editing, art, and graphics and use them to the advantage of Acme Electrical Corporation where applicable.

Personal Requirements

B.A. in technical communication desired.

Minimum of three years professional writing experience with a general knowledge of graphics and production.

Minimum of two years management experience with a knowledge of the general principles of management.

Must be conversant with the needs of support people, technical people, and customers.

FIGURE 1 Job Description

job search

To begin an effective search for a job, you must know essentially three things: (1) yourself (your specific skills, your goals, and so forth), (2) the job you are searching for (geographical area, likely industries, likely companies, and so on), and (3) where to get all the information you need to conduct an effective job search.

Yourself

Do some homework on yourself—some real soul searching. Decide exactly what your goals are. What would you like to do in the immediate future? When you have answered that, think about the kind of work you would like to be doing two years from now. Then think seriously about what you would like to be doing five years from now. When you have pinned down these answers, you have established your goals.

You must also know exactly what your strongest skills are, for they are your most salable products. You must know your appropriate level of entry into a job requiring those skills (entry level for the inexperienced, intermediate level for the moderately experienced, and senior level for the very experienced). Then you must know where you would be most likely to find a need for those skills (geographical location, professional vocation, which industries, which companies).

Determine how you are different, or better qualified, for a job than the other people who are likely to apply for it. Are you expert in your field? Are you especially painstaking with details? Are you particularly good with people? Are you persistent and determined in solving problems? Do you have any particular skills or talents, such as writing, public speaking, or analytical ability? If you know what makes you stand out and can communicate it during an interview, you greatly increase your chances.

The Job

Start with the geographical area that would most favor your job search or in which you prefer to live. Then narrow the geographical area by determining the field in which you prefer to work. Then narrow the field by identifying those specializations within that field that interest you most. Finally, pinpoint those companies within your selected specializations that need your skills and for which you feel you could work comfortably. Be sure to include small and even very small companies, since collectively, they create far more jobs than do large companies.

Thoroughly **research** each company. Visit the library to find information about the company in sources such as annual reports, the *Standard and Poor's Directory,* and *Thomas' Register of American Manufacturers* (see also **library research**). Ask family, friends, and even casual acquaintances about the companies. One of the key things to learn is who in the company has the authority to hire you. For a recent graduate, this may be the personnel manager; for a highly experienced professional, it is more likely to be the manager of the department doing the hiring. Your goal should be to get a face-to-face meeting with that person.

Sources of Information

The information you need to help you land a job can come from a number of sources: classified ads, **inquiry letters, Internet** resources, trade and professional journals, school placement services, employment agencies, friends already employed within your area of interest, and others.

Classified Ads. Many employers advertise in the classified sections of newspapers. For the widest selection, look in the Sunday editions of local and big city newspapers. Although reading want ads can be tedious, an item-by-item check is necessary if you are to make a thorough search. The job you are looking for might be listed in the classified ads under various titles. For example, an accountant seeking a job might find the specialty listed under "General Accountant," "Cost Analyst," or "Budget Analyst." So play it safe—read all the ads.

Occasionally, newspapers print special employment supplements that provide valuable information on many facets of the job market. Watch for these. As you read the ads, note such factors as salary ranges, job locations, and job duties and responsibilities.

Watch out for fake ads. These are ads for jobs that don't exist, run by placement firms to fatten their "résumé bank" for future clout with employers. They usually tell you nothing more than a job title and a salary (usually inflated). Blind ads (no company name, just a box number) rarely prove worthwhile. Legitimate advertisers have nothing to hide. Do not place a "job wanted" ad yourself, because potential employers don't read them—but placement services do (generally the less reputable ones).

Internet Resources. The **Internet** offers a variety of job search possibilities. For example, Internet discussion groups on topics in your job specialty can provide a useful way to keep up with trends and general employment conditions over the long term. More immediately, of course, some discussion groups post job openings that may be appropriate.

If you have access to the World Wide Web, you can search for job listings at various Web sites. Understand, however, that finding an employer on the Web involves searching thousands of entries in numerous different databanks, many of which will be irrelevant to your search. Cyberspace can be as crowded as other settings for finding employment.

Another option is establishing your own Web site and uploading your **résumé** in order to attract potential employers. If you post your résumé on the Web, keep the following points in mind:

> Keep your résumé as simple and clear as possible.
>
> Format the résumé as an ASCII text file so it will scroll correctly and be universally readable in any application.
>
> Use hyperlinks so prospective employers can simply click on your email address to contact you.
>
> Just below your name, provide a series of **hypertext** "jumps" for such important categories as "experience" and "education."
>
> Use a counter to keep track of the number of times your résumé Web page has been visited.

J

If you do not have a Web site or simply don't feel confident in adapting your résumé for the Web, you might consider using a résumé-writing company that will put your résumé online. Before employing such services, however, browse the Web to learn what type of employers (such as the computer industry or marketing companies) look at résumés on the Web. Be aware, too, that such companies often charge a monthly fee for their posting services, in addition to various setup and sign-on fees.

Inquiry Letters. If you are interested in a particular company, write and ask whether it has an opening for someone with your skills and qualifications. This is essentially an **application letter,** except that you are not responding to an advertisement for a job. Normally you should send the letter either to the director of personnel or to the specific department head; for a small company, however, you can write to the president.

Trade and Professional Journals. Many occupations have associations that publish periodicals containing listings of current job opportunities. If you are seeking a job in forestry, for example, you could check the job listings in the *Journal of Forestry,* published by the Society of American Foresters. To learn about the trade or professional associations for your occupation, consult the following reference books:

> *Encyclopedia of Associations*
> *Encyclopedia of Business Information Sources*
> *National Directory of Employment Services*

School Placement Services. Check with the career counselors in your school's job-placement office (you can sometimes use the job-placement services of other schools as well). Government, business, and industry recruiters often visit job-placement offices to interview prospective employees; the recruiters also keep college placement offices aware of their company's current employment needs. While you are in the placement office, ask to see a current issue of the *College Placement Annual.* This publication lists the occupational requirements and addresses of over a thousand industry, business, and government employers.

State Employment Agencies. Most states have free employment agencies that function specifically to match applicants and jobs. If your state has one, register with the local employment office. It may have just the job you want; if not, it will keep your résumé on file and call you if such a job comes along.

J

Private Employment Agencies. Private employment agencies are profit-making organizations that are in business to help people find jobs—for a fee. Reputable private employment agencies give you job leads and help you organize your campaign for the job you want. They may also provide useful information on the companies doing the hiring. Choose a private employment agency carefully. Some are well-established and quite reputable, but others have questionable reputations. Check with your local Better Business Bureau as well as with friends and acquaintances before signing an agreement with a private employment agency.

Who will pay the fee if you are offered a job through a private agency? Sometimes the employer will pay the agency's fee; if not, you must pay either a set fee or a percentage of your first month's salary. In general, if the employer pays the fee, the job is a prized position that is hard to fill, and the fact that the employer is paying the fee is a good indication that the employment agency is reputable. Before signing a contract, be sure you understand who is paying the fee and, if *you* are, how much you are agreeing to pay. As with any written agreement, read the fine print carefully.

Do not give a private employment agency "exclusive handling." If you do, and you find *yourself* a job independent of the agency, you may still have to pay it a fee.

Other Sources. When searching for a job, consult with people whose judgment you respect. Use all available resources. Recruit family members and friends to help, and alert as many people as possible to your search. Talk especially with people who are already working in your chosen field.

Local, state, and federal government agencies offer many employment opportunities. Local government agencies are listed in the white pages of your telephone directory under the name of your city, county, or state. For information about jobs with the federal government, contact the U.S. Office of Personnel Management or the Federal Job Information Center; both have branches in most major cities. The offices are listed in the blue pages of the telephone directory under "U.S. Government."

If you are a veteran, local and campus Veterans' Administration offices can provide material on special placement programs for veterans. Such agencies will supply you with the necessary information about the particular requirements or entrance tests for your occupation.

Keep a loose-leaf notebook with *dated* job ads, copies of letters of applications and résumés, notes sent in regard to interviews, and the names of important contacts. This notebook can act as a future resource and as a tickler file. (See also **résumé, interviewing for a job,** and **acceptance letters.**)

journal articles (see trade journal articles)

judicial/judicious

Judicial is a term that pertains only to law.

   EXAMPLE   The *judicial* branch is one of the three branches of the United States government.

Judicious refers to careful or wise judgment.

   EXAMPLE   We intend to convert to the new refining process by a series of *judicious* steps.

kind of/sort of

Kind of and *sort of* should be used in writing only to refer to a class or type of things.

K

> EXAMPLE They used a special *kind of* metal in the process.

Do not use *kind of* or *sort of* to mean "rather," "somewhat," or "somehow."

> CHANGE It was *kind of* a bad year for the firm.
> TO It was a bad year for the firm.

know-how

An informal term for "special competence or knowledge," *know-how* should be avoided in **formal writing.**

> CHANGE He has great technical *know-how.*
> TO He has great technical *skill.*

L

laboratory reports

A *laboratory report* communicates information acquired from a laboratory test or investigation. (Simpler tests use the less formal **test report** form.) Laboratory reports should state the reason the laboratory investigation was conducted, the equipment and procedures used, any problems encountered, any conclusions reached, and any recommendations based on the conclusions.

A laboratory report often places special **emphasis** on the equipment and the procedures used in the investigation because these two factors can be critical in determining the accuracy of the **data** obtained.

Present the results of the laboratory investigation clearly and concisely. (Use the Checklist of the Writing Process to write and revise your **report.**) Read the entries on **graphs** and **tables** if your report requires graphic or tabular presentation of data.

Each laboratory usually establishes the **organization** of its laboratory reports. Figure 1 on pages 313–315 is a typical example of a laboratory report.

lay/lie

Lay is a transitive **verb** (a verb that has a direct **object** to complete its meaning) and means "place" or "put." Its present **tense** form is *lay.*

EXAMPLE   We *lay* the foundation of the building one section at a time.

The past tense form of *lay* is *laid.*

PCB Exposure from Oil Combustion
Wayne County Professional Fire Fighters

Submitted to:
Mr. Philip Landowe
President, Wayne County Professional Fire-Fighters Association
Wandell, IN 45602

Submitted by:
Analytical Laboratories, Incorporated
Mr. Arnold Thomas
Certified Industrial Hygienist
Mr. Gary Seabolm
Laboratory Manager
Environmental Analytical Services
1220 Pfeiffer Parkway
Indianapolis, IN 46223

February 28, 19—

Introduction

Waste oil used to train fire fighters was suspected of containing polychlorinated biphenyls (PCB). According to information provided by Mr. Philip Landowe, President of the Wayne County Professional Fire-Fighters Association, it has been standard practice in training fire fighters to burn 20–100 gallons of oil in a diked area of approximately 25–50M^3. Fire fighters would then extinguish the fire at close range. Exposure would last several minutes, and the exercise would be repeated two or three times each day for one week.

Oil samples were collected from three holding tanks near the training area in Englewood Park on November 11, 19—. To determine potential fire-fighter exposure to PCB, bulk oil analyses were conducted on each of the samples. In addition, the oil was heated and burned to determine the degree to which PCB is volatized from the oil, thus increasing the potential for fire-fighter exposure via inhalation.

Testing Procedures

Bulk oil samples were diluted with hexane, put through a cleanup step, and analyzed in electron-capture gas chromatography. The oil from the underground tank that contained PCB was then exposed to temperatures of 100° C without ignition and 200° C with ignition. Air was passed over the enclosed sample during heating, and volatized PCB was trapped in an absorbing medium. The absorbing medium was then extracted and analyzed for PCB released from the sample.

Results

Bulk oil analyses are presented in Table 1. Only the sample from the underground tank contained detectable amounts of PCB. Aroclor 1260, containing 60 percent chlorine, was found to be present in this sample at 18 μg/g. Concentrations of 50 μg/g PCB in oil are considered hazardous. Stringent storage and disposal techniques are required for oil with PCB concentrations at these levels.

FIGURE 1 Laboratory Report

Results for the PCB volatization study are presented in Table 2. At 100° C. 1 μg PCB from a total of 18 μg PCB (5.6 percent) was released to the air. Lower levels were released at 200° C and during ignition, probably as a result of decomposition. PCB is a mixture of chlorinated compounds varying in molecular weight; lightweight PCBs were released at all temperatures to greater degree than the high molecular weight fractions.

TABLE 1. Bulk Oil Analysis

Source	Sample #	PCB Content (μg/g)
Underground Tank (11' deep)	# 6062	18*
Circle Tank (3' deep)	6063a	<1
	6063b	<1
Square Pool (3' deep)	6064a	<1
	6064b	<1

*Aroclor 1260 is the PCB type. This sample was taken for volatilization study.

TABLE 2. PCB Volatilization Study
for the 11-Foot Deep Underground Tank*

Outgassing Temp. (°C)	Outgassing Time (Min)	Sample Outgassed (g)	PCB Total (μg)	PCG Outgassed (μg)
100	30	1	18	1
200	30	1	18	0.6
200 with ignition	30	1	18	0.2

*Bulk analysis of 18 μg/g PCB

Discussion and Conclusions

At a concentration of 18 μg/g, 100 gallons of oil would contain approximately 5.5 grams of PCB. Of the 5.5 grams of PCB, about 0.3 grams would be released to the atmosphere under the worst conditions.

The American Conference of Governmental Industrial Hygienists has established a TLV* of 0.5 μg/M^3 air for a PCB containing 54 percent Cl as a time-weighted average over an 8-hour workshift and has stipulated that exposure over a 15-minute period should not exceed 1 mg/M^3. The 0.3 gram of released PCB would have to be diluted to 600 M^3 air to result in a concentration of 0.5 mg/M^3 or less. Since the combustion of oil lasted several minutes, a dilution to more than 600 M^3 is likely; thus exposure would be less than 0.5 mg/M^3. Since an important factor in determining exposure is time and the fire fighters were exposed only for several minutes at intermittent intervals, adverse effects from long-term exposure to low-level concentrations of PCB should not be expected.

*The Threshhold Limit Value (TLV) is the safe average concentration that most individuals can be exposed to in an 8-hour day.

FIGURE 1   Laboratory Report *(continued)*

It should be stressed, however, that these conclusions are based solely on oil containing 18 μg/g PCB. If, on previous occasions, the PCB content of the oil was much higher, greater exposure could have occurred. PCB is a known liver toxin and has also been classified as a suspected carcinogen. Although the primary route of entry into the body is by inhalation, PCB can be absorbed through the skin. PCB can cause a skin condition known as chloracne which results from a clogging of the pores. This condition, which is often associated with a secondary infection, should not occur at the PCB concentration found in this oil. A clinical test exists for determining if PCB has been absorbed by the liver.

In summary, because exposure to this oil was limited and because PCB concentrations in the oil were low, it is unlikely that exposure from inhalation would be sufficient to cause adverse health effects. However, we cannot rule out the possibility that excessive exposure may have occurred under certain circumstances, based on factors such as excessive skin contact and the possibility that higher-level PCB concentrations in the oil could have been used earlier. The practice of using this oil should be terminated.

FIGURE 1   Laboratory Report *(continued)*

L

> **EXAMPLE**   We *laid* the first section of the foundation on the 27th of June.

The perfect tense form of *lay* is also *laid*.

> **EXAMPLE**   Since June we *have laid* all but two sections of the foundation.

Lay is frequently confused with *lie,* which is an intransitive verb (a verb that does not require an object to complete its meaning) meaning "recline" or "remain." Its present tense form is *lie.*

> **EXAMPLE**   Injured employees should *lie* down and remain still until the doctor arrives.

The past tense form of *lie* is *lay.* (This form causes the confusion between *lie* and *lay.*)

> **EXAMPLE**   The injured employee *lay* still for approximately five minutes.

The past perfect tense form of *lie* is *lain.*

> **EXAMPLE**   The injured employee *had lain* still for approximately five minutes when the doctor arrived.

layout and design (see design and layout)

leave/let

As a **verb,** *leave* should never be used in the sense of "allow" or "permit."

> CHANGE   *Leave* me do it my way.
>   TO   *Let* me do it my way.

As a **noun,** however, *leave* can mean "permission granted."

> EXAMPLE   Employees are granted a *leave* of absence if they have a chronic illness.

lend/loan

Lend or *loan* each may be used as a **verb,** but *lend* is more common.

> EXAMPLES   You can *loan* them the money if you wish.
>   You can *lend* them money if you wish.

Unlike *lend,* *loan* can be a **noun.**

> EXAMPLE   We made arrangements at the bank for a *loan.*

L

letters (see **correspondence**)

libel/liable/likely

The term *libel* refers to "anything circulated in writing or pictures that injures someone's good reputation." When someone's reputation is injured in speech, however, the term is *slander.*

> EXAMPLES   When an editorial charged our board of directors with bribing a representative, the board sued the newspaper for *libel.*
>
>   If the mayor supports the bribery charge in his speech tonight, we will accuse him of *slander.*

The term *liable* means "legally subject to" or "responsible for."

> EXAMPLE   Employers are held *liable* for their employees' decisions.

In technical writing, *liable* should retain its legal meaning. Where a condition of probability is intended, use *likely.*

> CHANGE   Rita is *liable* to be promoted.
>   TO   Rita is *likely* to be promoted.

library research

To help you find the advice you need, the following entry is divided into the following sections:

L

The key tools for conducting library research include the library's catalog of holdings, periodical indexes, bibliographies, and subject-specific and general background reference works. The most efficient way to use these tools is through your library's computer-search system. Public access computer terminals provide online access to the library's card catalog, subject indexes, and many reference works. These systems have largely replaced paper card catalogs in academic and most other libraries. Although libraries provide printed instructions about how to conduct online searches, ask a librarian if you have questions about such searches or about any other facet of the library.

Online Card Catalog Searches

Using the computer to search the vast resources of a library can seem daunting at first, but your efficiency will improve with practice. The following example shows a step-by-step approach to an online search at a university library. Although search methods for online catalogs vary somewhat among libraries, the search screens shown are typical of most systems. Before beginning an online search, read the available instructions or consult a librarian.

A hypothetical search might begin with the first screen for the Colorado Alliance of Research Libraries system (called CARL). The screen (Figure 1) shows the broad categories of information available.

```
PRESS <RETURN> TO START THE PROGRAM (use //EXIT to QUIT)>>>

        The Colorado Alliance of Research Libraries
          offers access to the following
          groups of databases:

        1. Library Catalogs
             (including Government Publications)

        2. Current Article Indexes and Access
             (including Uncover and ERIC)

        3. Information Databases
             (including Encyclopedia)

        4. Other Library Systems

        5. Library and System News

Type //EXIT to leave this system
Enter the NUMBER of your choice, and press the <RETURN> key >> 1 <return>
```

FIGURE 1 Scope of Databases Available in CARL System

The researcher is looking for a book and so chooses *1. Library Catalogs*. The second screen (Figure 2) displays a list of the libraries whose catalogs are available on the CARL system.

```
1. Libraries  2. Articles  3. Information  4. Other Systems  5. News

                        LIBRARY CATALOGS

  6. Auraria Library                17. Regis University
  7. Colorado School of Mines       18. Luther College Network (IA)
  8. Univ Colo at Boulder           19. Northwest College (WY)
  9. Univ Colo Health Sciences Center  20. State Department of Education
 10. Univ Colo Law Library III System  21. Bemis Public Library (Littleton)
 11. Denver Public Library          22. Government Publications
 12. Denver University              23. Univ Colo Film/Video - Stadium
 13. Denver University Law Library   24. CCLINK -- Community Colleges
 14. University of Northern Colorado  25. MedConnect--Medical Libraries (CO)
 15. University of Wyoming          26. High Plains Regional Libraries
 16. Colorado State University      27. Teikyo Loretto Heights University
                                    92. Colo Dept of Health & Rocky Flats

 Enter the NUMBER of your choice, and press the <RETURN> key >> 15 <return>
 carl                                                              14:41
```

FIGURE 2 Library Catalogs Available in CARL System

The user then selects *15. University of Wyoming.*

This screen (Figure 3) tells users that they can search the University of Wyoming collection by title, series (for periodicals), and call number. (The call number is the library classification system number used to

```
              SELECTED DATABASE: University of Wyoming

  ENTER COMMAND (?H FOR HELP) >> b

  The BROWSE search allows you to view the collection
  in order by TITLE, SERIES, or CALL NUMBER.

  Would you like <T>ITLE, <S>ERIES, or <C>ALL NUMBER order? t <return>


  TITLE BROWSE allows you to view the collection in TITLE order.
  You should enter a title as exactly as you know it, such as

        THE WIND IN THE WILLOWS
  or    IN SEARCH OF EXCELLENCE

  ENTER TITLE TO START: van nostrand's scientific encyclopedia
  ■ carl                                                    14:57
```

FIGURE 3 Scope of Searches Available at the University of Wyoming Library

identify each work for shelving and retrieval; it is usually based on the Library of Congress cataloging system.) The user then selects *t* for *title* and the system displays suggestions for how to search for titles at the bottom of the screen. The user then enters the title, *van nostrand's scientific encyclopedia*. (Capital letters are unnecessary.)

The next screen (Figure 4) displays the library's holdings for that publication (the 1976, 1989, and 1995 editions), as well as several other

```
  1                                          UW   SCIREF    1976
     Van nostrand's scientific encyclopedia   Q 121 .V3 1976
  2                                          UW   see record 1989
     Van nostrand's scientific encyclopedia   Q 121 .V3 1989
  3                                          UW   SCIREF    1995
     Van nostrand's scientific encyclopedia   Q121 .V3 1995
  4 Liddle Ralph Alexand                      UW   GEO       1936
     Van oil field, van zandt county, texas   553.282 L619v
  5                                          UW   SCI       --
     Van overbeck: control of plant growth   FOLDER NO.6 BIOL 2020
  6                                          UW   SCI       --
     Van overbeck: control of plant growth   FOLDER NO.5 BIOL 2020
  7 Nierop Henk F K va                        UW   COE       1984
     Van ridders tot regenten : de hollandse adel in DJ 152 .N54 1984
  <RETURN> To continue display.
  Enter <Line number(s)> To Display Full Records (Number + B for Brief)
  <P>revious For Previous Page OR <Q>uit For New Search      3 <return>
  ■ carl                                                    14:58
```

FIGURE 4 Library Holdings for Selected Book Titles

publications with titles that begin with "van," because titles are alphabetized by the first important word in the title (excluding *a, an,* and *the*).

The user selects listing 3 (the 1995 edition) because it is the most up to date. The system next displays a full bibliographic citation for the book (Figure 5). This information was formerly located on a paper catalog card.

```
-----------------------------------------University of Wyoming------
TITLE(s):      Van Nostrand's scientific encyclopedia / Douglas M.
                Considine, editor ; Glenn D. Considine, managing editor.
               8th ed.

               New York : Van Nostrand Reinhold, c1995.
               2 v. (xvi, 3455 p.) : ill., maps ; 30 cm.
               "Animal life, biosciences, chemistry, earth and atmospheric
                sciences, energy sources and power technology, mathematics
                and information sciences, materials and engineering
                sciences, medicine, anatomy, and physiology, physics, plant
                sciences, space and planetary sciences."
               Includes bibliographical references and index.

Contents:      [1] A-I -- [2] J-Z.

OTHER ENTRIES: Science Encyclopedias.
               Engineering Encyclopedias.
               Considine, Douglas M.
               Considine, Glenn D.
               Scientific encyclopedia.

more follows -- press <RETURN> (Q to quit)
 ■ carl                                                           14:59
```

FIGURE 5   Full Bibliographic Citation for a Book

The citation includes the title, the publisher, the editor, the date of publication, the number of pages, a notation that it contains illustrations (ill.) and maps, and a listing of the subjects covered in the work. It also notes that the encyclopedia includes a bibliography and an index. In addition, the screen displays subjects and names that can be used to locate this or related works in the library's collection.

In the preceding search, the researcher knew the title of the encyclopedia and needed only to find its call number in order to locate it in the library. However, a user unfamiliar with the title, but searching for a scientific encyclopedia, could conduct a subject search. The subject search screen for the University of Wyoming system (Figure 6) indicates that users can search through its collection by name, word, title, call number, or series. The user enters "w" to begin a word (that is, subject) search.

```
Type S to try your search in another database, or

      R to repeat your search in University of Wyoming

      H to see a list of your recent searches, or

      <RETURN> for a new search:
3/27/96
3:05 P.M.       SELECTED DATABASE:  University of Wyoming

        Enter N for NAME search
              W for WORD search
              B to BROWSE by title, call number, or series
              S to STOP or SWITCH to another database

There is also a quick search -- type QS for details or

   Type ? for new information about searching

                 SELECTED DATABASE:  University of Wyoming

ENTER COMMAND (?H FOR HELP) >> w
■ carl                                              15:01
```

FIGURE 6 Subject Search Options

The next screen (Figure 7) suggests several ways to think about word searches and lists several word-search samples. The user then enters "science encyclopedias" when prompted at the bottom of the screen.

```
ENTER COMMAND (?H FOR HELP) >> W

                SELECTED DATABASE: University of Wyoming

REMEMBER -- WORDS can be words from the title, or can be subjects,
concepts, ideas, dates, etc.

        for example -- GONE WITH THE WIND
                       SILVER MINING COLORADO
                       BEHAVIOR MODIFICATION

Enter word or words (no more than one line, please)
separated by spaces and press <RETURN>.




>>science encyclopedias
■ carl                                              15:04
```

FIGURE 7 Word Search Suggestions

```
>>science encyclopedias
WORKING...
I am changing the order of your words for a quicker search......
ENCYCLOPEDIAS 476 ITEMS  University of Wyoming
patience -- SCIENCE is a long one...
ENCYCLOPEDIAS + SCIENCE  33 ITEMS

ENCYCLOPEDIAS + SCIENCE  33 ITEMS  University of Wyoming

You may make your search more specific (and reduce the size of the list)
by adding another word to your search. The result will be items in your
current list that also contain the new word.

to ADD a new word, enter it,

<D>ISPLAY to see the current list, or

<Q>UIT for a new search:

NEW WORD(S): d
■ carl                                                         15:04
```

FIGURE 8 Initial Results of Word Search

Word searches are not as precise as title searches, so the system displays the results of its initital findings: 476 items with the term *encyclopedia*, but only 33 items with *encyclopedia* and *science* used together (Figure 8).

After the user enters "d" for display, the next screen (Figure 9) lists the first 7 of the 33 works located, the fourth of which is the 1995 edition of the Van Nostrand *Scientific Encyclopedia*.

```
1                                          UW  COEREF READY 1995
    Worldmark encyclopedia of the nations  G63 W67 1995
2                                          UW                1995
  Encyclopedia of operations research and manageme
3                                          UW  COEREF        1995
    The encyclopedia of police science     HV7901 .E53 1995
4                                          UW  SCIREF        1995
    Van nostrand's scientific encyclopedia Q121 .V3 1995
5                                          UW  COEREF        1994
    Native America in the twentieth century : an en E76.2 .N36 1994
6                                          UW                1994
  Encyclopedia of social history
7                                          UW  COEREF        1994
    Brassey's encyclopedia of military history and b U24 .B73 1994

<RETURN> To continue display.

Enter <Line number(s)> To Display Full Records (Number + B for Brief)

<Q>uit For New Search                              4 <return>
■ carl                                                     15:05
```

FIGURE 9 Specific Titles Retrieved from Word Search

By selecting item 4, the screen displays the full bibliographic citation shown in Figure 5, including the call number, so that the user can locate the book in the library.

Reference Works

The library maintains collections of many other kinds of information organized systematically. The remainder of this entry presents a concise overview of the variety of information available, guidance on computer and other types of search strategies for how to obtain it, and tips on focusing your research to obtain the information you need.

Bibliographies and Periodical Indexes. Bibliographies list books, periodicals, and other research materials published in a particular subject area—for example, in business or engineering. Periodical indexes list journal, newspaper, and magazine articles organized by type of publication *(New York Times Index)* or by subject area *(Applied Science and Technology Index).* (Periodicals are publications that are issued at regular intervals—daily, weekly, monthly, and so on.) A selected list of bibliographies, periodical indexes, and other reference sources appears as the last section of this entry.

Encyclopedias. Encyclopedias—alphabetically arranged collections of articles—are often illustrated and usually published in multivolume sets. Some encyclopedias cover a wide range of general subjects, and others specialize in a particular subject. General encyclopedias provide the kind of overview that can be helpful to someone new to a particular subject. The articles in general encyclopedias are useful sources of background information and can be especially helpful in defining the terminology essential for understanding the subject. Some articles contain bibliographies, which list additional sources. A selected list of encyclopedias appears at the end of this entry.

Specialized encyclopedias provide detailed information on a particular field of knowledge. Their treatment of a subject is rather thorough, and so the researcher ought to have some background information on the subject in order to use the encyclopedia's information to full advantage. Specialized encyclopedias contain comprehensive bibliographies.

Dictionaries. Dictionaries—alphabetical arrangements of words with information about their forms, pronunciations, origins, meanings, and uses—are essential to all writers. For a list of major abridged and unabridged dictionaries, look in the **dictionary** entry of this book. Specialized dictionaries define those terms used in a particular field—

for example, *computers, architecture, geology.* Their definitions are generally more complete and are more likely to be current. The following specialized dictionaries are typical:

> Harris, C. M., ed. *Dictionary of Architecture and Construction*
> *McGraw-Hill Dictionary of Scientific and Technical Terms*
> Monkhouse, F. J., ed. *A Dictionary of Geography*

Handbooks and Manuals. Compilations of frequently used information in a particular field of knowledge, handbooks and manuals are usually single volumes that provide such information as brief definitions of terms or concepts, explanations and details about particular organizations, numerical data in **graphs** and **tables, maps,** and the like. Handbooks and manuals are useful sources of fundamental information on a particular subject, although they are most valuable for someone with a basic knowledge of the topic, particularly in scientific or technical fields. Every field has its own handbook or manual; the following are typical:

> *United States Government Printing Office Style Manual*
> *CRC Handbook of Chemistry and Physics*
> *Environmental Regulation Handbook*
> Merritt, F. S., ed. *Building Construction Handbook*

Statistical Sources. Collections of numerical data, these are the best sources for such information as the height of the Washington Monument; the population of Boise, Idaho; the cost of living in Aspen, Colorado; and the annual number of motorcycle fatalities in the United States. The answers to many statistical questions can be found in almanacs and encyclopedias. The answers to difficult or comprehensive questions, however, are most likely to be found in works devoted exclusively to statistical data, a selection of which follows:

> *American Statistics Index* (1978– with monthly, quarterly, and annual supplements. Lists and abstracts all statistical publications issued by agencies of the U.S. government, including periodicals, reports, special surveys, and pamphlets.)
> U.S. Bureau of the Census. *County and City Data Book* (1952– issued every five years. Includes a variety of data from cities, counties, metropolitan areas, and the like. Arranged by geographic and political areas, it covers such topics as climate, dwellings, population characteristics, school districts, employment, and city finances.)

U.S. Bureau of the Census. *Statistical Abstract of the United States* (1979– annual. Includes statistics on the U.S. social, political, and economic condition and covers broad topics such as population, education, public land, and vital statistics. Some state and regional data.)

Atlases. Collections of maps, atlases are classified into two categories based on the type of information their maps present: general atlases show physical and political boundaries; and thematic atlases give special information, such as climate, population, natural resources, or agricultural products. A selected list of atlases appears at the end of this entry.

Online Searches of Reference Works

Virtually all academic, industrial, and medical libraries offer computer-search access to bibliographies, periodical indexes, and other reference works, just as they do for the library card catalog. Computer-searchable data bases are available in medicine, business, psychology, biology, management, engineering, environmental studies, and many other subjects. Online searches can be used to prepare a bibliography of source material for your subject or can provide access to the full text of these sources. Be aware, however, that full text versions of many sources go back to the mid 1980s, before which the information is available in printed form only. There may be a charge for some online searches, but the cost is relatively low when the speed and convenience of the search are taken into account.

Computer-search services are usually located in the reference section of the library. (Searches are often conducted by reference librarians.) At some large universities, these services are located in satellite libraries. The reference staff can provide information about the range of subjects in the data base, the types and cost of search services, and procedures for using the service.

It is generally necessary to set up an appointment with a reference librarian to arrange for the search. Some libraries provide a form to fill out in advance that asks for a brief statement of the research topic and for any pertinent information necessary or helpful for conducting the search.

After completing the search, the librarian generally meets with the researcher to go over the printout. He or she can help you evaluate the citations for their relevance to your project and can also assist you in finding them. Some libraries provide with the printout a brief guide that indicates where the information on the list is located in the library.

Compact Disk/Read Only Memory (CD-ROM)

Libraries increasingly make reference materials available that use CD-ROM technology. The major benefit of this medium is its enormous storage capacity. Each disk can store approximately 275,000 pages. CD-ROM technology also permits full-text subject searches. Every word in a text (except **conjunctions, prepositions,** and articles) is indexed, thereby simplifying and greatly expanding the user's search capabilities compared with a manual search. Once the information is found, it can be printed on demand for reading ease and reference. The disks can store graphics, voice, and music, as well as text. For example, a "talking" version of *Webster's Ninth New Collegiate Dictionary* is now available, with the voice of a professional announcer reading each of the dictionary's 160,000 entries.

With their capacity to store, index, and permit interaction among text, graphics, music, and voice media, CD-ROMs have virtually unlimited potential for the storage of reference materials, including almanacs, encyclopedias, atlases, general and specialized dictionaries, books of quotations, thesauruses, and numerous economic, demographic, political, legal, and statistical data collections. The U.S. Bureau of the Census alone makes scores of data collections available on CD-ROMs in agriculture, housing, population, transportation, construction, retail and wholesale trade, and many more topics.

Check with a reference librarian to learn the kinds of CD-ROM databases available and for information about how to access the information.

Microforms

Some library source materials may require you to use microforms. Reduced photographic images of printed pages, microforms are used by many libraries for storing magazines, newspapers, and other materials. Because they are reduced, microforms must be magnified by machines called microreaders in order to be read. The most common kinds of microforms are *microfilm* and *microfiche.*

Microfilm. Rolls of 35-millimeter film, usually with four printed pages per frame, microfilm must be read on manually or electrically operated machines that advance the roll, frame by frame, for viewing.

Microfiche. A flat sheet of film, usually 4 × 6 inches, that can contain as many as 98 pages, microfiche is read on a microreader, where it can be moved horizontally or vertically from frame to frame for viewing.

If you are unsure about where to get microform materials or how to use microreaders, ask a librarian. Microform materials and microreaders are usually located in a special section of the library. Some microreaders are equipped with photocopying devices that for a fee permit you to make paper copies of microforms.

Tips for Doing Library Research

After you have located the information relevant to your research topic, you need to evaluate it and then document where it was obtained.

When you obtain the book, magazine, search screen, or microform, there are certain shortcuts that can help you decide quickly whether it has information that will be useful to your research. In a book it is best to start by looking over the table of contents and then by reading the **introductions, conclusions,** and summaries.

Does the book have an **index** or a **bibliography?** For a magazine article, it helps first to scan the **headings** to get an idea of its major topics. Is the magazine an authoritative one, or is it brief and topical in its scope? With microforms, you can quickly flip through the text, also looking for major topics. No matter what form your source takes, always consider its date: How current is the information? Timeliness is important in any research; you don't want to use out-of-date information.

To document sources, you can print out the search screen containing the full bibliographic citation, as in Figure 5. You can even download the bibliographic citation for all pertinent information from a full-text data base from the library terminal to your own disk for fuller evaluation outside the library. For paper sources, you can photocopy the relevant information and highlight important passages with a read-through marking pen.

Listing of Selected Reference Works

The following list provides many essential sources for library research. Most of these works are available both online and in CD-ROM. Ask a reference librarian about their availability before you begin your search.

Abstracts

Agricultural Engineering Abstracts, 1976–. (monthly)
American Statistics Index (ASI), 1974–. (monthly) (index and abstracts of government publications containing statistics)
Astronomy and Astrophysics Abstracts, 1969–. (semiannually)
Biological Abstracts, 1926–. (every two weeks)

Biological Abstracts/RRM, 1962–. (monthly) (formerly *Bioresearch Index*)

Chemical Abstracts (CA), 1907–. (every two weeks)

Computer Abstracts, 1967–. (monthly) (Section C of *Science Abstracts.*)

Cumulative Index to Nursing and Allied Health Literature, 1977–.

Dissertation Abstracts International, 1938–. (monthly) Part B contains science and engineering; Part C contains European abstracts.

Ecological Abstracts, 1974–. (every other month)

Electrical and Electronic Abstracts, 1898–. (monthly) (Section B of Science Abstracts.)

Energy Information Abstracts, 1976–. (monthly)

Energy Research Abstracts, 1976–. (every two weeks)

Engineering Index, 1906–, (monthly)

Environment Abstracts, 1971–. (monthly)

Geographical Abstracts, 1972–.

INIS Atomindex, 1970–. (every two weeks) (formerly *Nuclear Science Abstracts*)

International Aerospace Abstracts, 1961–. (every two weeks)

Key Abstracts: Electrical Measurement and Instrumentation, 1976–. (monthly)

Key Abstracts: Physical Measurement and Instrumentation, 1976–. (monthly)

Key Abstracts: Solid State Devices, 1978–. (monthly)

Mathematical Reviews, 1940–. (monthly)

Metals Abstracts, 1968–. (monthly)

Oceanic Abstracts, 1966–. (every other month)

Physics Abstracts (PA), 1898–. (every two weeks) (Section A of *Science Abstracts.*)

Pollution Abstracts, 1970–. (every two months)

Psychological Abstracts, 1927–.

Review of Plant Pathology, 1922-. (monthly) (Formerly *Review of Applied Mycology.*)

Atlases

Moore, Patrick. *The Atlas of the Universe.*

National Geographic Atlas of the World.

Rand McNally Commercial Atlas and Marketing Guide. (updated annually)

The Rand McNally Atlas of the Oceans.

Times Atlas of the World: Comprehensive Edition.

United States Geological Survey. *National Atlas of the United States of America.*

Bibliographies

Alred, Gerald J., Diana C. Reep, and Mohan R. Limaye. *Business and Technical Writing: An Annotated Bibliography of Books,* 1880–1980.
Besterman, Theodore. *A World Bibliography of Bibliographies.*
Bibliographic Index: A Cumulative Bibliography of Bibliographies, 1937–. (updated three times annually)
Hernes, Helga. *The Multinational Corporation: A Guide to Information Sources.*
Sheehy, Eugene P. *Guide to Reference Books.*
Walford, A. J. *Walford's Guides to Reference Material.* (Volume 1 contains science and technology.)

Biographies

American Men and Women of Science.
Biographical Encyclopedia of Scientists, 2nd ed., Institute of Physics Pub.
Biography Index, 1946–1973.
Current Biography, 1940-. (monthly)
Dictionary of Scientific Biography, 1970–1980.
Ireland, Norma Olin. *Index to Scientists of the World from Ancient to Modern Times: Biographies and Portraits.*
Prominent Scientists: an Index to Collective Biographies, 3rd ed.
Who's Who in America, 1899-. (updated every two years)
Who's Who in Science and Engineering.

Book Guides

Books in Print. (issued annually) (lists author, title, publication data, edition, and price)
Cumulative Book Index, 1928-. (lists all books published in English, with author, title, and subject listings; issued annually with monthly supplements)
New Technical Books, 1915-. (annotated selections, arranged by subject and updated every two months)
Proceedings in Print, 1964-. (announcements of conference proceedings with citation, subject, agency, and price; updated every two months)

Book Reviews

Book Review Digest, 1905-. (monthly, with annual accumulation)
Current Book Review Citations, 1976-.
Technical Book Review Index, 1935-. (guide to reviews in scientific, technical, and trade journals; issued monthly except for July and August)

Commercial Guides and Professional Directories

Encyclopedia of Associations. (annual guide to professional groups and publications with a quarterly supplement, New Associations and Projects)

MacRae's Blue Book (annually updated buying directory for industrial material and equipment)

National Trade and Professional Associations of the United States and Canada and Labor Unions, 1966–. (annual)

Thomas Register of American Manufacturers, 1905–. (monthly)

Current Awareness Services

Chemical Titles, 1960–. (lists current journal articles, indexed by keyword and author, issued every two weeks)

Conference Papers Index, 1973–. (monthly)

Current Contents, 1961–. (weekly listing of current journals' tables of contents; sections include agriculture, biology, and environment services; engineering and technology; physical, chemical, and earth sciences; life sciences)

Current Papers in Electrical and Electronics Engineering, 1969–. (monthly)

Current Papers in Physics, 1966–. (every two weeks)

Current Papers on Computer and Control, 1969–. (monthly)

Environmental Periodicals Bibliography, 1972–. (monthly)

Index to Scientific and Technical Proceedings, 1978–. (monthly)

Dictionaries

For unabridged dictionaries and desk dictionaries, see **dictionaries.**

Dissertations

Dissertation Abstracts International, 1938–. (monthly) (since 1966, Part B has been devoted to science and engineering)

Masters Abstracts, 1962–. (quarterly)

Encyclopedias

The American Medical Association Encyclopedia of Medicine.

Encyclopedia Americana.

Encyclopedia of Bioethics.

Encyclopedia of Human Biology.

Encyclopedia of Computer Science and Technology.

Encyclopedia of Physics.

International Encyclopedia of Statistics.

McGraw-Hill Encyclopedia of Economics.

McGraw-Hill Encyclopedia of Engineering.

McGraw-Hill Encyclopedia of the Geological Sciences.

McGraw-Hill Encyclopedia of Science and Technology.
The New Encyclopaedia Britannica.
The Prentice-Hall Encyclopedia of Mathematics.
Sills, David L., ed. *International Encyclopedia of the Social Sciences.*
Van Nostrand's Scientific Encyclopedia.

Fact Books and Almanacs

The Budget in Brief. (a condensed version of the annual U.S. federal budget)
Demographic Yearbook, 1979–. (annual U.N. collection of information on world economics and trade)
Facts on File, 1974–. (weekly digest of world news, collected annually)
Johnson, Otto T., et al. *Information Please Almanac; Atlas Yearbook.*
Statistical Abstract of the United States (annual publication of the U.S. Bureau of the Census that summarizes data on industrial, political, social, and economic organizations in the United States)

Government Publications

Government Reports Announcements (GRA), 1946–. (twice-monthly listing of federally sponsored research reports arranged by subject with abstracts)
Government Reports Index, 1965-. (semimonthly index of reports abstracted in GRA and arranged by subject, author, and report number)
Index to U.S. Government Periodicals, 1970–. (quarterly)
Monthly Catalog of U.S. Government Publications, 1895–. (lists non-classified publications of all federal agencies by subject, author, title, and report number; also gives ordering instructions)
Monthly Checklist of State Publications, 1940–. (state government publications arranged by state, with an annual subject and title index)
Schmeckebier, Laurance F., and Roy B. Eastin. *Government Publications and Their Use.* (guide to government publications)

Graphics

American Society for Testing and Materials. *Illustrations for Publication and Projection.*
Arkin, Herbert, and Raymond R. Colton. *Statistical Methods.*
Cardamone, Tom. *Chart and Graph Preparation Skills.*
Enrick, Norbert Lloyd. *Handbook of Effective Graphic and Tabular Communication.*
Field, Ron M. *A Guide to Micropublishing.*
Holmes, Nigel. *Designer's Guide to Creating Charts and Diagrams.*
Kepler, Harold B. *Basic Graphical Kinematics.*

Language Usage

Bernstein, Theodore M. *The Careful Writer: A Guide to English Usage.*
Copperud, Roy H. *American Usage and Style: The Consensus.*
Morris, William, and Mary Morris. *Harper Dictionary of Contemporary Usage.*
Nicholson, Margaret. *A Dictionary of American English Usage.*
The Reader's Digest Association. *Success with Words: A Guide to the American Language.*

Logic

Beardsley, Monroe C. *Thinking Straight: Principles of Reasoning for Readers and Writers.*
Brennan, Joseph G. *A Handbook of Logic.*
Copi, Irving M. *Introduction to Logic.*
Weddle, Perry. *Argument: A Guide to Critical Thinking.*

Parliamentary Procedure

Deschler, Lewis. *Deschler's Rules of Order.*
Robert, Henry M., et al. *Robert's Rules of Order.*

L

Periodical Indexes and Directories

Applied Science and Technology Index, 1958–. (alphabetical subject list of more than 50,000 periodicals; issued monthly)
Bibliography of Agriculture, 1942–. (every two months)
Engineering Index, 1906–.
General Science Index, 1978–. (monthly)
Index Medicus, 1880–. (monthly)
New York Times Index, 1851–. (every two weeks)
Readers' Guide to Periodical Literature, 1900–. (every two weeks)
Science Citation Index, 1962–. (quarterly)
Ulrich's International Periodicals Directory: A Classified Guide to Current Periodicals, Foreign and Domestic (annually)
Wildlife Review, 1935–. (quarterly)
Zoological Record, 1864–. (annually)

Speech and Group Discussion

Capp, Glen R., and G. Capp, Jr., *Basic Oral Communication.*
Harnack, R. Victor, et al. *Group Discussion: Theory and Technique.*
Maier, Norman. *Problem-solving Discussions and Conferences: Leadership Methods and Skills.*
Wilcox, Roger P. *Oral Reporting in Business and Industry.*

(See also **oral presentations.**)

Style Manuals and Guides

For a list of style manuals and guides, see **documenting sources.**

Subject Guides

Chen, Ching-chih. *Scientific and Technical Information Sources.*
Geologic Reference Sources.
Malinowsky, Harold R. *Science and Engineering Literature: A Guide to Reference Sources.*
Mount, Ellis. *Guide to Basic Information Sources in Engineering.*
Owen, Dolores B. *Abstracts and Indexes in Science and Technology: A Descriptive Guide.*

Synonyms

Rodale, J. L. *The Synonym Finder.*
Roget, Peter M. *Roget's International Thesaurus.*
Webster's Collegiate Thesaurus.

-like

The **suffix** *-like* is sometimes added to **nouns** to make them into **adjectives.** The resulting **compound word** is hyphenated only if it is unusual or might not immediately be clear.

> **EXAMPLES**   childlike, lifelike, dictionary-like, computer-like
> Her new assistant works with machine-*like* efficiency.

like/as

To avoid confusion between *like* and *as,* remember that *like* is a **preposition** and *as* is a **conjunction.** Use *like* with a **noun** or **pronoun** that is not followed by a **verb.**

> **EXAMPLE**   The new supervisor behaves *like a novice.*

Use *as* before **clauses** (which contain verbs).

> **EXAMPLES**   He acted *as though he owned the company.*
> He responded *as we expected he would.*

Like may be used in elliptical constructions that omit the verb.

> **EXAMPLE**   She took to architecture *like a bird to nest building.*

If the omitted portions of the elliptical construction were restored, however, *as* would be used.

> **EXAMPLE**   She took to architecture *as a bird takes to nest building.*

linking verbs (see verbs)

listening

Listening is a learned skill; it is not automatic or instinctive. Listening requires concentration and mental effort. But it is worth the effort it takes, for good listening facilitates the exchange of ideas, opinions, and information.

The first thing to understand is that there are a number of natural human barriers to effective listening. One of them is that if the speaker says something that is contrary to your own prejudice, you may immediately start to prepare a rebuttal—and stop listening. Another is that your personal feelings about the speaker can get in your way; for example, if you dislike the speaker, your feelings may convince your subconscious mind that he or she will not have anything to say that is worth listening to—and your mind is more likely to stray. Physical barriers can also get in the way of effective listening. If you are hungry or tired, for example, you will certainly find concentrating more difficult. And finally, you may be distracted by noises and sights. The important thing is to be aware of these potential distractions and try to prevent them or, if you cannot, to overcome them.

The following guidelines may help you improve your listening skills:

1. Make a conscious effort to listen carefully, to stay involved, and to react to what is being said. It may help to try to put yourself in the speaker's place.
2. Learn to use the spare time that results from your mind's absorbing information faster than the speaker can present it to listen more effectively and efficiently. For example, you might try to find the speaker's pattern of organization. Identify the introduction, the method being used to develop the topic, the transition (so that you will be alert for the next point), and the conclusion. Or evaluate the evidence the speaker uses to substantiate his or her points. Listen "between the lines" for meaning that may not be in the words. Or review the talk to this point. These techniques will help you not only probe the meaning of the subject but also remember key points.
3. Don't allow yourself to be distracted by the speaker's personality or mannerisms. Respond thoughtfully to the speaker's words, and avoid making judgments too quickly.
4. Be prepared. Think about the subject of a meeting or workshop ahead of time. This will help you understand and remember the material.
5. Take notes. If you are following a speaker's presentation in order to take notes, your mind will be less likely to wander or be distracted. However, your note taking must enhance your listening and not detract from it. Also, if you have taken notes, you are much more likely

to review what you have heard—and reviewing your notes can remedy weaknesses in your listening skills.

The best test of listening well is the ability to respond well. Get into the habit of summarizing and paraphrasing in your own words what has been said. Doing so will both reduce the probability of listening mistakes and increase retention.

lists

Lists can save **readers** time by allowing them to see at a glance specific items, questions, or directions. Lists also help readers by breaking up complex statements that include figures and by allowing key ideas to stand out.

> EXAMPLE   Before we agree to hold the convention at the Brent Hotel, we should make sure the hotel facilities meet the following criteria:
>
> 1. At least eight meeting rooms that can accommodate 25 people each.
> 2. Ballroom and dining facilities for 250 people.
> 3. Duplicating facilities that are adequate for the conference committee.
> 4. Overhead projectors, flip charts, and screens that are sufficient for eight simultaneous sessions.
> 5. Ground-floor exhibit area that can provide room for thirty 8 × 15-foot booths.
>
> To confirm that the Brent Hotel is our best choice, perhaps we should take a look at its rooms and facilities during our stay in Kansas City.

Notice that all the items in this example have a **parallel structure.** In addition, all the items are balanced—that is, all points are relatively equal in importance and are of the same general length.

Consider, however, the following example and its revision:

> CHANGE   I believe we should consider several important items at the meeting.
>
> 1. Our 19— Production Schedule
> 2. The Five-Year Corporate Plan
> 3. We should also develop an agenda for the next meeting that includes a discussion of criteria for relocating the Westdale Division. Perhaps you can give me some tentative ideas at this meeting.
> 4. The draft of our Year-End Financial Report

TO   Please bring to the meeting the following items:

1. Our 19— Production Schedule
2. The Five-Year Corporate Plan
3. The draft of our Year-End Financial Report

In addition, you might give me some tentative ideas about criteria for relocating the Westdale Division. Since we must formally discuss this problem soon, we should include such a discussion in the agenda for the committee's next meeting.

In the original version, item 3 is neither parallel in structure nor balanced in importance with the other items. In the revision, item 3 is discussed separately, since it deserves a more developed treatment.

In an attempt to avoid writing **paragraphs,** some writers tend to overuse lists, however. A **memorandum** or **report** that consists almost entirely of lists, for example, can be difficult to understand, for the reader is forced to connect the separate items and mentally to provide coherence. Do not expect a reader to deal with unexplained lists of ideas.

To ensure that the reader understands how a list fits with the surrounding sentences, always provide adequate **transitions** before and after any lists. If you do not wish to indicate rank or sequence, which numbered lists suggest, you can use bullets, as shown in the list of tips that follows. In typography a bullet is a small *o* that is filled in with ink.

Tips for Using Lists

- List only comparable items.
- Use parallel structure throughout.
- Use only words, phrases, or short sentences.
- Provide adequate transitions before and after lists.
- Use bullets when rank or sequence is not important.
- Do not overuse lists.

literature

In technical contexts, the word *literature* applies to a body of writing pertaining to a specific field, such as finance, insurance, or computers. We commonly speak, for example, of medical literature or campaign literature.

EXAMPLE   Please send me any available *literature* on computerized translations of foreign languages.

literature reviews

A *literature review* is a summary **report** on the **literature,** or printed material, that is available on a particular subject over a specified period of time. For example, a literature review might describe all material published within the past five years on a special technique for debugging computers, or it might describe all reports written in the past fifteen years on efforts to improve quality control at a particular company. A literature review tells the **reader** what is available on a particular subject and gives him or her an idea of what should be read in full.

Some **trade journal articles** or theses begin with a brief literature review to bring the reader up to date on current research in the area. The author can then use the review as background for his or her own discussion of the subject. As an example, Figure 1 shows a one-paragraph literature review that serves as an **introduction** for a medical article on clefts in newborns. On the other hand, a literature review could be a whole document, as illustrated by the article shown (abridged) in Figure 2. This article reviews publications from 1980–1995 on the treatment of acute pain in order to make recommendations to the medical community. Such fully developed literature reviews serve as good starting points for detailed **research.** Managers in busi-

L

Assessment of attractiveness of newborn infants with unrepaired facial clefts by different groups of individuals was studied to investigate the validity of a surgeons' rule of thumb system for this based upon cleft severity.

Recent studies have attempted to construct objective rating scales to quantify the severity of facial disfigurement in children and young people (Eliason et al., 1991; Poole et al., 1991; Roberts-Harry et al., 1992; Tobiasen and Hiebert, 1993). The medical community increasingly recognizes the need for a holistic approach to congenital disfigurement, including such aspects as parents' reactions, feelings, and attitudes (Bull and Rumsey, 1988). Results of these studies suggest that there may be a mistaken assumption of consensus between plastic surgeons and people unfamiliar with facial clefts in how each group respectively rates attractiveness of facial appearance. The studies all refer, however, to repaired facial clefts and other anomalies; none has assessed unrepaired facial clefts in neonates. Tobiasen (1991) has concluded that it is premature to suggest that surgeons rate impairment differently from others. Surgeons commonly use a rule of thumb system to classify severity on the basis of dichotomous categorizations (i.e., bilateral or unilateral clefting along the axis of symmetry, complete or incomplete extension to the nasal alar, and with or without palate involvement). The proposed system used by plastic surgeons and derived from Freedlander et al. (1990) is shown in Table 1.

Source: Slade, Pauline. "Relationships between Cleft Severity and Attractiveness of Newborns with Unrepaired Clefts." *Cleft Palate-Craniofacial Journal,* vol. 32, no. 4 (July 1995): 318–322.

FIGURE 1   Literature Review in Introductory Paragraph

Quality Improvement Guidelines for the Treatment of Acute Pain and Cancer Pain

American Pain Society Quality of Care Committee

UNDERTREATMENT of acute pain and chronic cancer pain persists despite decades of efforts to provide clinicians with information about analgesics. A high prevalence of unrelieved pain has been documented in a variety of clinical settings, including general medical[1-4] and surgical units,[1,2,4-6] oncology wards and clinics,[7-11] burn units,[12] emergency departments,[13] and pediatric wards.[14] In response to this problem, clinicians have identified factors that contribute to poor treatment outcomes and have designed corrective programs.[15] The barriers to pain relief include gaps in physicians' and nurses' undergraduate and graduate education about pain treatment,[16-24] concerns of clinicians[25] and patients[26-30] about the risk of addiction to opioids, state and federal regulation of the prescribing of opioid analgesics,[31,32] and reimbursement policies for analgesic treatments.[33]

During the decade following the article by Marks and Sachar[1] that called attention to undertreatment of pain, most recommendations stressed the need to educate individual clinicians and patients, imparting knowledge about methods of relieving pain and the low risks of addiction.[1,5,7,34-36] Although experts agree that such educational approaches are essential, several studies have focused on the problem that pain may frequently go unrecognized by clinicians.[2,37,38] Donovan et al[2] showed that among 454 randomly selected patients on the medical and surgical units of a midwestern academic hospital, 78% reported having experienced pain during hospitalization and 45% reported having had excruciating pain. Of the patients with pain, only 45% recalled a nurse discussing their pain with them, and in only 49% of charts was there a progress note mentioning pain. Grossman et al[37] asked the responsible nurse, house officer, and oncology fellow to estimate each of 104 cancer patients' pain using a 10-cm visual analog scale. For the 15 patients who rated their pain in the most severe range (>7 of a possible 10), only one of the nurses, three of the house officers, and four of the oncology fellows estimated the patient's pain in that range.

The widespread failure to recognize the presence of pain even in institutions with active analgesic education programs suggests a flaw in the design of local systems for care[39-41] rather than lapses by individual clinicians. The same clinicians almost never fail to recognize and take action on an elevated temperature in a neutropenic cancer patient, because temperature is routinely measured and recorded at the patient's chart and at the nursing station, and because standard practice holds that a high fever compels immediate action. In contrast, reports of unrelieved pain do not invariably result in corrective measures; pain may not be visible at the coordinating centers of the ward, and physicians and nurses have not traditionally been held accountable for providing titrated analgesia. A clinician can provide excellent analgesic care if he or she constantly remembers to seek out and relieve pain, but when other priorities require attention, lapses may occur. These considerations have convinced many pain clinicians that traditional educational approaches must be complemented by interventions in health care systems that more directly influence the routine behaviors of clinicians and patients,[39-43] a perspective that has long been advocated by the quality improvement (QI) movement.[44-47]

FIGURE 2 Full Article Literature Review (abridged)

Preparation of American Pain Society Guidelines

Recognizing that the growth of QI approaches in health care organizations offered promise for pain management, the American Pain Society (APS, Glenview, Ill), a multidisciplinary scientific society devoted to pain research, treatment, and education, established a committee in 1988 to develop guidelines for such efforts. Three physicians, two psychologists, and one nurse experienced in pain treatment, consultation, and research drafted a prototype of the current guidelines in 1989. The guidelines were applicable to inpatient settings, and the range of pain conditions addressed in the guidelines was narrowed to types of pain that usually respond to appropriate doses of opioid and nonopioid analgesics, ie, acute pain[6] and cancer pain.[8] The targeted outcome was that each patient would receive timely and optimal doses of analgesic drugs. The guidelines did not discuss which treatment should be used for which clinical circumstance but focused on general issues of pain assessment and institutional commitment to patient satisfaction.

The draft guidelines were circulated to the APS membership in late 1989, which resulted in a number of revisions and the addition of members to the committee. The resulting prototype of the guidelines, published in 1991,[48] was evaluated at about 20 medical centers, and three have published results.[3,4,10] These findings have been presented and QI approaches discussed at annual workshops at the APS scientific meetings. All society members have been invited to participate in the annual meetings of the Quality of Care Committee. Many of the suggestions in the draft APS QI guidelines were incorporated into the clinical practice guidelines on acute pain and cancer pain treatment prepared by the Agency for Health Care Policy and Research (AHCPR), which also discussed in detail many types of treatments for pain.[6,8]

Development of the current version of the APS guidelines began in November 1993. Based on research and experience of the committee members with the guidelines and a systematic literature review, the committee chair prepared four successive drafts. Comments and revisions were invited from 50 pain clinicians and researchers who had indicated an interest in QI approaches and from the APS Board of Directors.

The APS Guidelines

The APS guidelines embody key elements for favorably influencing behaviors of patients and clinicians. The guidelines are intended for settings in which conventional analgesic methods (eg, intermittent parenteral or oral analgesics) are used exclusively and those using the most modern technology for pain management. The quality of pain relief can be enhanced by a dedicated pain management approach by a multidisciplinary group of clinicians who acquire special training in pain management. While this guideline focuses on the assessment of pain and its treatment with analgesic drugs, which we consider to be the two most important components of the treatment of acute pain and caner pain, nonpharmacological measures can also contribute to effective therapy.[6,8]

I. Recognize and Treat Pain Promptly

IA. Chart and Display Patients' Self-report of Pain.—A measure of pain intensity should be recorded in a way that makes it highly visible and facilitates regular review by members of the health care team. This information should be incorporated in the patient's permanent record. The data can be recorded on a vital sign sheet at the

FIGURE 2 Full Article Literature Review (abridged) *(continued)*

patient's bedside, a page at the front of the patient's record, or a chart in the nursing station or outpatient clinic, depending on the routine work flow of the health care team. Unrelieved pain should be a "red flag" that promptly turns attention to this problem.

IA1. The intensity of pain and discomfort should be assessed and documented during the initial evaluation of the patient, after any known pain-producing procedure, with each new report of pain, and at regular intervals that depend on the severity of pain and the clinical situation. A simple, valid measure of intensity should be selected by each clinical unit. A 0 to 10 numerical scale has been the most widely used; other well-validated scales include a 100-mm visual analog scale with "no pain" and "worst possible pain" at each end and a category scale consisting of the words none, slight, moderate, severe, and excruciating. For children, age-appropriate measures should be used.[8] For patients with cognitive impairment, it may be necessary to use the simpler scales, such as the category scale. For patients who do not speak English, pain scales written in other languages may need to be developed.

IA2. The degree of reduction in pain intensity should be determined after each pain management intervention after sufficient time has elapsed for the treatment to reach peak effect.

IB. Commit to Continuous Improvement of One or Several Outcome Variables.—Each clinical unit should identify the outcome variables of pain treatment that will be targeted for improvement (eg, average pain intensity scores in cancer patients and time to remediation in postoperative patients). An interdisciplinary team including at least physicians and nurses will develop a plan to improve these outcomes, implement this plan, and review and revise approaches based on ongoing review of clinical data. A common starting point is to select a pain intensity rating that will elicit an immediate review of treatment, such as a value of 5 or greater on a 0 to 10 numerical scale, and to develop clinical algorithms to address common problems encountered with pain treatment. As the general quality of treatment improves, the clinical unit should upgrade its standards to encourage a continuous process of improvement. In a small outpatient clinic, one or two individuals rather than an interdisciplinary team may need to carry out these activities.

IC. Document Outcomes Based on Data and Provide Prompt Feedback.—At regular intervals defined by the clinical unit and the continuous quality improvement (CQI) task force (described herein), each clinical unit will assess a randomly selected sample of patients. These patients need not have been previously identified as likely to have pain. Every effort should be made to include patients at high risk for undertreatment of pain, such as children, adults with mental retardation, intensive care unit patients, non-English speaking patients, and elderly patients.

Patients should be asked whether they have had pain during the past 24 hours. If they have had paid, they should be asked about (1) current pain intensity; (2) intensity of the worst pain experienced in the past 24 hours (or other interval selected by the clinical unit); (3) effects of pain on sleep and functioning; (4) satisfaction with responsiveness of the staff to the patient's reports of pain; (5) satisfaction with relief provided; (6) if there is persistent pain despite treatment, whether the patient would like more intensive analgesic treatment, and if not, the reasons why; (7) (for inpatients) maximum time the patient waited after a request for additional pain medication or for a change in regimen; and (8) (for outpatients) adequacy of instructions

FIGURE 2   Full Article Literature Review (abridged) *(continued)*

regarding the pain treatment plan, reasons for any need to contact their physician or nurse, and ability to contact them when needed.

Prompt feedback of this information is essential if clinicians are to recognize and correct practices that contribute to suboptimal pain relief. The presentation need not be elaborate; for example, a brief weekly presentation may consist of the worst and average pain scores for 10 randomly selected patients from the previous week. Integration of this feedback into existing activities of the health care team, such as conferences that include both physicians and nurses, may be most useful.

II. Make Information About Analgesics Readily Available

Information about analgesics and other methods of pain management, such as charts of relative potencies of analgesics, the AHCPR practice guidelines on acute pain[6] and cancer pain,[8] the APS *Principles of Analgesic Use,*[50] and computer programs to guide analgesic dosing,[6] should be available to clinicians involved in writing orders for pain treatment in a way that facilitates writing and interpeting orders. Appropriate training in pain management should be included in orientation of all health care professionals and in continuing education activities. Consultants should be available within and outside the institution to handle difficult problems related to pain management.

III. Promise Patients Attentive Analgesic Care

The patient at risk for pain should be informed verbally and in an electronic or printed format at the time of the initial evaluation that effective pain relief is an important part of treatment, that his or her report of unrelieved pain is essential, and that the staff will respond quickly to the patient's requests for pain treatment. This presentation should make clear that the institution and health care professionals are making themselves accountable to the patient for promptly addressing their concerns about pain, even though it is not always possible to provide complete pain relief. Pediatric patients and their parents should receive materials appropriate to the age of the patient. Special efforts should be made to reach other high-risk groups, such as using materials designed for comprehension by these patients.[6,8] Patients and their clinicians should agree on individualized goals of pain management, eg, using one of the standard pain assessment scales to define the maximum level of pain that would not interfere with important patient activities.

IV. Define Explicit Policies for Use of Advanced Analgesic Technologies

Advanced pain control techniques, including intermittent intraspinal opioid administration, systemic or intraspinal patient-controlled opioid infusion or continuous opioid infusion, local anesthetic infusion, and inhalational analgesia, must be governed by policies and standard procedures that define the acceptable level of monitoring of patients and define appropriate roles, accountability, and limits of practice for all groups of health care providers involved.[51] Such policies and procedures should include definitions of staff competencies, define a program for the certification of staff skills in the provision of care by the advanced technologies that are used, and set forth criteria for identifying patients suitable for particular methods. A program for the periodic updating and recertification of skills necessary to administer these technologies should be addressed in such policies. Because these advanced technologies can be associated with serious unwanted effects of therapy, the regular documentation of adverse effects and complications of therapy is essential. A QI program must

FIGURE 2 Full Article Literature Review (abridged) *(continued)*

include ongoing monitoring and documentation of adverse effects and complications of the advanced analgesic technologies used in the clinical setting.

V. Examine the Process and Outcomes of Pain Management With the Goal of Continuous Improvement

VA. Each health care organization should have an interdisciplinary committee to study systematically the processes involved in pain management and the reasons for breakdowns in care, and to evaluate changes in these processes to improve service to their patients and families and to the care givers who need supportive services for treating pain. This committee, often termed a *CQI task force,* should include physicians, pharmacists, nurses, and, when appropriate, clinicians from other appropriate disciplines (eg, physical therapists). The committee should assess the implementation of guidelines for pain assessment and management, monitoring of outcomes, and feedback to clinicians and should make recommendations to improve the process. In a small outpatient clinic, nursing home or very small hospital in which an interdisciplinary CQI task force is not feasible, one or several individuals may fulfill this role, using consultation as needed to interpret the data and to explore alternatives for improvement.

VB. At least one person on the CQI task force should have experience working with issues related to effective pain management. In facilities that have a pain management consulting group, the CQI task force should include a representative of that group.

VC. The CQI task force should meet as needed but at least every 3 months to review process and outcomes related to pain management. This APS guideline emphasizes the direct assessment of patient pain report and patient function and satisfaction as the most important types of outcome data. The CQI task force may also choose to monitor and evaluate other questions, such as "Are pain assessments being performed and recorded in a way that assures recognition of pain?" and "Are the prescribed doses and administered doses of opioid analgesics within the range recommended by the AHCPR clinical practice guidelines on acute and cancer pain management?"

VD. In clinical settings using advanced analgesic technologies, the task force should assess whether the rates of complications and adverse effects are appropriate given the types of pain problems that are being addressed.

VE. The task force should interact with clinical units to establish procedures for improving pain management where necessary and review the results of these changes within 3 months of their implementation.

VF. The task force should provide regular reports to administration and to the medical, nursing, and pharmacy staffs about its activities, results, and recommendations.

VG. Larger health care organizations that establish a pain management CQI program are also advised to create an executive CQI committee on pain, comprising individuals in positions of authority in each relevant division to provide support needed for institutional change. The CQI task force should call on this group when needed. The guidelines are accompanied by a patient outcome questionnaire (Table 1).

Experience With the APS Quality Improvement Guidelines

Three articles[3,4,10] describe experience with the APS patient outcome questionnaire, and one study[10] describes the results of comprehensive implementation of the APS guidelines in a large cancer hospital.

The studies focusing on the patient outcome questionnaire in patients with acute pain and cancer pain showed that the tool was easily used in either an interview[4,10] or

FIGURE 2 Full Article Literature Review (abridged) *(continued)*

in a patient self-report format.[3] In each study, the average score for "worst pain over 24 hours" was approximately 7 on a scale of 0 to 10. Serlin et al[52] have shown that this level of pain is associated with significant impairment of sleep, activity, and mood, and these results reinforce previous reports that pain tends to be under-treated.[1,2,5-9,11-14] In two of the studies of the APS questionnaire,[3,4] 24% and 8% of the patients had to wait more than 30 minutes for pain medication after requesting it. The most surprising findings were the high satisfaction ratings of patients despite high levels of pain, suggesting that patients do not expect consistent pain relief and that the use of patient satisfaction questions without other questions about pain ratings may overlook suboptimal pain relief. All three studies[3,4,10] revealed modest but statistically significant correlations between patients' degree of dissatisfaction and measures of pain intensity or waiting time for medication.

Bookbinder et al[10] studied the impact of the first year of implementing the APS guidelines in an academic cancer hospital. The program included routine monitoring of pain, staff education, and focus groups to identify organizational obstacles to effective pain treatment. During the first year, patient satisfaction increased significantly but the "worst pain levels over past 24 hours" remained unchanged (at 7.7 on a scale of 0 to 10). Preliminary results from the program's second and third years suggest considerable reduction of pain intensity on targeted hospital units. Major change did not occur, however, until pain assessment had become routine and the resulting data had convinced physicians to take part in the programs (Bookbinder et al, unpublished data, November 1995). Similar findings were reported by Dietrick-Gallagher et al,[53] who implemented the QI recommendations that were adapted from the APS and AHCPR guidelines, although they did not use the APS patient satisfaction questionnaire. On a 320-bed surgical nursing service consisting of 10 services, nurses first implemented routine pain assessment, then worked with nurses and physicians on the services where patients had the most severe pain to develop new practices, including giving greater authority for altering patient-controlled analgesia orders to nurses well trained in pain management, and orienting new residents to the pain control initiative. Substantial and progressive improvement of pain relief was achieved on the targeted units during the 15 months described in the report.[53]

Questions for Future Research

Application of QI methods to pain treatment has just begun. The guidelines herein are based on inferences from our experience as pain treatment consultants, the literature on pain treatment and QI, a modest amount of experience with QI approaches to pain treatment, and on three studies related to the guidelines.[3,4,10] We encourage readers to adapt these methods to the needs of their clinical setting, learn what is effective, and let their colleagues know about their findings and insights. Additional research is needed, and specific research questions that might be addressed include the following:

What Are the Essential Items for a Brief Patient Outcome Questionnaire for General Clinical Use? Brevity is a great advantage for surveys used in QI programs. The determination of a few items that can capture the adequacy of pain relief would encourage more widespread use.

What Determines Patient Satisfaction With Pain Treatment? Many pain clinicians think that patients are too readily satisfied with partial and intermittent relief of pain[3,42] and that encouraging patients to demand more complete relief would im-

FIGURE 2 Full Article Literature Review (abridged) *(continued)*

prove overall outcomes. Examination of the determinants of patient satisfaction requires sophisticated research approaches, including questionnaires that would be more extensive than the APS questionnaire.

How Can Organizations Integrate a QI Program for Pain Treatment With Other Interventions, Such as Clinician Education and the Use of Tools to Guide Prescribing? Pilot projects implementing the APS guidelines have not solved their institutions' problems with pain management in the first year.[10,53,58] Recognition of pain must be improved before understanding the best local corrective approaches. Therefore, we recommend that research projects into comprehensive QI-based approaches be conceived and funded for a period of at least 3 years and that they integrate a variety of approaches to changing clinician practices.[59] It is likely that specifically tailored informational approaches, particularly those using computer applications to guide medication ordering and dispensing[60-62] will prove to be an essential complement to regular pain assessment.

Are QI Approaches to Pain Treatment Effective and Cost-effective? This question can only be answered by carefully controlled studies. Quality improvement programs require considerable work and some expense. In addition to the benefit of lessening patients' suffering, QI programs that improve pain control might reduce costs both by preventing or decreasing the length of hospital stays[63] and by preventing medication errors.

Conclusion

The QI approach to pain treatment is based on the assumption that although clinicians are concerned with patient comfort, habits and procedures of practice do not support them in achieving effective pain relief. For the drug-responsive acute pain and cancer pain snydromes addressed by these guidelines, clinicians should be able to relieve pain effectively. Although pain has received the most study, some experts believe that many common symptoms of medical illness are neglected because patterns of medical practice and accountability have evolved out of a focus on structural disease rather than on patient complaints.[40,64-66] As the public demands that medical practice and research justify their costs, we must heed the concerns of our patients, well represented by Montaigne[67] in 1589:

> For heaven's sake, let medicine some day give me some good and perceptible relief and you will see how I shall cry out in good earnest: "At last I yield to an efficient science."

Note: The literature review is accompanied by a numbered list of references.
Source: JAMA, vol. 274, no. 23 (December 20, 1995): 1874–1880.

FIGURE 2 Full Article Literature Review (abridged) *(continued)*

ness and industry also use literature reviews to keep specialists informed of the latest developments and trends in their fields.

To prepare a literature review, you must begin with **library research** of published material on your **topic.** Since your reader may begin research on the basis of your literature review, be especially careful to cite accurately all bibliographic information. As you review each source, note the scope of the book or article and judge its value to the reader. Save all printouts of computer-assisted searches—you may wish to incorporate the sources in the **bibliography** of the final version of your literature review.

Begin a literature review by defining the area to be covered and the types of works to be reviewed. For example, a literature review may be limited to articles and reports and not include any books. You can arrange your discussion chronologically, beginning with a description of the earliest relevant literature and progressing to the most recent. You can also subdivide the topic, discussing works on various subcategories of the topic.

Annotated Bibliographies

Related to literature reviews are *annotated bibliographies,* which also give readers information about published material. Rather than discussing the literature in **paragraphs,** however, an annotated bibliography lists each item in a standard form (see **documenting sources**) and then briefly describes it. This description (or *annotation*) may include the purpose of the book, its scope, the main topics covered, its historical importance, and anything else the writer thinks the reader should know. For example, see the annotation in **bibliography.**

logic

Logic, or correct reasoning, is essential to convincing your **reader** that your conclusion is valid. Errors in logic can quickly destroy your credibility with your reader. Although a detailed discussion of logic is beyond the scope of this book, the following discussion points out some common errors in logic that you should watch out for in your writing.

Lack of Reason

When a statement violates the reader's common sense, that statement is not reasonable. If, for example, you stated "New York City is a small town," your reader might immediately question your logic. Common sense would suggest that a city of over eight million people is not a small town. If, however, you stated "Although New York's population is over eight million, it is a city composed of small towns," your reader could probably accept the statement as reasonable—that is, if you then demonstrated the truth of the statement with reasonable examples. Always be sure that your statements are sensible and that they are supported by sound examples.

Sweeping Generalizations

When the scope of an opinion is unlimited, the opinion is called a *sweeping generalization*. Such statements, though at times tempting to make, should be qualified during revision. Consider the following:

> **EXAMPLES**   Computer instructions are always confusing.
> Engineers are poor writers.

These statements ignore any possibility that some computer instructions are clear or that there might be an engineer who writes superbly. Moreover, one person's definition of "clear" or "confusing" or "poor writing" may be different from another's. No matter how certain you are of the general applicability of an opinion, use such all-inclusive terms as *anyone, everyone, no one, all, always, never,* and *in all cases* with caution. Otherwise, your opinions are likely to be judged irresponsible.

Non Sequitur

A statement that does not logically follow from a previous statement or has little bearing on the previous statement is called a *non sequitur.*

> **EXAMPLE**   I arrived at work early today, so the weather is calm.

In this example, common sense tells us that arriving early for work does not produce calm weather, or any other kind. Thus, the second part of the sentence does not follow logically from the first. Notice how much more logical the following statement is. It makes sense because each idea is logically linked to the other.

> **EXAMPLE**   The snow slowed traffic in the city, so although I left for work at the usual time, I arrived a bit late.

Of course, most non sequiturs are not so obvious as the first example. They often occur when a writer neglects to express adequately the logical links in a chain of thought.

> **EXAMPLE**   Last year our laboratory increased its efficiency by 50 percent by using part-time help. I suggest that we hire two full-time laboratory assistants.

Although both the statements are about the same subject—additional laboratory staff—a logical connection is missing. The reader is forced to guess the meaning. Does the writer mean that full-time assistants will be more efficient? Are part-time employees unavailable? Such non sequiturs can cause the reader to misunderstand what he or she is reading or to assume that it is illogical. In your own writing be careful that all points stand logically connected; non sequiturs cause difficulties for reader and writer alike.

Post Hoc, Ergo Propter Hoc

This term means literally "after this, therefore because of this," and it refers to the logical fallacy that because one event happened after another event, the first somehow caused the second.

> **EXAMPLE**   I didn't bring my umbrella today. No wonder it is now raining.

Many superstitions are based on this error in logic. In on-the-job writing, this error in reasoning usually results when the writer hastily concludes that events are related, without examining the logical connection between the two events.

> **EXAMPLE**   We issued the new police uniforms on September 2. Arrests then increased 40 percent.

In the example, it is not logical to assume that new uniforms would produce such a dramatic increase in arrests. To demonstrate such a claim logically would require an explanation of the special circumstances that produced the result. For example, if the new uniforms resulted in a union settlement with police officers, which in turn caused the officers to end a work slowdown, then the conclusion would be reasonable. Even in that case, however, the new uniforms per se did not increase arrests; rather, a complete chain of events produced the result.

Biased or Suppressed Evidence

A **conclusion** reached as a result of self-serving data, questionable sources, or purposely incomplete facts is illogical—and probably dishonest. If you were asked to prepare a **report** on the acceptance of a new policy among employees and you distributed **questionnaires** only to those who thought the policy was effective, the resulting evidence would be biased. If you purposely ignored employees who did not believe the policy was effective, you would also be suppressing evidence. Any writer must be concerned about the fair presentation of evidence—especially in reports that recommend or justify actions.

Fact versus Opinion

Distinguish between fact and opinion. Facts include verifiable data or statements, whereas opinions are personal conclusions that may or may not be based on facts. For example, it is a verifiable fact that distilled water boils at 100°C; that distilled water tastes better than tap water is an opinion. In order to prove a point, some writers mingle facts with opinions, thereby obscuring the differences between them. This is of course unfair to the reader. Be sure always to distinguish your facts from your opinions in your writing so that your reader can clearly understand and judge your conclusions.

> **EXAMPLES**   The new milling machines produce parts that are within 2 percent of specification. (This sentence is stated as a fact and can be verified by measurement.)
>
> The milling machine operators believe the new models are safer than the old ones. (The word *believe* identifies the statement as an opinion—later statistics on the accident rates may or may not verify the opinion as a fact.)

L

The opinion of experts, or their professional judgment, in their area of expertise is often accepted as valid evidence in court. In writing, too, such testimony can help you convince your reader of the logic of your conclusion. Be on guard against flawed testimonials, however. For example, a chemist may well make a statement about the safety of a product for use on humans, but the chemist's opinion would not be as valid as that of a medical researcher who has actually studied the effects of that product on humans. When you quote the opinion of an authority, be sure that his or her expertise is appropriate to your point. Make sure also that the opinion is a current one and that the expert is indeed highly respected.

Loading

When you include an opinion in a statement and then reach conclusions that are based on that statement, you are *loading the argument.* Consider the following opening for a **memorandum:**

> **EXAMPLE**   I have several suggestions to improve the *poorly written* policy manual. First, we should change . . .

By opening with the assumption that the manual is poorly written, the writer has loaded the statement to get readers to accept his arguments and conclusions. Yet, he has said nothing to establish how the manual is poorly written. In fact, it may be that only parts of the manual are poorly written. And if so, what are the exact problems? Be careful not to load arguments in your writing; conclusions reached with loaded statements are weak and ultimately unconvincing.

long variants

Guard against inflating plain words beyond their normal value by adding extra **prefixes** or **suffixes,** a practice that creates *long variants.* The following is a **list** of some normal words followed by their inflated counterparts:

analysis	analyzation
connect	interconnect
finish	finalize
orient	orientate
priority	prioritization (or prioritize)
use	utilize (see also **utilize**)

(See also **gobbledygook** and **word choice.**)

loose/lose

L

Loose is an **adjective** meaning "not fastened" or "unrestrained."

> EXAMPLE   He found a *loose* wire.

Lose is a **verb** meaning "be deprived of" or "fail to win."

> EXAMPLE   Did you *lose* the operating instructions?

lowercase and uppercase letters

Lowercase letters are small letters, as distinguished from **capital letters** (known as *uppercase* letters). The terms were coined in the early history of printing when printers kept the small letters in a "case," or tray, below the tray where they kept the capital letters.

Avoid using all capital letters ("all caps") to emphasize important information. Setting an entire passage in all caps makes it hard to read, because the shapes of capital letters are not as distinctive as are the shapes of lowercase letters or a combination of uppercase and lower case letters. Uniformity of shape deprives readers of important visual clues about the identities of letters and thus slows them down. Instead, request boldface or **italic** for such passages, which makes the words stand out without affecting the unique shapes of the letters. (See also **design and layout.**)

malapropisms

A *malapropism* is a word that sounds similar to the one intended but that is ludicrously wrong in the context.

> **CHANGE** The service technician cleaned the printer's *plankton.*
> **TO** The service technician cleaned the printer's *platen.*

Intentional *malapropisms* are sometimes used in humorous writing; unintentional malapropisms can embarrass a writer. (See also **antonyms** and **synonyms.**)

male

The term *male* is usually restricted to scientific, legal, or medical contexts (a *male* patient or suspect). Keep in mind that this term sounds cold and impersonal. The terms *boy, man,* and *gentleman* are acceptable substitutes in other contexts; however, be aware that these substitute words also have **connotations** involving age, dignity, and social position. (See also **female.**)

maps

Maps can be used to show specific geographic features (roads, mountains, rivers, and the like) or to show information according to geographic distribution (population, housing, manufacturing centers, and so forth). Bear these points in mind in creating and using maps:

1. Label the map clearly.
2. Assign the map a figure number if you are using enough **illustrations** to justify use of figure numbers. (See Figure 1.)

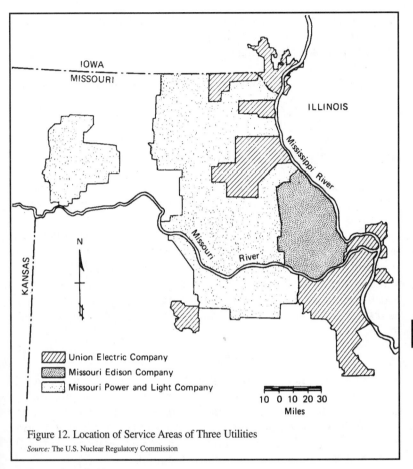

Figure 12. Location of Service Areas of Three Utilities

Source: The U.S. Nuclear Regulatory Commission

FIGURE 1 Sample Map

3. Make sure all boundaries within the map are clearly identified. Eliminate unnecessary boundaries.
4. Eliminate unnecessary information from your map (if population is important, do not include mountains, roads, rivers, and the like).
5. Include a scale of miles or feet to give your **reader** an indication of the map's proportions.
6. Indicate which direction is north.
7. Show the features you want emphasized by using shading, dots, cross-hatching, or appropriate symbols when color reproduction cannot be used.

8. If you use only one color, remember that only three shades of a single color will show up satisfactorily.

9. Include a key or legend telling what the different colors, shadings, or symbols represent.

10. Place maps as close as possible to the portion of the text that refers to them.

mathematical equations

Material with mathematical equations can be accurately prepared and easily read by following a few standard guidelines on displaying equations.

Short and simple equations, such as $x(y) = y^2 + 3y + 2$, should be set as part of the running text rather than being displayed or set on a separate line. (However, an equation or mathematical symbol should never appear at the beginning of a sentence.) If a document contains equations that are referred to in the text, even short equations should be displayed and identified with a number, as the following example shows:

$$x(y) = y^2 + 3y + 2 \qquad\qquad (1)$$

Equations are usually numbered consecutively throughout the work. Place the equation number, in parentheses, at the right margin, on the same line as the equation (or on the first line if the equation runs longer than one line). Leave at least four spaces between the equation number and the equation. Number displayed equations consecutively, and refer to them in the text as "Equation 1" or "Eq. 1."

Positioning Displayed Equations

Equations that are set off from the text need to be surrounded by space. Triple-space between displayed equations and the normal text. Double-space between one equation and another and between the lines of multiline equations. Count space above the equation from the uppermost character in the equation; count space below from the lowermost character.

Type displayed equations either at the left margin or indented five spaces from the left margin, depending on their length. When a series of short equations is displayed in sequence, however, align them on their equal signs.

$$p(x,y) = \sin(x + y) \tag{2}$$
$$p(x,y) = \sin x \cos y + \cos x \sin y$$
$$p(x_0, y_0) = \sin x_0 \cos y_0 + \cos x_0 \sin y_0$$

$$q(x,y) = \cos(x + y)$$
$$= \cos x \cos y - \sin x \sin y$$
$$q(x_0, y_0) = \cos x_0 \cos y_0 - \sin x_0 \sin y_0$$

Break an equation requiring two lines at the equal sign, carrying the equal sign over to the second portion of the equation.

$$_0\!\int^1 (f_n - \tfrac{n}{r}f_n)^2\, r\, dr + 2n\, _0\!\int^1 f_n f_n dr \tag{3}$$
$$= _0\!\int^1 (f_n - \tfrac{n}{r}f_n)^2\, r\, dr + nf_n^2(1)$$

If you cannot break it at the equal sign, break it at a plus sign or minus sign that is not within **parentheses** or **brackets.** Bring the plus sign or minus sign to the next line of the equation, which should be positioned to end near the right margin.

$$\emptyset(x,y,z) = (x^2 + y^2 + z^2)^{1/2}\, (x - y + z)\, (x + y - z)^2 \tag{4}$$
$$- [f(x,y,z) - 3x^2]$$

The next best place to break an equation is between parentheses or brackets that indicate multiplication of two major elements.

For equations requiring more than two lines, start the first line at the left margin, end the last line at the right margin, and center intermediate lines inside the margins. Whenever possible, break equations at operational signs, parentheses, or brackets.

Omit punctuation after displayed equations, even when they end a sentence and even when a key list defining terms follows (e.g., P = pressure, psf; V = volume, cu ft; T = temperature, °). Punctuation may be used before an equation, however, depending on the grammatical construction.

The symbol $(n)_r$ may be written in a more familiar way by using the following algebraic device:

$$(n)_r = \frac{(n)(n-1)(n-2)\ldots(n-r+1)(n-r)(n-r-1)\ldots 3\cdot 2\cdot 1}{(n-r)(n-r-1)\ldots 3\cdot 2\cdot 1} \tag{5}$$
$$= \frac{n!}{(n-r)!}$$

Expressing Mathematical Equations

Type mathematical equations rather than writing them out. Many word-processing software programs can produce a comprehensive range of mathematical and scientific letters and symbols and even permit you to adjust their size to suit the complexity of the equation.

maybe/may be

Maybe (one word) is an **adverb** meaning "perhaps."

> EXAMPLE   *Maybe* the legal staff can resolve this issue.

May be (two words) is a **verb** phrase.

> EXAMPLE   It *may be* necessary to ask for an outside specialist.

media/medium

Media is the plural of *medium* and should always be used with a plural **verb.**

> EXAMPLES   The *media are* a powerful influence in presidential elections.
> The most influential *medium is* television.

memorandums

M

The *memorandum* is one of the most frequently used forms of communication among members of the same organization. Called memos for short, memorandums are routinely used for internal communications of all kinds—from short notes to small **reports** and internal **proposals.** (See also **email.**) Among their many uses, memos announce policies, confirm conversations, exchange information, delegate responsibilities, request information, transmit documents, instruct employees, and report results. As this partial list illustrates, memos provide a record of decisions made and actions taken in an organization. For this reason, clear and effective memos are essential to the success of any organization. The slogan "Put it in writing" reflects the importance of memos and the care with which they should be written. The result of a carelessly prepared memo is a garbled message that could baffle **readers,** waste valuable time, produce costly errors, or irritate employees with an offensive **tone.**

Effective managers use memos to keep employees informed about company goals, motivate them to achieve these goals, and keep their morale high in the process. To achieve these ends, managers must be clear and accurate in their memos in order to gain respect and maintain credibility among their subordinates. Consider the unintended secondary messages the following notice conveys:

> EXAMPLE   It has been decided that the office will be open the day after Thanksgiving.

The first part of the sentence ("It has been decided") not only sounds impersonal but also communicates an authoritarian management-versus-employee tone: *somebody* decides that *you* work. The passive **voice** also suggests that the decision maker does not want to say, "I have decided" and thus be identified. One solution, of course, is to remove the first part of the sentence.

> EXAMPLE   The office will open the day after Thanksgiving.

But even this statement sounds impersonal. A better solution would be to suggest both that the decision is good for the company and that employees should be privy to (if not part of) the decision-making process.

> EXAMPLE   Because we must meet the December 15 deadline to be eligible for the government contract, the office will be open the day after Thanksgiving.

This version is forthright and informative and helps the employees understand the decision.

Organizing Memos

Outline your memos, even if that means simply jotting down the points to be covered and then ranking them in a logical **method of development.** This will not only help you be logical, but it will also help you keep within the scope of your memo.

Memos should ordinarily deal with only one subject; if you need to cover two subjects, you may wish to write two memos, since multi-subject memos are difficult to file.

Memo Openings

A memo should normally begin with a statement of its main idea.

> EXAMPLE   Because of our inability to serve our present and future clients as efficiently as in the past, I recommend we hire an additional claims representative and a part-time receptionist. (main idea)

The only exceptions to stating the main point first are (1) when the reader is likely to be highly skeptical and (2) when you are disagreeing with persons in positions of higher authority. In such cases, a more persuasive tactic is to state the problem first (rather than *your* solution) and then present the specific points that will support your final recommendation. (See also **persuasion.**)

> EXAMPLE   Following our meeting, I met with several of our most important clients and learned that our problem involves both communication and service.

As with other writing, if your reader is not familiar with the subject or the background of a problem, provide an introductory background **paragraph.** A brief background is especially important in memos that serve as records that can provide crucial information months (or even years) later. Generally, longer memos or those dealing with complex subjects benefit most from developed **introductions.** However, even when you are writing a short memo and the recipient is familiar with the situation, you need to remind your reader of the context. Readers have so much crossing their desks that they need a quick orientation. (Words that provide context are shown below in *italics.*)

> **EXAMPLES** *As we decided after yesterday's meeting,* we need to set new guidelines for . . .
>
> *As Jane recommended,* I reviewed the office reorganization plan. I like most of the features; however, the location of the receptionist and word processor . . .

Lists and Headings

It is often a good idea to use **lists** to emphasize your points in a memo. If you are trying to convince a skeptical reader, a list of your points—from most to least persuasive—will stand out rather than being lost in a lengthy paragraph. Be careful, however, not to overuse lists. A memo that consists almost entirely of lists is difficult for readers to understand because they are forced to connect the separate items for themselves and mentally provide coherence. Lists lose their impact when they are overused.

Another useful device, particularly in long memos, is **headings.** Headings have a number of advantages:

1. They divide material into manageable segments.
2. They call attention to main topics.
3. They signal a shift in topic.

Headings also function like the parts of a **formal report** in that readers interested only in one section of a memo can easily identify that section.

Writing Style

Like all **correspondence,** the level of formality in memos depends entirely on your reader and **purpose.** For example, is your reader a peer, superior, or subordinate? A memo to an associate who is of equal rank and is a friend is likely to be informal and personal. However, a memo written as an internal proposal to several readers or a memo to someone two or three levels higher in your organization is likely to use the more formal **style** of a report. Consider the following original and revised versions of a subordinate's statement to a superior:

CHANGE   I can't agree with your plan because I think it poses logistical problems. (informal, personal, and forceful)

TO   The logistics of moving the department may pose serious problems. (formal, impersonal, and cautious)

A memo giving **instructions** to a subordinate will also be relatively formal and impersonal—unless you are trying to reassure or praise. When writing to subordinates, also remember that managing does not mean ordering. A dictatorial tone will prevent a memo from being an effective management tool.

Format and Parts

Memo format and customs vary greatly from organization to organization. The following are some of the styles used:

preprinted half sheets (8 ½″ × 5 ½″)
message-and-reply forms ("speed messages")
regular 8 ½″ × 11″ letter stationery
special 8 ½″ × 11″ forms with company name or logo

Although there is no single, standard form, Figure 1 shows a typical memorandum format. When memos exceed one page, use a second page heading like the one shown in **correspondence.** Some firms also print short purpose statements on the side or bottom of memos, with space for the writer to make a check mark.

EXAMPLES   ☐ For your information
☐ For your action
☐ For your reply

You can also use this quick-response system for memos sent to numerous readers whose responses you need to tabulate.

EXAMPLE   I can meet at 1 p.m. ☐
2 p.m. ☐
3 p.m. ☐

One of the most important parts of a memo is the subject line because it serves as an important orientation when the reader first sees the memo. Therefore, although titles in the subject line should be very brief, they must be as specific as possible.

CHANGE   Subject: Tuition Reimbursement

TO   Subject: Tuition Reimbursement for Time Management Seminar

---

PROFESSIONAL PUBLISHING SERVICES
MEMORANDUM

DATE: April 14, 19—

TO: Hazel Smith, Publications Manager

FROM: Herbert Kaufman *HK*

SUBJECT: Schedule for Acme Electronics Brochure

Acme Electronics has asked us to prepare a comprehensive brochure for their Milwaukee office by August 9, 19—.

We have worked with electronic firms in the past, so this job should be relatively easy. My guess is that it will take nearly two months. Ted Harris has requested time and cost estimates. Fred Moore in accounting will prepare the cost estimates, and I would like you to prepare a schedule for the estimated time.

Additional Personnel
In preparing the schedule, check the following:

1. Production schedule for all staff writers.
2. Available free-lance writers.
3. Dependable graphic designers.

Ordinarily, we would not need to depend on outside personnel; however, since our bid for the *Wall Street Journal* special project is still under consideration, we could be pressed in June and July. We have to keep in mind staff vacations that have already been approved.

Time Estimates
Please give me time estimates by April 19. A successful job done on time will give us a good chance to obtain the contract to do Acme's publications for their annual stockholders' meeting this fall.

I know your staff can do the job.

lcs
Copies: Ted Harris, Senior Vice-President
            Fred Moore, Accounting Manager

FIGURE 1 Typical Memo Format

Subject lines function much like the titles of reports, aiding in filing and later retrieval. However, *subject lines should not substitute for an opening that provides a context for the message.*

Capitalize all major words in the title of a subject line. Do not capitalize an article, **conjunction,** or **preposition** of fewer than four letters unless it is the first or last word.

The final step is signing or initialing a memo, which lets the reader know that you approve of its contents, especially if you did not type it. Where you sign or initial the memo depends on the practice of your organization: Some writers sign at the end, and others sign their initials next to their typed name.

metaphor (see figures of speech)

methods of development

After you have completed your **research,** but before beginning your **outline,** ask yourself how you can most effectively "unfold" your **topic** for your **reader.** An appropriate method of development will make it easy for your reader to understand your topic and will move the topic smoothly and logically from an **introduction** (or **opening**) to a **conclusion.** There are several common methods of development, each best suited to particular purposes.

If you are writing a set of **instructions,** for example, you know that your readers need the instructions in the order that will enable them to perform some task. Therefore, you should use a **sequential method of development.** If you wished to emphasize the time element of a sequence, however, you could follow a **chronological method of development.**

If writing about a new topic that is in many ways similar to another, more familiar topic, it is sometimes useful to develop the new topic by comparing it to the old one, thereby enabling your readers to make certain broad assumptions about the new topic, based on their understanding of the familiar topic. By doing so, you are using a **comparison method of development.**

When describing a mechanical device, you may divide it into its component parts and explain each part's function as well as how all the parts work together. In this case, you are using a **division-and-classification method of development.** Or you may use a **spatial method of development** to describe the physical appearance of the device from top to bottom, from inside to outside, from front to back, and so on.

If you are writing a report for a government agency explaining an airplane crash, you might begin with the crash and trace backwards to its cause, or you could begin with the cause of the crash (for example, a structural defect) and show the sequence of events that led to the crash. Either way is the **cause-and-effect method of development.**

You can also use this approach to develop a report dealing with the solution to a problem, beginning with the problem and moving on to the solution, or vice versa.

If you are writing about the software for a new computer system, you might begin with a general statement of the function of the total software package, then explain the functions of the larger routines within the software package, and finally deal with the functions of the various subroutines within the larger routines. You would be using a **general-to-specific method of development.** (In another situation, you might use the **specific-to-general method of development.**)

To explain the functions of the departments in a company, you could present them in a sequence that reflects their importance within the company: the executive department first and the custodial department last, with all other departments (sales, engineering, accounting, and so on) arranged in the relative order of importance they are given in that company. You would be using the **decreasing-order-of-importance method of development.** (In another situation, you might use the **increasing-order-of-importance method of development.**)

Methods of development often overlap, of course. Rarely does a writer rely on only one method of development in a written work. The important thing is to select one primary method of development and then to base your outline on it. For example, in describing the organization of a company, you would actually use elements from three methods of development: you would *divide* the larger topic (the company) into departments, present the departments *sequentially,* and arrange the departments by their *order of importance* within the company.

minutes of meetings

Organizations and committees keep official records of their meetings; such records are known as *minutes.* If you attend many business-related meetings, you may be asked to serve as recording secretary and write and distribute the minutes of a meeting. At each meeting, the minutes of the previous meeting are usually read aloud if printed copies of the minutes were not distributed to the members beforehand; the group then votes to accept the minutes as prepared or to revise or clarify specific items. The bylaws or policies and procedures of your organization may specify what must be included in your minutes of meetings. As a general guide, *Robert's Rules of Order* recommends that the minutes of meetings include the following:

1. Name of the group or committee holding the meeting.
2. Place, time, and date of the meeting.
3. Kind of meeting (a regular meeting or a special meeting called to discuss a specific subject or problem).
4. Number of members present and, for committees or boards of ten or fewer members, their names.
5. A statement that the chairperson and the secretary were present, or the names of any substitutes.
6. A statement that the minutes of the previous meeting were approved, revised, or not read.
7. A list of any reports that were read and approved.
8. All the main motions that were made, with statements as to whether they were carried, defeated, or tabled (vote postponed), and the names of those who made and seconded the motions (motions that were withdrawn are not mentioned).
9. A full description of resolutions that were adopted and a simple statement of any that were rejected.
10. A record of all ballots with the number of votes cast for and against.
11. The time that the meeting was adjourned (officially ended) and the place, time, and date of the next meeting.
12. The recording secretary's signature and typed name and, if desired, the signature of the chairperson.

M

Except for recording motions, which must be transcribed word for word, summarize what occurs and paraphrase discussions.

Since minutes are often used to settle disputes, they must be accurate, complete, and clear. When approved, minutes of meetings are official and can be used as evidence in legal proceedings.

Keep your minutes brief and to the point. Give complete information on each **topic,** but do not ramble—conclude the topic and go on to the next one. Following a set **format** will help you keep the minutes concise. You might, for example, use the heading *TOPIC,* followed by the subheadings *Discussion* and *Action Taken,* for each major point discussed.

Avoid abstractions and generalities; always be specific. If you are referring to a nursing station on the second floor of a hospital, write "the nursing station on the second floor" or "the second-floor nursing station," not just "the second floor."

Be especially specific when referring to people. Avoid using titles (the chief of the Word Processing Unit) in favor of names and titles (Ms. Florence Johnson, chief of the Word Processing Unit). And be consistent in the way you refer to people. Do not call one person *Mr.*

Minutes of the Regular Meeting of the Credentials Committee

DATE: April 18, 19—

PRESENT: M. Valden (Chairperson), R. Baron, M. Frank, J. Guern, L. Kingson, L. Kinslow (Secretary), S. Perry, B. Roman, J. Sorder, F. Sugihana

Dr. Mary Valden called the meeting to order at 8:40 p.m. The minutes of the previous meeting were unanimously approved, with the following correction: the name of the secretary of the Department of Medicine is to be changed from Dr. Juanita Alvarez to Dr. Barbara Golden.

Old Business
None.

New Business
The request by Dr. Henry Russell for staff privileges in the Department of Medicine was discussed. Dr. James Guern made a motion that Dr. Russell be granted staff privileges. Dr. Martin Frank seconded the motion, which passed unanimously.

Similar requests by Dr. Ernest Hiram and Dr. Helen Redlands were discussed. Dr. Fred Sugihana made a motion that both physicians be granted all staff privileges except respiratory-care privileges because the two doctors had not had a sufficient number of respiratory cases. Dr. Steven Perry seconded the motion, which was passed unanimously.

Dr. John Sorder and Dr. Barry Roman asked for a clarification of general duties for active staff members with respiratory-care privileges. Dr. Richard Baron stated that he would present a clarification at the next scheduled staff meeting on May 15.

Dr. Baron asked for a volunteer to fill the existing vacancy for Emergency Room duty. Dr. Guern volunteered. He and Dr. Baron will arrange a duty schedule.

There being no further business, the meeting was adjourned at 9:15 p.m. The next regular meeting is scheduled for May 15, at 8:40 p.m.

Respectfully submitted,

Leslie Kinslow

Leslie Kinslow
Medical Staff Secretary

Mary Valden M.D.

Mary Valden, M.D.
Chairperson

FIGURE 1 Sample of a Set of Minutes

Jarrel and another *Janet* Wilson. It may be unintentional, but a lack of consistency in titles or names may reveal a deference to one person at the expense of another. Avoid **adjectives** and **adverbs** that suggest either good or bad qualities, as in "Mr. Sturgess's *capable* assistant read the *extremely comprehensive* report to the subcommittee." Minutes should always be objective and impartial.

If a member of the committee is to follow up on something and report back to the committee at its next meeting, state clearly the person's name and the responsibility he or she has accepted.

When assigned to take the minutes at a meeting, be prepared. Bring more than one pen and plenty of paper. If convenient, you may bring a tape recorder as a backup to your notes. Have ready the minutes of the previous meeting and any other material that you may need. If you do not know shorthand, take memory-jogging notes during the meeting and then expand them with the appropriate details immediately after the meeting. Remember that minutes are primarily a record of specific actions taken. (See also **note taking**.)

Figure 1 is a sample set of minutes.

misplaced modifiers (see modifiers)

mixed constructions

A mixed construction occurs when a sentence contains grammatical forms that are improperly combined. These constructions are improper because the grammatical forms are inconsistent with one another. The most common types of mixed constructions result from the following causes:

Tense

CHANGE   The pilot *lowered* the landing gear and *is approaching* the runway. (shift from past to present tense)

TO   The pilot *lowered* the landing gear and *approached* the runway.

OR   The pilot *has lowered* the landing gear and *is approaching* the runway.

Person

CHANGE   The *technician* should take care in choosing *your* equipment. (shift from third to second person)

TO   The *technician* should take care in choosing *his or her* equipment.

Number

CHANGE   My *car*, though not as fast as the others, *operate* on regular gasoline. (singular subject with plural verb form)

TO   My *car*, though not as fast as the others, *operates* on regular gasoline.

Voice

CHANGE   I will *check* your report, and then *it will be returned* to you. (shift from active to passive voice)

TO   I will *check* your report, and then *I will return* it to you.

(See also **parallel structure** and **agreement.**)

modifiers

Modifiers are words, **phrases,** or **clauses** that expand, limit, or make more precise the meaning of other elements in a sentence. Although we can create sentences without modifiers, we often need the detail and clarification they provide.

EXAMPLES   Production decreased. (without modifiers)
Automobile production decreased *rapidly.* (with modifiers)

Most modifiers function as **adjectives** or **adverbs.** An adjective makes the meaning of a **noun** or **pronoun** more precise by pointing out one of its qualities or by imposing boundaries on it.

EXAMPLES   *ten* automobiles   *this* crane
an educated person   *loud* machinery

An adverb modifies an adjective, another adverb, a **verb,** or an entire clause.

EXAMPLES   Under test conditions, the brake pad showed *much* less wear than it did under actual conditions. (modifying the adjective *less*)

The wear was *very* much less than under actual conditions. (modifying another adverb, *much*)

The recording head hit the surface of the disc *hard.* (modifying the verb *hit*)

Surprisingly, the machine failed even after all the tests that it had passed. (modifying a clause)

Adverbs become **intensifiers** when they increase the impact of adjectives (*very* fine, *too* high) or adverbs (*rather* quickly, *very* slowly). As a rule, be cautious in using intensifiers; their overuse can lead to exaggeration and hence to inaccuracies.

Jammed Modifiers

Some writing is unclear or difficult to read because it contains jammed modifiers, or strings of modifiers preceding nouns.

CHANGE   Your *staffing level authorization reassessment* plan should result in a major improvement.

In this sentence the noun *plan* is preceded by four modifiers; this string of modifiers slows the **reader** down and makes the sentence awkward and clumsy. Jammed modifiers often result from an overuse of **jargon** or **vogue words.** Occasionally, they occur when writers mistakenly attempt to be concise by eliminating short **prepositions** or connectives—exactly the words that help to make sentences clear and readable. See how breaking up the jammed modifiers makes the previous example easier to read.

TO   Your plan for the reassessment of staffing-level authorizations should result in a major improvement.

OR   Your plan to reassess authorizations for staffing levels should result in a major improvement.

Misplaced Modifiers

A modifier is misplaced when it modifies, or appears to modify, the wrong word or phrase. It differs from a **dangling modifier** in that a dangling modifier cannot *logically* modify any word in the sentence because its intended referent is missing. The best general rule for avoiding misplaced modifiers is to place modifiers as close as possible to the words they are intended to modify. A misplaced modifier can be a word, a **phrase,** or a **clause.**

M

Misplaced Words. Adverbs are especially likely to be misplaced because they can appear in several positions within a sentence.

EXAMPLES   We *almost* lost all of the parts.
We lost *almost* all of the parts.

The first sentence means that all of the parts were *almost* lost (but they were not), and the second sentence means that a majority of the parts (*almost all*) were in fact lost. Possible confusion in sentences of this type can be avoided by placing the adverb immediately before the word it is intended to modify.

CHANGE   *All* navigators are *not* talented in mathematics. (The implication is that no navigator is talented in mathematics.)
TO   *Not all* navigators are talented in mathematics.

Misplaced Phrases. To avoid confusion, place **phrases** near the words they modify. Note the two meanings possible when the phrase is shifted in the following sentences:

EXAMPLES The equipment *without the accessories* sold the best. (Different types of equipment were available, some with and some without accessories.)

The equipment sold the best *without the accessories*. (One type of equipment was available, and the accessories were optional.)

Misplaced Clauses. To avoid confusion, **clauses** should be placed as close as possible to the words they modify.

CHANGE We sent the brochure to four local firms *that had three-color art.*
TO We sent the brochure *that had three-color art* to four local firms.

Squinting Modifiers. A modifier "squints" when it can be interpreted as modifying either of two sentence elements simultaneously, so that the reader is confused about which is intended.

EXAMPLE We agreed *on the next day* to make the adjustments.

The meaning of the preceding example is unclear; it could have either of the following two senses:

EXAMPLES We agreed to *make the adjustments on the next day.*
On the next day we agreed to make the adjustments.

A squinting modifier can sometimes be corrected simply by changing its position, but often it is better to recast the sentence:

EXAMPLES We agreed that *on the next day* we would make the adjustments. (The adjustments were to be made on the next day.)

On the next day we agreed that we would make the adjustments. (The agreement was made on the next day.)

(See also **dangling modifiers.**)

mood

The grammatical term *mood* refers to the verb functions—and sometimes form changes—that indicate whether the verb is intended to (1) make a statement or ask a question (indicative mood), (2) give a command (imperative mood), or (3) express a hypothetical possibility (subjunctive mood).

The *indicative mood* refers to an action or a state that is conceived as fact.

EXAMPLES *Is* the setting correct?
The setting *is* correct.

The *imperative mood* expresses a command, suggestion, request, or entreaty. In the imperative mood, the implied subject "you" is not expressed.

> EXAMPLES *Install* the wire today.
> Please *let* me know if I can help.

The *subjunctive mood* expresses something that is contrary to fact, that is conditional, hypothetical, or purely imaginative; it can also express a wish, a doubt, or a possibility. The subjunctive mood may change the form of the verb, but the verb *be* is the only one in English that preserves many such distinctions.

> EXAMPLES The senior partner insisted that he (I, you, we, they) *be* in charge of the project.
>
> If we *were* to close the sale today, we would meet our monthly quota.
>
> If I *were* you, I would postpone the trip.

The advantage of the subjunctive mood is that it enables us to express clearly whether or not we consider a condition contrary to fact. If so, we use the subjunctive; if not, we use the indicative.

> EXAMPLES If I *were* president of the firm, I would change several personnel policies. (subjunctive)
>
> Although I *am* president of the firm, I don't feel that I control every aspect of its policies. (indicative)

Be careful not to shift from one mood to another within a sentence; to do so makes the sentence not only ungrammatical but unbalanced as well.

> CHANGE *Put* the clutch in first (imperative); then you *can* put the truck in gear. (indicative)
> TO *Put* the clutch in first (imperative); then *put* the truck in gear. (imperative)

Ms./Miss/Mrs.

Ms. is a convenient form of addressing a woman, regardless of her marital status, and it is now almost universally accepted. *Miss* is used to refer to an unmarried woman, and *Mrs.* is used to refer to a married woman. Some women indicate a preference for *Miss* or *Mrs.*, and such a preference should be honored. An academic or professional title *(Doctor, Professor, Captain)* should take preference over *Ms.*, *Miss*, or *Mrs.*

MS/MSS

MS (or *ms*) is the **abbreviation** for *manuscript.* The plural form is *MSS* (or *mss*). Do not use this abbreviation (or any abbreviation) unless you are certain that your **reader** is familiar with it.

mutual/in common

When two or more persons (or things) have something in common, they share it or possess it jointly.

> **EXAMPLES**   What we have *in common* is our desire to make the company profitable.
>
> The fore and aft guidance assemblies have a *common* power source.

Mutual may also mean "shared" (as in *mutual* friends), but it usually implies something given and received reciprocally and is used with reference to only two persons or parties.

> **EXAMPLES**   Smith mistrusts Jones, and I am afraid the mistrust is *mutual.* (Jones also mistrusts Smith.)
>
> Both business and government consider the problem to be of *mutual* concern.

N

narration

Narration is the presentation of a series of events in a prescribed (usually chronological) sequence. Much narrative writing explains how something happened: a laboratory or field study, a site visit, an accident, the events and decisions in an important meeting.

Effective narration rests on two key writing techniques: the careful, accurate sequencing of events and a consistent **point of view** on the part of the narrator. Narrative sequence and essential shifts in the sequence are signaled in three ways: chronology (clock and calendar time), transitional words pertaining to time *(before, after, next, first, while, then)*, and verb tenses that indicate whether something has happened (past **tense**) or is under way (present tense). The point of view indicates the writer's relation to the information being narrated as reflected in the use of **person.** Narration usually expresses either a first- or third-person point of view. First-person narration indicates that the writer is a participant ("this happened to me"), and third-person narration indicates that the writer is writing about what happened to someone or something else ("this happened to her, to them, or to it").

The narrative presented in Figure 1 reconstructs the final hours of the flight of a small aircraft that attempted to land at the airport in Hailey, Idaho. Instead, the plane crashed into the side of a mountain, killing the pilot and copilot. Because of the disastrous outcome of the flight, the investigators needed to "tell the story" in detail so that any lessons learned could be made available to other flyers. To do so, they had to recount the events as closely as possible. Thus, the chronology of the events is specified throughout the narrative in local (Hailey, Idaho) time.

To "tell the story," the narrator used the third-person point of view throughout, except for the first-person reports from witnesses. The

History of the Flight*

At 0613 m.s.t.[1] on January 3, 19—, N805C, a Canadair Challenger owned and operated by the A.E. Staley Company departed Decatur, Illinois, on a flight to Friedman Memorial Airport, Hailey, Idaho. The route of the flight was via Capitol, Illinois, Omaha, Nebraska, Scotts Bluff, Nebraska, Riverton, Wyoming, Idaho Falls, Idaho, direct 43°30′ north latitude, 114°17′ west longitude.

The en route portion of the flight was uneventful, and about 35 nmi east of Idaho Falls N805C was cleared by the Salt Lake City, Utah, Air Route Traffic Control Center (ARTCCS) to descend from 39,000 feet to 22,000 feet. N805C descended to 22,000 feet and the flightcrew then requested a descent to 17,000 feet. About 35 nmi east of Sun Valley Airport, after being cleared, N805C descended to 17,000 feet. About 0901, N805C's flightcrew cancelled their flight plan and, shortly thereafter, changed the transponder from 1311, the assigned discrete code, to 1200, the visual flight rules (VFR) code. At 0901:07, the data analysis reduction tool (DART) radar data showed a 1311 beacon code at 17,000 feet about 11 nmi east of the Sun Valley Airport. At 0901:37, the DART radar data showed a 1200 VFR transponder beacon code with no altitude readout about 2 nmi west of the 1311 beacon code that was recorded at 0901:07.

At 0904:10, DART radar data recorded a 1200 code target at 13,500 feet almost directly over the Sun Valley Airport. According to an employee of the airport's fixed base operator, N805C's flightcrew called on the airport's UNICOM[2] frequency and requested a landing advisory and asked if a food order had been placed. The flightcrew then stated that there would "be a quick-turn," and placed a fuel request. This was the last transmission heard from N805C. The employee said that she provided the latest altimeter setting to N805C, and "since we did not have the cloud conditions in the area, I was glad when other pilots were able to give reports as they saw things from the air."

The flightcrew of Cessna Citation, N13BT, which had landed at Sun Valley about 0903, also heard N805C report "over the field." According to N13BT's pilot, N805C reported over the field "sometime during our final approach or landing." According to the pilot, the weather at the airport when he landed "was 800 (feet) overcast with 10 miles visibility. The tops of the overcast or fog bank was about 6,800 (feet) m.s.l." He said that the overcast was "solid northwest up the valley. Visibility appeared lower (to the) northwest."

About 0908, Trans Western Flight 1301, a Convair 580, landed at Sun Valley Airport. Flight 1301 had descended through a hole in the overcast about 15 nmi southwest of Bellevue, Idaho, which is about 3 nmi southeast of the Sun Valley Airport. The first officer said that he gave position reports to the Sun Valley UNICOM when

---

*National Transportation Safety Board, "Aircraft Accident Report: A. E. Staley Manufacturing Co., Inc., Canadair Challenger CL-600, N805C, Hailey, Idaho, January 3, 1983. National Technical Information Service, Springfield, Va., 1983.

[1]All times, unless otherwise noted, are mountain standard time based on the 24-hour clock.
[2]UNICOM. The Non-government air/ground radio communications facility which may provide airport advisory information at certain airports. The Sun Valley UNICOM did not record, nor was it required to record or log, the time of radio communications.

FIGURE 1 Example of a Narration

the flight was 15 nmi from the airport, when it was 10 nmi from the airport over Bellevue turning on final approach for runway 31, and when it was 1 mile from the runway. The first officer said that he could see the visual approach slope indicator (VASI) lights for runway 30 during the landing approach. The captain and the first officer said that they neither saw N805C nor heard radio transmissions from N805C.

About 0900, a man who was driving his truck north on the highway between Bellevue and Hailey, Idaho, saw a twin engine, cream-colored jet, break through the clouds when he was about 2.5 miles north of Bellevue. He saw that the landing gear was down but he did not see any lights on the airplane. When the airplane appeared, "it was about 300 to 500 yards from the west hills adjacent to the airport and about 1,000 feet from the valley floor." The airplane was in a noseup attitude. The witness said that after the airplane descended below the clouds "and (the pilot) saw how close to the hills he was, he then started a sharp right turn." The airplane disappeared from his view into "low hanging clouds" over the northwest side of the hangar at the airport.

Between 0900 and 0930, another man, who was in the yard of his home in northeast Hailey, saw a jet airplane east of his home. The airplane was "white or silver with a blue tint." (N805C was painted white with blue and gold stripes along the length of the fuselage and tops of the wings.) The airplane was below the clouds, and he had "a good view of the airplane for about 10 to 15 seconds." He said that the airplane had a noseup attitude and "the wings were rocking up and down about 20°." The witness said that the clouds obscured all but the lower peaks of the mountains to the east and that after he lost sight of the airplane he thought it was "odd that the aircraft was under the cloud cover."

Shortly after 0900, a woman who was located in an apartment in southeast Hailey, heard a jet airplane fly over "in a northerly direction." She thought that this was "odd because jets don't go over us heading north from the airport The engines sounded very loud. . . ."

A fourth witness said that, between 0900 and 0940, she heard a jet airplane overfly her house in northeast Hailey. The woman was in the living room of her house when she heard the airplane and thought that "it must be low because of the loudness of the (engine) noise," and that "the sound of the jet did not trail off as they do as they fly farther away from you. The sound stopped less than 30 seconds from the time I first heard it." At the time, the clouds were resting on and hiding the tops of the mountains to the east.

About 1030, the chief pilot of the A. E. Staley Manufacturing Company, who was to board N805C at Sun Valley, arrived at the airport. Since the airplane was overdue, he instituted inquiries to several nearby airports to determine where the airplane had landed. At 1300, he asked air traffic control to make a full communications search. At 1400, after being told that the airplane had not been found, he requested an air search. While waiting for search and rescue teams to arrive, the chief pilot rented an airplane and at about 1700, found the accident site. The impact site, elevation about 6,510 feet, was about 2.2 nmi north of Sun Valley Airport at coordinates 43°32'50" N latitude, 114°17'35" W longitude.

FIGURE 1   Example of a Narration *(continued)*

words of each witness were chosen carefully and quoted for their bearing on what happened, from that person's vantage point.

The events were sequenced as precisely as possible from the beginning of the flight until the crash site was located by referencing verified clock times and approximating those that could not be verified. The verb tenses throughout indicated a past action: *departed, canceled, cleared, called, reported.*

Important but secondary information was either mentioned in a footnote or inserted, like the color of the plane, in parentheses. Information about the color of the plane was important to the narration only because it verified a sighting by an eyewitness. Otherwise, this information would have been unnecessary.

Once a narrative is under way, it should not be interrupted by lengthy explanations or analysis. Explain only what is necessary for readers to follow the action. In the plane-crash narrative, numerous pertinent issues might have been explained or evaluated in more detail. The weather, the condition of the plane's radar equipment and engines, the pilot's professional and medical history, the airport and its personnel and equipment are all pertinent, but each was assessed after the narrative in a separate section of the report. To have addressed each in the course of the narrative would have made the reader's task of finding out what happened unnecessarily difficult.

nature

Nature, used to mean "kind" or "sort," can—like those words—often be vague. Avoid the word in your writing. Say exactly what you mean.

> **CHANGE**   The *nature* of the engine caused the problem.
> **TO**   The compression ratio of the engine caused the problem.

needless to say

Although the **phrase** *needless to say* sometimes occurs in speech and writing, it is redundant because it always precedes a remark that is stated despite the inference that nothing further need be said.

> **CHANGE**   *Needless to say,* departmental cutbacks have meant decreased efficiency.
> **TO**   Departmental cutbacks have meant decreased efficiency.
> **OR**   Understandably, departmental cutbacks have meant decreased efficiency.

neo-

The **prefix** *neo-* is derived from a Greek word meaning "new." It is hyphenated when used with a proper **noun,** and it may be hyphenated when used with a noun beginning with the vowel o.

   EXAMPLES   neo-Fascism, neo-Darwinism, neo-orthodoxy
                        neologism, neonatal, neoplasm

new words

New words (also called *neologisms*) continually find their way into the language from a variety of sources. Some come from other languages.

   EXAMPLES   discotheque (French), skiing (Norwegian), whiskey (Gaelic)

Some come from technology.

   EXAMPLES   software, vinyl, nylon

Some come from scientific research.

   EXAMPLES   berkelium, transistor

Some come from brand or trade names.

   EXAMPLES   Jello, Teflon, Kleenex

Some, called **blend words,** are formed by combining existing words.

   EXAMPLE   smoke + fog = smog

Some come from **acronyms.**

   EXAMPLES   scuba, radar, laser

Business and technology are responsible for many new words. Some are necessary and unavoidable; however, it is best to avoid creating a new word if an existing word will do. If you do use what you believe to be a new word or expression, be sure to define it the first time you use it; otherwise, your **reader** will be confused. (See also **long variants** and **vogue words.**)

newsletter articles

A *newsletter* is a company publication or an employee publication that is produced by the employer. Its primary purpose is to keep employees informed about the company and its operations and policies.

Many different types of newsletters are found in business and industry—from the gossip sheet to the more sophisticated instruments of
company policy that increase the employee's understanding of the
company, its purpose, and the business in which it is engaged. If your
company publishes a newsletter of the latter type, you may be asked
to contribute an article on a subject that you are especially qualified to
write about. The editor may offer some general advice but can give you
little more help until you submit a draft.

Before beginning to write, consider the traditional *who, what, where,
when,* and *why* of journalism (who did it? what was done? where was
it done? when was it done? and why was it done?)—and then add *how,*
since the how of your subject may be of as much interest to your fellow employees as any of the five *w's.*

Before going any further, determine whether there is an official
company policy or position on your subject. If so, adhere to it as you
prepare your article. If there is no company policy on your subject,
determine as nearly as you can what management's attitude is toward
your subject.

Gather several fairly recent issues of your company newsletter, and
study the **style** and **tone** of the writing and the company's approach
to various kinds of subjects in past issues. Try to emulate these as you
work on your own article. Ask yourself the following questions about
your subject: What is its significance to the company? What is its significance to employees? The answers to these questions should help
you establish the style, tone, and approach for your article. These
answers should also heavily influence the conclusion you write for
your article.

Research for a newsletter article frequently consists of **interviewing.** Interview everyone concerned with your subject. Get all available
information and all points of view. (Be sure to give maximum credit
to the maximum number of people.)

Writing a newsletter article requires a little more imagination than
does writing **reports.** With a newsletter, you do not have a captive
audience; therefore, you may want to provide four helpful ingredients
to ensure a successful article: (1) an intriguing **title** to catch the **readers'** attention (a rhetorical question is often effective); (2) eye-catching
photographs or **illustrations** that entice them to read your lead
paragraph to see what the subject is really about; (3) a lead, or first
paragraph, that is designed to encourage further reading (this paragraph
generally makes the **transition** from the title to the down-to-earth
treatment of the subject); and (4) a well-developed presentation of
your subject to hold the readers' interest all the way to the **conclusion.**

Author! Author!

It was a number of centuries back that the Roman satirist Persius said, "Your knowing is nothing, unless others know you know."

Today at Allen-Bradley some of our engineers are subscribing to Persius's philosophy. They contend it takes more than the selling of a product for a corporation to enjoy continuing success.

"We must sell our knowledge as well," stressed Don Fitzpatrick, Commercial Chief Engineer. "We have to get our customers to think of us as knowledge experts."

Fitzpatrick's point is well made. In today's high technology business climate, it is essential to present a vanguard image as well as produce quality products. A corporation can be considered a "knowledge expert" when it "not only has a product for sale, but is the acknowledged leader in the application of that product," Fitzpatrick explained.

How does a corporation acquire an image as a "knowledge expert?"

Customer Roundtables, distributor schools, and participation by our engineers in professional societies, such as the Institute of Electrical and Electronics Engineers (IEEE), have helped Allen-Bradley to enhance its image in the industrial control market, noted Fitzpatrick. Yet, one subliminal selling tool that Fitzpatrick believes A-B engineers could sharpen to foster more influence is the writing and presentation of technical papers.

"I'm talking about technical writing. Not the kind of technical writing that explains our products or how to use them safely, but the kind of writing that is done as a method of sharing knowledge among companies who have similar concerns," Fitzpatrick explained. Such writings discuss new methods being used in the industry, or a company's difficulties working with an electrical phenomenon. They introduce state-of-the-art technology, new theories, or are tutorials on the developments in the industry to date, he added.

Fitzpatrick's Paper Cited

Fitzpatrick knows about technical papers. The former Purdue University professor has been preparing and presenting topics for Allen-Bradley Roundtables and distributor schools for several years. He also has been active in the presentation of papers for IEEE conferences. This past September in Houston, Fitzpatrick received a second-place award for a paper he presented at a Petrochemical Industrial Conference (PCIC) in San Diego. (PCIC is an organization of the Industry Application Society, which is part of the IEEE.) The award-winning paper, entitled "Transient Phenomena in the Motor Control Distribution Systems," also was published in IEEE's renowned annual publication "IAS Transactions."

Only a few papers are accepted annually for presentation at professional conferences, noted Fitzpatrick. In turn, just a few of those presentations are published in IEEE-related "Transactions" catalogs. Occasionally a paper's topic also may generate an article in a trade publication. For these reasons preparing such papers can be plums—for both company recognition and the author's esteem among colleagues.

Cultivating peer recognition is as important as nurturing an expert image for your company in today's technological world, said Steven Bomba, Director of Advanced Technology Development. "In the area of advanced technology the only way to test

FIGURE 1 Newsletter Article

if your work is current is to share it with people outside of the company," he said. "When you publish, you don't reveal proprietary information. You publish information that will be economically beneficial to your organization. Our employees should be anxious to test their sword against their competitors."

Growing Fraternity Here

Bomba and Fitzpatrick noted that there is a "new spirit for Allen-Bradley" in the area of technical writing and presentation. The number of engineers participating as guest speakers at national engineering, application, and technical conferences is a small but growing fraternity. Participation by A-B personnel in these professional conferences spurs a "psyche tune-up" and is critical to "innovation, staying current, and invention. Without it, a company lacks a reference point. Encouraging your people to write these papers provides a measure of value and stimulation," Bomba contends.

Technical papers are always noncommercial. They don't mention specific products or the company's name. The papers are bylined, however, usually with an editor's note containing information about the writer's background and employment. "If you word a paper right, you can sell your company. You can make the paper work as a subliminal public relations tool for you. One payoff for the extra time it often takes to prepare a paper is the direct influence in the public eye that is gained. It is impossible to measure the exact influence, but the exposure generally generates positive reaction," Fitzpatrick notes.

Contacts at Conferences

"Participation in conferences through writing of technical papers results in an employee becoming more active in professional organizations. That in turn leads to more business contacts. We're trying to broaden our market base, enter new industries. Roundtables, publishing, and conference programs are by-products, or spin-offs, that work for us. They help us to sell our ability," said Fitzpatrick.

They also work for the engineers and technicians who get involved in the paper's preparation. Often a paper may be co-authored or researched as a team. Engineers or technicians who take the time to research a possible topic, prepare an abstract for query, and finally write a paper upon abstract acceptance, are taking those small steps of knowledge that lead to leaps in industry technology.

"Getting published places the author in the elite 'Invisible College of Knowledge Workers,' which is a collection of technical people who transcend the boundaries of industry, cities, and nations," said Bomba. "Only 1 percent of degreed engineers write technical papers that are published," he added.

"Peer fear" may be the greatest barrier for an engineer to pick up the pen—or word processor, as the case may be today. Walt Maslowski, Principal Engineer, Dept. 756, was an engineer for 12 years before he decided to give technical writing a whirl three years ago. He joined A-B in 1979. This past September he presented his second paper as an A-B employee at an Industry Application Society conference in Cincinnati. His latest paper, which addressed the techniques in controlling three-phase induction motor drives, received a second-place award from IAS. The article will be published in IEEE's next issue of IAS Transactions. "There was a lot of positive response to the paper, and motor drives recognition is an area that A-B has been pursuing for some time," Maslowski noted.

FIGURE 1 Newsletter Article *(continued)*

Many Benefits Intangible

"The fear of possibly writing substandard papers held me back from writing for years," Maslowski continued. "What I've discovered, though, is that when you do write a qualified paper, your technical expertise becomes known. Many of your personal gains are intangible, but you meet people, open contacts, and absorb more knowledge."

How does one decide to write a paper? "First you should review enough of a field to get a pulse for it. Read key papers. Review key results that people are talking about. Analyze the unsolved areas—or holes—of these writings and then figure out how those holes might relate to your job," suggested Dave Linton, a project engineer in Dept. 756. Linton presented a tutorial on computer software design at a Digital Equipment Computer User Society symposium in early November.

Another barrier for some engineers is getting an abstract accepted, since all papers are reviewed by published professionals, or "referees." "Making that jump from the inside to the outside is hard without some middle ground to test your ideas. Some companies active in publishing have internal review systems where your co-workers give you some input and encouragement. It would be nice if Allen-Bradley had a similar system of internal publishing," Linton suggested. Bomba, Fitzpatrick, and others involved in technical writing here, concurred.

"Industry does not make the 'publish or perish' threat that is prevalent in the academic world. At A-B we really encourage our employees to contribute to our industry: to get involved," said Bomba.

Being recognized in your field builds self-confidence and generates a sense of pride and accomplishment. Writing technical papers is a good way to let "others know you know."

FIGURE 1   Newsletter Article *(continued)*

N

Your conclusion should emphasize the significance of your subject to your company and its employees. Because of its strategic location to achieve **emphasis,** your conclusion should include the thoughts that you want your readers to retain about your subject.

In preparing your newsletter article you will find it helpful to refer to the Checklist of the Writing Process and to follow the steps listed there. Figure 1 (pp. 375–377) is an article written for a newsletter. The original included photographs and boxed quotations.

no doubt but

In the **phrase** *no doubt but,* the word *but* is redundant.

> CHANGE   There is *no doubt but* that he will be promoted.
>   TO   There is *no doubt that* he will be promoted.

nominalizations

We have a natural tendency to want to make our on-the-job writing sound formal, even "impressive." One practice that contributes to both is the use of *nominalizations,* or a weak **verb** (*make, do, conduct, perform,* and so forth) and a **noun,** when the verb form of the noun would communicate the same idea more effectively in fewer words.

> CHANGE   The Legal Department will *conduct an investigation* of the purchase.
> TO   The Legal Department will *investigate* the purchase.
>
> CHANGE   The quality assurance team will *perform an evaluation* of the new software.
> TO   The quality assurance team will *evaluate* the new software.

You may occasionally find a legitimate use for a nominalization. You might, for example, use a nominalization to slow down the **pace** of your writing. But if you use nominalizations thoughtlessly or carelessly or just to make your writing sound more formal, the result will be **affectation.**

none

None may be considered either a singular or plural **pronoun,** depending on the context.

> EXAMPLES   *None* of the material *has* been ordered. (always use a singular **verb** with a singular noun—in this case, "material")
>
> *None* of the clients *has* been called yet. (singular even with reference to a plural noun *[clients]* if the intended emphasis is on the idea of *not one*)
>
> *None* of the clients *have* been called yet. (plural with a plural noun)

For **emphasis,** substitute *no one* or *not one* for *none* and use a singular verb.

> EXAMPLE   I paid the full retail price for three of your firm's machines, *no one* of which was worth the money.

(See also **agreement.**)

nor/or

Nor always follows *neither* in sentences with continuing negation.

> EXAMPLE   They will *neither* support *nor* approve the plan.

Likewise, *or* follows *either* in sentences.

> **EXAMPLE**   The firm will accept *either* a short-term *or* a long-term loan.

Two or more singular subjects joined by *or* or *nor* usually take a singular **verb.** But when one subject is singular and one is plural, the verb agrees with the subject nearer to it.

> **EXAMPLES**   *Neither* the manager *nor* the secretary *was* happy with the new filing system. (singular)
>
> *Neither* the manager *nor* the secretaries *were* happy with the new filing system. (plural)
>
> *Neither* the secretaries *nor* the manager *was* happy with the new filing system. (singular)

(See also **conjunctions.**)

notable/noticeable

Notable, meaning "worthy of notice," is sometimes confused with *noticeable,* meaning "readily observed."

> **EXAMPLES**   His accomplishments are *notable.*
>
> The construction crew is making *noticeable* progress on the new building.

N

note taking

The purpose of *note taking* is to summarize and record the information that you extract from your **research** material. The great challenge in taking notes is to condense another's thoughts in your own words without distorting the original thinking. Meeting this challenge involves careful reading on your part because to compress someone else's ideas accurately, you must first understand them.

When taking notes on abstract ideas, as opposed to factual data, be careful not to sacrifice **clarity** for brevity. You can be brief—if you are accurate—with statistics, but notes expressing concepts can lose their meaning if they are too brief. The critical test of a note is whether after a week has passed, you will still know what the note means and from it be able to recall the significant ideas of the passage. If you are in doubt about whether or not to take a note, take it—it is much easier to discard a note you don't need than to find the source again if the note is needed.

As you extract information, be guided by the **purpose** of your writing and by what you know about your **readers.** (How much do they know about your subject? What are their needs?)

Resist the temptation to copy your source word for word as you take notes. Paraphrase the author's idea or concept in your own words. But don't just change a few words in the original passage; you will be guilty of **plagiarism.** On occasion, when your source concisely sums up a great deal of information or points to a trend or development important to your subject, you are justified in quoting it verbatim and then incorporating that **quotation** into your paper. As a general rule, you will rarely need to quote anything longer than a **paragraph.** (See also **paraphrasing.**)

If a note is copied word for word from your source, be certain to enclose it in **quotation marks** so that you will know later that it is a direct quotation. In your finished writing, be certain to give the source of your quotation; otherwise, you will be guilty of plagiarism. (See also **documenting sources.**)

The mechanics of note taking are simple and, if followed conscientiously, can save you much unnecessary work. The following guidelines may help you take notes more effectively:

N

1. First, don't try to write down everything. Select the most important ideas and concepts to record as notes.
2. Mark any notes you don't fully understand or think you might need to pursue further.
3. Be sure to record all vital names, dates, and definitions.
4. Create your own shorthand. Many words can be indicated by symbols or shortened forms like & for *and,* + for *plus,* w for *with,* vwl for *vowel,* and 7 for *seven.*
5. Photocopy pages for passages that you intend to copy word for word, marking key sections with a highlighter.
6. Use file folders to store copied pages from library sources. (Remember, however, that the photocopy machine cannot replace careful note taking.)
7. Take notes from your own documents on your computer by cutting and pasting them to a "notes" file.
8. Print out key sections from CD-ROM disks or download the information to a "notes" file.
9. Check your notes for accuracy against the printed material before leaving it.

Consider the information in the following paragraph:

> Long before the existence of bacteria was suspected, techniques were in use for combating their influence in, for instance, the decomposition of meat. Salt and heat were known to be effective, and these do in fact kill bacteria or prevent them from multiplying. Salt acts by the osmotic effect of extracting water from the bacterial cell fluid. Bacteria are less easily destroyed by osmotic action than are animal cells because their cell walls are constructed in a totally different way, which makes them very much less permeable.

The paragraph says essentially three things:

1. Before the discovery of bacteria, salt and heat were used in combating bacteria.
2. Salt kills bacteria by extracting water from their cells by osmosis, hence its use in curing meat.
3. Bacteria are less affected by the osmotic effect of salt than are animal cells because bacterial cell walls are less permeable.

If your readers' needs and your objective involved tracing the origin of the bacterial theory of disease, you might want to note that salt was traditionally used to kill bacteria long before people realized what caused meat to spoil, though it might not be necessary to your topic to say anything about the relative permeability of bacterial cell walls.

So that you will give proper credit when you incorporate your notes into your writing, be sure to include the following with the first note taken from a book: author, title, publisher, place and date of publication, and page number. (On subsequent notes from the same book, you will need to include only the author and the page number.)

Figures 1 and 2 are examples of a first and a subsequent note taken from the paragraph on the use of salt as a preservative.

N

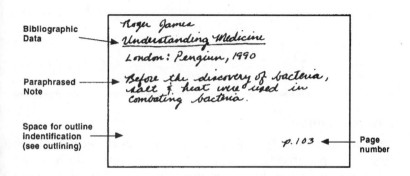

Bibliographic Data
Paraphrased Note
Space for outline indentification (see outlining)
Page number

FIGURE 1 First Note from a Source

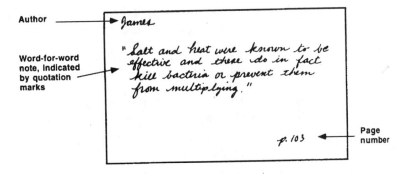

FIGURE 2 Subsequent Note from a Source

Although you should record notes in the way that you find efficient, nothing replaces 3 × 5 cards (as shown in Figures 1 and 2) in some applications. They are especially useful, for example, for **outlining** a complex project.

nouns

N

A *noun* names a person, place, thing, concept, action, or quality. The two basic types of nouns are *proper nouns* and *common nouns*.

Proper Nouns

A *proper noun* names a specific person, place, thing, concept, action, or quality and therefore is always capitalized.

Proper Nouns	Common Nouns
Jane Jones	person
Chicago	city
General Electric	company
Tuesday	day
Declaration of Independence	document

Common Nouns

Common nouns name general classes or categories of persons, places, things, concepts, actions, or qualities. Common nouns, which are not capitalized unless they begin a sentence, include all types of nouns except proper nouns.

Common Nouns	Proper Nouns
boy	Toby Wilson
city	Chicago
company	General Electric
day	Tuesday
document	Declaration of Independence

Some nouns may be both common and proper.

EXAMPLES   turkey/Turkey, china/China

Abstract nouns are common nouns that refer to something intangible that cannot be discerned by the five senses.

EXAMPLES   love, loyalty, pride, valor, peace, devotion, harmony

Concrete nouns are common nouns used to identify those things that can be discerned by the five senses.

EXAMPLES   house, carrot, ice, tar, straw, grease

A *count noun* is a type of concrete noun that identifies things that can be separated into countable units.

EXAMPLES   desks, chisels, envelopes, engines, pencils
There were four *calculators* in the office.

A *mass noun* is a type of concrete noun that identifies things that comprise a mass, rather than individual units, and cannot be separated into countable units.

EXAMPLES   electricity, water, sand, wood, air, uranium, gold, oil, wheat, cement
The price of *silver* increased this year.

Collective nouns are common nouns that name a group or collection of persons, places, things, concepts, actions, or qualities.

EXAMPLES   army, committee, crowd, team, public, class, jury, humanity

When a collective noun refers to a group as a whole, it takes a singular **verb** and **pronoun.**

EXAMPLE   The staff *was* divided on the issue and could not reach *its* decision until May 15.

When a collective noun refers to individuals within a group, it takes a plural verb and pronoun.

EXAMPLE   The staff *returned* to *their* offices after the conference.

A better way to emphasize the individuals on the staff would be to use the **phrase** *members of the staff.*

> **EXAMPLE**   The members of the staff returned to *their* offices after the conference.

Treat organization names and titles as singular.

> **EXAMPLE**   LRM Company *has* grown 200 percent in the last three years; *it* will move to a new facility in January.

Some collective nouns regularly take singular verbs *(crowd)*; others do not *(people)*.

> **EXAMPLES**   The crowd *was* growing impatient.
> Many people *were* able to watch the first space shuttle land safely.

Some collective nouns have regular plural forms *(team, teams)*; others do not *(sheep)*.

Noun Function

Nouns may function as subjects of verbs, as **objects** of verbs and **prepositions,** as **complements,** or as **appositives.**

> **EXAMPLES**   The *metal* bent as *pressure* was applied to it. (subjects)
> The bricklayer cemented the *blocks* efficiently. (direct object)
> The company awarded our *department* a plaque for safety. (indirect object)
> The event occurred within the *year.* (object of a preposition)
> An equestrian is a *horseman.* (subjective complement)
> We elected the sales manager *chairperson.* (objective complement)
> George Thomas, the *treasurer,* gave his report last. (appositive)

Words usually used as nouns may also be used as **adjectives** and **adverbs.**

> **EXAMPLES**   It is *company* policy. (adjective)
> He went *home.* (adverb)

Using Nouns

Forming the Possessive. Nouns form the possessive most often by adding *'s* to the names of living things and by adding an *of* phrase to the names of inanimate objects.

> **EXAMPLES**   The *president's* statement was forceful.
> The keys *of the calculator* were sticking.

However, either form may be used.

> **EXAMPLES**   The *table's* mahogany finish was scratched.
> Two personal friends *of the chairperson* were on the committee.

Plural nouns ending in *s* need only to add an **apostrophe** to form the **possessive case.**

> **EXAMPLE**   The *architects'* design manual contains many illustrations.

Plural nouns that do not end in *s* require both the apostrophe and the *s.*

> **EXAMPLE**   The installation of the plumbing is finished except in the *men's* room.

With group words and compound nouns, add the *'s* to the last noun.

> **EXAMPLES**   The *chairman of the board's* report was distributed.
> My *son-in-law's* address was on the envelope.

To show individual possession with coordinate nouns, use the possessive with both.

> **EXAMPLES**   Both the *Senate's* and *House's* galleries were packed for the hearings.
> *Mary's* and *John's* presentations were the most effective.

To show joint possession with coordinate nouns, use the possessive with only the last.

> **EXAMPLES**   The *Senate and House's* joint committee worked out a satisfactory compromise.
> *Mary and John's* presentation was the most effective.

Forming the Plural.   Nouns normally form the plural by simply adding *s* to their singular forms.

> **EXAMPLES**   Many new business *ventures* have failed in the past ten years.
> *Partners* are not always personal friends.

Nouns ending in *s, z, x, ch,* and *sh* form the plural by adding *es.*

> **EXAMPLES**   Partners in successful *businesses* are not always friends.
> How many size *sixes* did we produce last month?
> The letter was sent to all area *churches.*
> Technology should not inhibit our individuality; it should fulfill our *wishes.*
> The young engineers were known throughout the company as *whizzes.*
> The instructor gave three *quizzes* during the course.

N

Nouns ending in a consonant plus *y* form the plural by changing *y* to *ies*.

EXAMPLE   The store advertises prompt delivery but places a limit on the number of *deliveries* in one day.

Some nouns ending in *o* add *es* to form the plural, but others add only *s*.

EXAMPLES   One tomato plant produced twelve *tomatoes*.
We installed two *dynamos* in the plant.

Some nouns ending in *f* or *fe* add *s* to form the plural; others change the *f* or *fe* to *ves*.

EXAMPLES   cliff/cliffs, fife/fifes, knife/knives

Some nouns require an internal change to form the plural.

EXAMPLES   woman/women, man/men, mouse/mice, goose/geese

Some nouns do not change in the plural form.

EXAMPLE   *Fish* swam lazily in the clear brook while a few wild *deer* mingled with the *sheep* in a nearby meadow.

Compound nouns form the plural in the main word.

EXAMPLES   sons-in-law, high schools

Compound nouns written as one word add s to the end.

EXAMPLE   Use seven *tablespoonfuls* of freshly ground coffee to make seven cups of coffee.

nowhere near

The **phrase** *nowhere near* is colloquial and should be avoided in writing.

CHANGE   His ability is *nowhere near* Jim's.
TO   His ability does *not approach* Jim's.
OR   His ability is *not comparable* to Jim's.
OR   His ability is *far inferior* to Jim's.

number

Number is the grammatical property of **nouns, pronouns,** and **verbs** that signifies whether one thing (singular) or more than one (plural) is being referred to. Nouns normally form the plural by simply adding *s* or *es* to their singular forms.

EXAMPLES   Many new business *ventures* have failed in the past ten years.
             Partners in successful *businesses* are not always personal friends.

But some nouns require an internal change to form the plural.

EXAMPLES   woman/women, man/men, goose/geese, mouse/mice

All **pronouns** except *you* change internally to form the plural.

EXAMPLES   I/we he, she, it/they

By adding an *s* or *es,* most verbs show the singular of the third **person,** present **tense,** indicative **mood.**

EXAMPLES   he *stands,* she *works,* it *goes*

The verb *be* normally changes form to indicate the plural.

EXAMPLES   I *am* ready to begin work. (singular)
             We *are* ready to begin work. (plural)

(See also **agreement.**)

numbers

General Guidelines

1. Write numbers from zero to ten as words and numbers above ten as figures.
2. Spell out approximate numbers.
3. In most writing, do not spell out ordinal numbers, which express degree or sequence (42nd), unless they are single words (tenth, sixteenth).
4. When several numbers appear in the same sentence or paragraph, express them alike, regardless of other rules and guidelines.

   EXAMPLE   The company owned 150 trucks, employed 271 people, and rented 9 warehouses.

5. Spell out numbers that begin a sentence, even if they would otherwise be written as figures.

   EXAMPLE   One hundred and fifty people attended the meeting.

   If spelling out such a number seems awkward, rewrite the sentence so that the number does not appear at the beginning.

CHANGE   Two hundred and seventy-three defective products were returned last month.
TO   Last month, 273 defective products were returned.

6. Do not follow a word representing a number with a figure in parentheses representing the same number.

CHANGE   Send five (5) copies of the report.
TO   Send five copies of the report.

Plurals

The plural of a written number is formed by adding *s* or *es* or by dropping *y* and adding *ies*, depending on the last letter, just as the plural of any other noun is formed. (See **nouns.**)

EXAMPLES   sixes, elevens, twenties

Use an **apostrophe** with a number or letter only if confusion would result without one.

EXAMPLES   5s, 12s, two 100s, seven *I*'s

Document Parts

In typed manuscript, page numbers are written as figures, but chapter or volume numbers may appear as figures or words.

EXAMPLES   Page 37
Chapter 2 or Chapter Two
Volume 1 or Volume One

Figure and table numbers are expressed as figures.

EXAMPLE   Figure 4 and Table 3

Measurements

Units of measurement are expressed in figures.

EXAMPLES   3 miles, 45 cubic feet, 9 meters, 27 cubic centimeters, 4 picas

When numbers appear run together in the same phrase, write one as a figure and the other as a word.

CHANGE   The order was for 12 6-inch pipes.
TO   The order was for twelve 6-inch pipes.

Percentages are normally given as figures, and the word *percent* is written out, except when the number is in a table.

EXAMPLE   Approximately 85 percent of the area has been sold.

Fractions

Fractions are expressed as figures when written with whole numbers.

EXAMPLES $27^1/_2$ inches, $4^1/_4$ miles

Fractions are spelled out when they are expressed without a whole number.

EXAMPLES   one-fourth, seven-eights

Numbers with decimals are always written as figures.

EXAMPLE   5.21 meters

Time

Hours and minutes are expressed as figures when a.m. and p.m. follow.

EXAMPLES   11:30 a.m., 7:30 p.m.

When not followed by a.m. or p.m., however, time should be spelled out.

EXAMPLES   four o'clock, eleven o'clock

Dates

The year and day of the month should be written as figures. Dates are usually written in a month-day-year sequence, in which the year may or may not be followed by a comma.

EXAMPLES   The August 26, 19— issue of *Computer World* announced the new system.

The August 26, 19—, issue of *Computer World* announced the new system.

In the day-month-year sequence, commas are not used.

EXAMPLE   The 26 August 19— issue of *Computer World* announced the new system.

The slash form of expressing dates (8/24/97) is used in informal writing only.

Addresses

Numbered streets from one to ten should be spelled out except when space is at a premium.

EXAMPLE   East Tenth Street

Building numbers are written as figures. The only exception is the building number *one*.

EXAMPLES   4862 East Monument Street
                 One East Tenth Street

Highways are written as figures.

EXAMPLES   U.S. 70, Ohio 271, I-94

numeral adjectives (see adjectives)

N

objective (see **purpose**)

objective complements (see **complements**)

objects

There are three kinds of objects: direct object, indirect object, and object of a **preposition**. All objects are **nouns** or noun equivalents (**pronoun,** gerund, infinitive, noun **phrase,** noun **clause**) and can be replaced by a pronoun in the objective **case.**

O

Direct Objects

The direct object answers the question "what?" or "whom?" about a **verb** and its subject.

> EXAMPLES   John built a *business.* (noun)
>
> I like *jogging.* (gerund)
>
> I like *to jog.* (infinitive)
>
> I like *it.* (pronoun)
>
> I like *what I saw.* (noun clause)
>
> Sheila designed a new *circuit.* (*Circuit,* the direct object, answers the question, "Sheila designed *what?*")
>
> George telephoned the *chief engineer.* (*Chief engineer,* the direct object, answers the question, "George telephoned *whom?*")

A verb whose meaning is completed by a direct object is called a transitive verb.

392 the indirect object answers the question

Indirect Objects

An indirect object is a noun or noun equivalent that occurs with a direct object after certain kinds of transitive verbs, such as *give, wish, cause, tell,* and their **synonyms** or **antonyms.** The indirect object is usually a person or persons and answers the question "to whom or what?" or "for whom or what?" The indirect object always precedes the direct object.

EXAMPLES   Wish *me* success.

It caused *him* pain.

Their attorney wrote *our firm* a follow-up letter.

Give *the car* a push.

We sent the *general manager* a full report. (*Report* is the direct object. The indirect object, *general manager,* answers the question, "We sent a full report *to whom?*")

The general manager gave the *report* careful consideration. (*Consideration* is the direct object. The indirect object, *report,* answers the question, "The general manager gave careful consideration *to what?*")

The purchasing department bought *Sheila* a new oscilloscope. (*Oscilloscope* is the direct object. The indirect object, *Sheila,* answers the question, "The purchasing department bought a new oscilloscope *for whom?*")

Objects of Prepositions

The object of a preposition, the word or phrase following the preposition, is always in the objective **case.**

observance/observation

An *observance* is the "performance of a duty, custom, or law"; it is sometimes confused with *observation,* which is the "act of noticing or recording something."

EXAMPLES   The *observance* of Veterans Day as a paid holiday varies from one organization to another.

The laboratory technician made careful *observations* during the experiment.

OK/okay

The expression *okay* (also spelled *OK* or *O.K.*) is common in informal writing but should be avoided in more formal **correspondence** and **reports.**

CHANGE   Mr. Sturgess gave his *okay* to the project.
TO   Mr. Sturgess *approved* the project.

CHANGE   The solution is *okay* with me.
TO   The solution is *acceptable* to me.

on account of

The **phrase** *on account of* should be avoided as a substitute for *because.*

CHANGE   He felt that he had lost his job *on account of* the company's switch to automated equipment.
TO   He felt that he had lost his job *because* the company switched to automated equipment.

on/onto

On is normally a **preposition** meaning "supported by," "attached to," or "located at."

EXAMPLE   Install the telephone *on* the wall.

Onto implies movement to a position on or movement up and on.

EXAMPLE   The union members surged *onto* the platform after their leader's defiant speech.

on the grounds that/of

The **phrase** *on the grounds that*—or *on the grounds of*—is a wordy substitute for *because.*

CHANGE   She left *on the grounds that* the NASA position offered a higher salary.
TO   She left *because* the NASA position offered a higher salary.

one

When used as an indefinite **pronoun,** *one* may help you avoid repeating a **noun.**

> EXAMPLE   We need a new plan, not an old *one*.

One is often redundant in **phrases** in which it restates the noun, and it may take the proper **emphasis** away from the **adjective.**

> CHANGE   The computer program was not a unique *one*.
> TO   The computer program was not unique.

One can also be used in place of a noun or personal pronoun in a statement such as the following:

> EXAMPLE   *One* cannot ignore *one's* physical condition.

Using *one* in this way is formal and impersonal; in any but the most formal writing you are better advised to address your **reader** directly and personally as *you.*

> CHANGE   *One* cannot be too careful about planning for leisure time. The cost of *one's* equipment can, for example, force *one* to work more and therefore reduce *one's* leisure time.
> TO   *You* cannot be too careful about planning for leisure time. The cost of *your* equipment can, for example, force *you* to work more and thus reduce *your* leisure time.

(See also **point of view.**)

one of those . . . who

A dependent **clause** beginning with *who* or *that* and preceded by *one of those* takes a plural **verb.**

> EXAMPLES   She is *one of those* executives *who are* concerned about their writing.
>
> This is *one of those* policies *that make* no sense when you examine them closely.

In the preceding two examples, *who* and *that* are subjects of dependent clauses and refer to plural antecedents *(executives* and *policies)* and thus take plural verbs, *are* and *make.*

Because people so often use singular verbs in such constructions in everyday speech and very informal writing, this rule confuses many writers. The principle behind the rule becomes clearer if *among* is substituted for *one of*. Compare the following examples with the preceding ones:

EXAMPLES   She is *among* those executives *who are* concerned about their writing. You would not write, "She is among those executives *who is* concerned . . .")

This is *among* those policies *that make* no sense when you examine them closely. (You would not write "This is among those policies *that makes* no sense . . .")

There is one exception to the plural-verb rule stated at the beginning of this entry: If the phrase *one of those* is preceded by *the only (the only one . . .),* the verb in the following dependent clause should be singular:

EXAMPLES   She is *the only one* of those executives *who is* concerned about her writing. (The verb is singular because its subject, *who*, refers to a singular antecedent, *one*. If the sentence were reversed, it would read, "Of those executives, she is the only *one who is* concerned about her writing.")

This is *the only one* of those policies *that makes* no sense when you examine it closely. (If the sentence were reversed, it would read, "Of those policies, this is the *only one that makes* no sense when you examine it closely.")

only

In writing, the word *only* should be placed immediately before the word or **phrase** it modifies.

CHANGE   We *only* lack financial backing; we have determination.
TO   We lack *only* financial backing; we have determination.

Incorrect placement of *only* can change the meaning of a sentence.

EXAMPLES   *Only* he said that he was tired. (He alone said that he was tired.)

He *only* said that he was tired. (He actually was not tired, although he said he was.)

He said *only* that he was tired. (He said nothing except that he was tired.)

He said that he was *only* tired. (He was nothing except tired.)

openings

If your **reader** is already familiar with your subject or if what you are writing is short, you may not need to begin your writing project with a full **introduction.** You may simply want to focus the reader's attention with a brief *opening*.

For many types of on-the-job writing, openings that simply get to the point are quite adequate, as shown in the following examples:

Correspondence

> Mr. George T. Whittier
> 1720 Old Line Road
> Thomasbury, WV 26401
>
> Dear Mr. Whittier:
>
> You will be happy to know that we have corrected the error in your bank balance. The new balance shows . . .

Progress Report Letter

> William Chang, M.D.
> Phelps Building
> 9003 Shaw Avenue
> Parksville, MD 21221
>
> Dear Dr. Chang:
>
> To date, 18 of the 20 specimens you submitted for analysis have been examined. Our preliminary analysis indicates . . .

Longer Progress Report

O

> PROGRESS REPORT ON REWIRING THE SPORTS ARENA
>
> The rewiring program at the Sports Arena is continuing on schedule. Although the costs of certain equipment are higher than our original bid had indicated, we expect to complete the project without exceeding our budget because the speed with which the project is being completed will save labor costs.
>
> *Work Completed*
>
> As of August 15, we have . . .

Memorandum

> To: Jane T. Meyers, Chief Budget Manager
> From: Charles Benson, Assistant to the Personnel Director
> Date: June 12, 19-
> Subject: Budget Estimates for Fiscal Year 19—
>
> As you requested, I am submitting the personnel budget estimates for fiscal year 19—.

You may also use an intriguing or interesting opening, either by itself or in conjunction with an **introduction,** to stimulate your reader's interest. Such openings have two purposes: to indicate the subject and to catch the interest of the **reader.**

Statement of the Problem

One way to give the reader the perspective of your report is to present a brief account of the problem that led to the study or project being reported.

> EXAMPLE Several weeks ago a brewmaster noticed a discoloration in the grain supplied by Acme Farms, Inc. He immediately reported his discovery to his supervisor. After an intensive investigation, we found that Acme . . .

Definition

Although a definition can be useful as an opening, do not define something with which the reader is familiar or provide a definition that is obviously a contrived opening (such as "Webster defines business as . . ."). A definition should be used as an opening only if it offers insight into what follows.

> EXAMPLE Risk is a loosely defined term. It is used here in the sense of physical risk as a qualitative combination of the probability of an event and the severity of the consequences of that event. Risk assessment is the process of estimating the probabilities and consequences of events and of establishing the accuracy of these estimates. Another necessary term is risk appraisal. This goes far beyond assessment and involves judgment about people's perception of risk and their reactions to this perception; it extends to the final process of making decisions. Risk appraisal is thus an essential part of the work of a licensing and regulatory body in formulating safety policy and applying this policy to individual plants. It is part of the everyday work of inspectors. However, only in recent years have attempts been made to quantify the appraisal aspects of risk.

Interesting Detail

Often an interesting detail of your subject can be used to gain the readers' attention and arouse their interest. This requires, of course, that you be aware of your readers' interests. The following opening, for example, might be especially interesting to members of a personnel department.

> EXAMPLE The small number of graduates applying for jobs at Acme Corporation is disappointing, particularly after many years of steadily increasing numbers of applicants. There are, I believe, several reasons for this development . . .

Sometimes it is possible to open with an interesting statistic.

EXAMPLE   From asbestos sheeting to zinc castings, from a chemical analysis of the water in Lake Maracaibo (to determine its suitability for use in steam injection units) to pistol blanks (for use in testing power charges), the purchasing department attends to the company's material needs. Approximately 15,000 requisitions, each containing from one to fourteen separate items, are processed each year by this department. Every item or service that is bought . . .

Anecdote

An anecdote can also be used to catch your reader's attention and interest.

EXAMPLE   In his poem "The Calf Path," Sam Walter Foss tells of a wandering, wobbly calf trying to find its way home at night through the lonesome woods. It made a crooked path, which was taken up the next day by a lone dog that passed that way. Then "a bellwether sheep pursued the trail over vale and steep, drawing behind him the flock, too, as all good bellwethers do." At last the path became a country road; then a lane that bent and turned and turned again. The lane became a village street, and at last the main highway of a flourishing city. The poet ends by saying, "A hundred thousand men were led by a calf, three centuries dead."

Many companies today follow a "calf path" because they react to events rather than planning . . .

Background

The background or history of a certain subject may be quite interesting and may even put the subject in perspective for your reader. Consider the following example from a newsletter article describing the process of oil drilling:

EXAMPLE   From the bamboo poles the Chinese used when the pyramids were young to today's giant rigs drilling in a hundred feet of water, there has been a lot of progress in the search for oil. But whether four thousands years ago or today, in ancient China or a modern city, in twenty fathoms of water or on top of a mountain, the object of drilling is and has always been the same—to manufacture a hole in the ground, inch by inch. The hole may be either for a development well . . .

This type of opening is easily overdone, however; use it only if the background information is of some value to your reader. Never use it just as a way to get started.

Quotation

Occasionally, you can use a **quotation** to stimulate interest in your subject. To be effective, however, the quotation must be pertinent—not some loosely related remark selected from a book of quotations.

> EXAMPLE   Richard Smith, president of P. R. Smith Corporation, recently said, "I believe that the Photon projector will revolutionize our industry." His statement represents a growing feeling among corporate . . .

Objective

In reporting on a project or activity of some kind, you may wish to open with a statement of the project or activity's objective. Such an opening gives the reader a basis for judging the actual results as they are presented.

> EXAMPLE   The primary objective of the project was to develop new techniques to measure heat transfer in a three-phase system. Our first step was to investigate . . .

Summary

You can provide a summary opening by greatly compressing the results, conclusions, or recommendations of your article or report. Do not start a summary, however, by writing "This report summarizes . . ."

> CHANGE   This report summarizes the advantages offered by the photon as a means of examining the structural features of the atom. The photon is a specially designed laser used for examining . . .
>
> TO   As a means of examining the structure of the atom, the photon offers several advantages. Since the photon is especially designed for examining . . .

Forecast

Sometimes you can use a forecast of a new development or trend to arouse the reader's interest.

> EXAMPLE   In the future, we may be able to have our bodies "beamed" to a distant location, like a radio signal, and reconstructed at the other end—but we'd better pray that we can eliminate any static between here and there!

Scope

At times you may want to present the **scope** of your document in your opening. By providing the parameters of your material—the limitations of the subject or the amount of detail to be presented—you enable your readers to determine whether they want or need to read your document.

> EXAMPLE   This pamphlet provides a review of the requirements for obtaining an FAA pilot's license. It is not designed as a textbook to prepare you to take the examination itself; rather, it gives you an idea of the steps you need to take and the costs involved . . .

oral presentations

Many of the steps required to prepare an effective oral presentation parallel the steps required to write a document. You must focus your presentation by determining its **purpose** and the audience to whom you will present it. You must find and gather the facts that will support your **point of view** and **proposal,** and you must then logically **organize** the information you have collected.

Focusing the Presentation

Focusing the presentation means (1) determining the specific **purpose** of the presentation and (2) determining your audience's needs. When your presentation is focused, the **scope** of coverage (the **topics** you need to include) fulfills your purpose in making the presentation and meets the needs of your audience.

Purpose. Every presentation has a purpose—even if it is only to share information. One of the most frequent presentation problems is that the presenter does not have a specific enough purpose. Following is an example of a general purpose statement that, given a couple of attempts, gets tightened to a good, direction-giving purpose statement.

> EXAMPLES   I would like to discuss productivity with you today. (vague)
>
> I would like to discuss how the productivity in Area B can be improved. (better)
>
> The productivity in Area B can be improved with a few simple, inexpensive ideas that I want to share with you. (specific)

To determine the purpose of your presentation, use one or more of the following questions as a guide.

- What do I want the audience to *know* when I've finished this presentation?

    EXAMPLE   When I've finished with this presentation, I want the audience to know that *Brand X copier is the best copier for us based on our finances, needs, and space.*

- What do I want the audience to *believe* when I've finished this presentation?

    EXAMPLE   When I've finished with this presentation, I want the audience to believe that *I've thoroughly investigated all the applicable copiers, using identical criteria for each. To ensure that the audience believes that, I will show the system I used to evaluate each copier.*

- What do I want the audience to *do* when I've finished with this presentation?

    EXAMPLE   When I've finished with this presentation, I want the audience to *approve the purchase of the Brand X copier. To get their approval, I need to assure them that the copier is within budget and that the money is being spent for a good, long-lasting product.*

You may need to answer all of these questions, or you may answer any one of them. Once you've answered them, write a succinct purpose statement for your presentation, based on the answers. This statement will most likely become a part of your presentation's **introduction.**

    EXAMPLE   The purpose of my presentation is to show the audience the copiers I compared and to make my recommendations so that the audience will approve funds to buy the copier I am recommending.

Audience Analysis.   The second step in focusing a presentation is to determine not only the type of information to give your audience, but also how to talk to your audience. To know both those things, you must analyze your audience by asking the following questions.

- What is my audience's level of knowledge about and experience with my topic before hearing my presentation?

    EXAMPLE   The experience or level of knowledge that my audience currently has about my topic is fairly basic. They know how to use a copier, and they know what they use it for. Beyond that, they have little knowledge. Therefore, I should begin with a basic description of how a copier works and how copiers differ so that they will understand how I made my comparisons.

- What is the general educational level of my audience?

    EXAMPLE   Since the general educational level of my audience is bachelor's degree and up, I don't need to worry about talking over their heads as far as general English vocabulary is concerned.

- What type of information will it take to move my audience to act on what I am presenting?

    EXAMPLE   The type of information I should provide this audience in order to achieve my objective is:

    - the basics of how a copier works;
    - the criteria used to compare copiers;
    - the benefits of Brand X over the other copiers in terms of quality, performance, how it fits our needs, and how it will be performing after 500,000 copies; and
    - how we can justify the extra expense up front—how that extra expense will save money in the long run.

Next, anticipate the questions that may come to the audience's mind during the presentation. Then incorporate the answers to those questions in your presentation.

    EXAMPLE   Some questions the audience might have during the presentation are:

    - Can Brand X handle the volume we do without breaking down during a project?
    - Is a maintenance contract available?
    - What is the response time for maintenance?
    - Will the vendor provide us with a loaner should our copier need to go out for repairs?
    - Does the vendor deliver supplies?
    - How soon will our copier be installed should we approve the purchase?
    - Will our people be trained, and if so what kind of schedule is proposed for training?

Gathering Information

Now you need to find the facts that will support your point of view or proposal. While you are gathering information, keep in mind that you should give the audience only the facts it will need to enable you to accomplish your presentation goals. Too much information tends to overburden the audience, and too little information leaves your audience with a sketchy understanding of your topic. Know exactly what information to include to achieve your purpose and meet the needs of your audience.

Ask yourself the following question to help guide you in gathering information: "Based on what I know about the members of my audience, what do I need to include to meet their needs?"

EXAMPLE   Based on what I know about the audience, I'll need to:

- explain how copiers work;
- provide an analysis of our needs in a copier;
- explain which copiers I chose to compare and why;
- go over the maintenance contract, comparing all brands (including cost per page for maintenance and supplies);
- show the long-term performance histories of all three copiers;
- show the quality of copies after 120,000 copies; and
- show the copier I recommend and why I am recommending it.

Organizing the Presentation

Next, **organize** your data in a way that is logical and easy to understand. Organizing the presentation includes deciding (1) how to develop the topic and (2) how to structure the presentation.

Developing the Topic.   You first need to determine the best method for unfolding (or developing) your topic for your audience. Choosing a method of development means determining the best way to sequence your information to make it easiest to understand or most persuasive. An appropriate **method of development** will move the topic smoothly and logically from your opening to your closing. Make certain that the method of development you select is appropriate to both your audience and your objective and that it is based on your audience's needs.

Structuring the Presentation.   A presentation has three distinct parts: an introduction, a body, and **closing.** It is often best to write the introduction last because you'll be able to write a better introduction after you have written the body of the document.

BODY

How to put your presentation together is going to be evident now because you know what your purpose is, you know what your audience needs, you know how to talk to your audience, you have collected the appropriate data, and you have decided how to develop your topic.

Now it is only a matter of listing your major points, in the most appropriate order. Continuing with the copier example, you could start with the following major divisions:

Introduction
How a Copier Works
Construction of Compared Copiers
Quality of Copies by Compared Copiers
Maintenance Contracts of Compared Copiers
Cost of Compared Copiers
Proposal
Closing

The next step is to add subdivisions under your major divisions and incorporate the data you have collected under the appropriate major and minor divisions.

I. Introduction
II. How a Copier Works
   A. Overhead with Drawing
III. Construction of Compared Copiers
   A. Brand X
   B. Brand Y
   C. Brand Z
IV. Quality of Copies by Compared Copiers
   A. Brand X
   B. Brand Y
   C. Brand Z
V. Maintenance Contracts of Compared Copiers
   A. Brand X
   B. Brand Y
   C. Brand Z
VI. Cost of Compared Copiers
   A. Brand X
   B. Brand Y
   C. Brand Z
VII. Proposal
   A. A Comparison Chart of All Three Copiers
VIII. Closing

INTRODUCTION

The **introduction** may include an **opening**—a catchy beginning that is designed to focus the audience's attention. In the copier presentation, you could use any of the following types of openings.

- *An attention-getting statement,* such as, "The Brand X copier exceeds the ability of Brands Y and Z in several areas by a wide margin."

- *A rhetorical question,* such as, "Are all copiers alike?"
- *A dramatic or entertaining story,* such as, "When I first began researching these three brands of copiers, I thought there really couldn't be much difference. Boy, was I wrong!"
- *A personal experience,* such as, "I first saw the Brand X copier at our branch office in Memphis. After using it a few times, I feel sure this copier is just what we need."
- *An appropriate quotation,* such as, "The bitterness of poor quality lingers long after the sweetness of low price is forgotten."
- *A historical event,* such as, "When the first copy machine came out, there weren't many options. There have been a log of changes since then, some better than others."
- *A reference to a current news story.*
- *A joke or humorous story* that ties directly to your topic. Just be sure your humor is related and is not off-color.

Following your opening, if you have one, create a formal introduction that sets the stage for our audience. The following questions will help you create your introduction.

- *What is the purpose of your presentation?* Not every presentation needs to announce its purpose in the introduction, but many do. If there is a good strategic reason to wait until the closing to announce your purpose, by all means do so.
- *What general information will your audience need* in order to understand the more detailed information in the body of your presentation?
- *What is your method of development?* It is sometimes beneficial to tell your audience how you are going to tell them what you have to say.

> EXAMPLE   The purpose of my presentation is to show you the copiers I compared and to recommend the one I believe will best meet our needs. To do that, I'm going to show you:
>
> - how a copier works;
> - the criteria I used to compare copiers;
> - the brands of copiers I compared, and why; and
> - which copier I propose we buy.

CLOSING

The closing should be designed to achieve the goals of your presentation. If your purpose is to motivate the audience to take action, ask them to do what you want them to do; if your purpose is to get your audience to think about something, summarize what you want them to think about. The mistake presenters make most frequently with closings is that they fail to close at all—they simply quit talking, shuffle papers, and ask, "Are there any questions?"

Since your closing is what your audience is most likely to remember, it's the time to be strong and persuasive. Let's return to our copier example and take a look at one possible closing.

> EXAMPLE Based on all the data, I have concluded that Brand X is the best copier for our needs. The fact that it is $600 higher in price than the other brands I looked at is unimportant compared to the value we would be getting. If you could allocate the money to buy this copier by the fifteenth of this month, we can not only be up and running by the first of next month, but we'll be well equipped to prepare for next quarter's customer presentations.

This closing brings the presentation full cycle and asks the audience to fulfill the purpose of the presentation—exactly what a closing should do.

Transitions. Planned **transition** should appear between the body and the closing. Transitions let the audience know you're moving from one topic to the next, and provides you, the speaker, with insurance that you know where you're going and how to get there.

Most delivery problems in a presentation happen at points of transition. When you are on a point that you know, it is easy to speak. The problems usually occur when you have to move from one point to the next. Most of a presenter's fidgeting and audible pauses ("uh," "okay?" and "you know") happen when the presenter can't figure out how to get off one point and on to the next. Transition neatly solves that problem.

> EXAMPLE Before getting into the specifics of each of the copiers I compared, I'd like to show how copiers in general work. That knowledge will provide you with the background you'll need to compare the different brands of copiers.

It is also a good idea to pause for a moment after you have delivered a transitional line between topics to let the audience shift gears with you. Remember, they don't know your plan.

Visual Aids. Well-planned visual aids can not only add interest and emphasis to your presentation, but they can also clarify and simplify your message because they communicate clearly, quickly, and vividly. Use visual aids, however, *only* if they clarify or emphasize a point. Don't try to use visual aids to flesh out a skimpy or weak presentation, and don't use them to deflect the attention of your audience away from you. Above all, don't use visual aids, especially overheads, simply because "everyone else does." Visual aids that don't aid communication just get in the way of communication.

It is a good idea to start planning your visual aids when you begin to gather information. For example, the bulleted **list** of the kinds of information needed for the sample presentation described under "Gathering Information" in this entry could have been altered as follows to include the kind of visuals that would most likely be needed.

EXAMPLE   Based on what I know about the audience, I'll need to:

- explain how copiers work—*good place for a drawing that shows the process as simply as possible*
- provide an analysis of our needs in a copier—*perhaps a bulleted list on an overhead transparency*
- explain which copiers I chose to compare and why—*perhaps a chart that lists the necessary criteria*
- go over the maintenance contracts, comparing all brands (including cost per page for maintenance and supplies)—*overhead transparency*
- show the long-term performance histories of all three copiers
- show the quality of copies after 120,000 copies
- state the copier I recommend and why I am recommending it—*perhaps on an overhead that lists the advantages of Brand X over the others*

Keep your visual aids simple and uncluttered. Keep the amount of information on each visual aid to a minimum (usually one idea per visual). When the visual contains text, space the lines of type so that the audience can read them easily and make the letters large enough to be read easily.

As you prepare your presentation, decide which visuals you will use and where each one will go. If you have a large number of them, use slides rather than overheads to avoid the distraction of constantly changing overheads. Have visuals made up far enough in advance so you can rehearse with them.

Preparing to Deliver a Presentation

When you have taken all the steps outlined so far, you are ready to think about delivery techniques. It is *only* after you have the presentation carefully created that you can begin practicing your delivery.

Practice. Begin by familiarizing yourself with the sequence of the material in your **outline.** Only then are you ready to practice your presentation. Remember the following points about practicing.

Practice on your feet and out loud. The reason for practicing out loud rather than just mentally rehearsing is that you can process the information in your mind many times faster than you can possibly speak it. Mental rehearsing will not tell you how long your presentation will take or if there are any problems, such as awkward transitions.

Videotape your practice session if possible. Seeing yourself will show you not only what you are doing wrong but also what you are doing right. If you can't videotape yourself, at least use an audiotape recorder to evaluate your vocal presentation.

Force yourself to exaggerate gesturing, physical movement, and vocal inflection in your practice sessions. Continue practicing those things until you don't feel awkward with any of them.

Practice with your visual aids. The more you actually handle your visuals and practice using them, the more smoothly you'll use them during the presentation. And even then things may still go wrong, but being prepared and practiced will give you the confidence and poise to go on.

Try to get a practice session in the room where you'll be giving the presentation. This will let you learn the idiosyncrasies of the room: acoustics, lighting, how the chairs will most likely be arranged, where electrical outlets and switches are located, and so forth.

Delivery Techniques. Delivery is both audible and visual. Your audience is affected by all of you, not just your words or message. Therefore, you need to breathe life into your presentation. To make an impact on your audience members and keep them mentally with you, you must be animated. Words will have more punch when they are delivered with physical and vocal animation. The audience must believe in your enthusiasm for your topic if you want them to share your point of view.

When using visuals, give your audience time to absorb the information before you comment on it. Pause—don't talk—while you are changing visuals or doing something physical.

During the practice sessions, point to the visual aid—with both gestures and words. Be sure you are talking to your audience, however, rather than to the visual aid, and be sure you do not block the visual aid with your body.

If possible, don't distribute handouts early in the presentation. Hold them until you are ready for your audience to look at them.

There are a number of techniques that you should use to give your presentation the animation it needs to make you look enthusiastic: eye contact, movement, gestures, and pace.

EYE CONTACT
Eye contact is the best way to establish rapport with your audience. The smaller the audience, the more important it is that you make eye

contact with as many people as possible. In a large audience, find people in different parts of the audience who are responding to you and talk directly to those people, one at a time. Look at one person and talk to that person for several seconds. Then go to the next person and do the same thing, and so on.

MOVEMENT

To get maximum interest from your audience, use physical movement to add animation to your presentation. Movement can mean taking a step or two to one side or the other after you have been talking a minute or so. The strategic points at which movement is most likely to be effective are between points, before transitionary words or **phrases,** after pauses, or after a point of **emphasis.** Be careful, however, to stay in one place for a minute or so—do not pace.

One way to get movement into your presentation is to walk to the screen and point. Touch the screen with the pointer, to hold it in place, and then turn back to the audience before beginning to speak (touch, turn, and talk). Or you can leave the screen and go to the overhead projector, pointing by placing the pointer on the overhead (the pointer will cast a shadow pointing to the appropriate item on the overhead).

GESTURES

Gestures not only help provide animation to your presentation, but they can help you communicate your message. Most people gesture naturally when they talk; however, nervousness can inhibit you from gesturing when you make a presentation. If you discipline yourself to always keep one hand free and above your waist during your presentation, you will almost certainly use that hand to gesture. Leave your other hand by your side or in your pocket. Be careful not to put both hands above your waist and then lock them together; there is nothing wrong with bringing them together for effect, but don't lock them up and keep them together because that inhibits gesturing. If you make yourself gesture during practice, you will automatically gesture at the appropriate time and in the appropriate way when you make your presentation.

VOICE

The effective use of your voice is another way to get animation into your presentation—and a very important one. Since much of your believability is projected through your vocal tone, your ability to use your voice effectively is of enormous importance. Hypnotists use a monotone speech pattern to hypnotize people, and as a speaker you can have the same effect on your audience if you speak in a monotone. It also allows you to emphasize and de-emphasize different points.

Your vocal delivery should sound conversational rather than contrived and formal. By using simple, everyday words, you will draw your audience into your subject more quickly and easily. When your delivery is conversational, each member of the audience feels as though he or she is the only one you are speaking to. Using the personal pronoun "you," speaking as though you are talking to only one person, can make this easier to do.

PROJECTION

Most speakers think they are projecting more loudly than they are. Just remember that if anyone in the audience cannot hear you, your presentation has been ineffective for that person—and that if the audience has to strain to hear you, they will usually give up.

INFLECTION

Vocal inflection is the rise and fall of your voice at different times, such as the way your voice naturally rises at the end of a question. It is like using a musical scale in your speaking. Vocal inflection helps you communicate your meaning more effectively.

PACE

Pace is the speed at which you deliver your presentation. If the pace is too fast, your words will run together, making it difficult for your audience to follow you. If your pace is too slow, the audience will get impatient and their minds are likely to wander to another topic.

Managing Nervousness

It is natural to be apprehensive, and that feeling will probably never go away. Stage fright shouldn't worry you. Perhaps the lack of it should, because nerves can be a helpful stimulant. But nervousness and fear are not the same thing. Fear is the result of a lack of confidence, while nervousness is the energy you feel when you are as well prepared as possible. If you are prepared, you can expect to be nervous; however, if you know what you are going to say and how you are going to say it, you will probably relax as you start to speak. You can use the power of positive imaging (imagining yourself making the presentation perfectly) to overcome your nervousness. You can also use the "as if" technique: act "as if" you are completely in control, or act "as if" you can't wait to get started with the presentation. You can also give yourself a pep talk to control nervousness: "My subject is important. I am ready. My listeners are here to listen to what I have to say."

oral/verbal

Oral refers to what is spoken.

> EXAMPLE   He made an *oral* commitment to the policy.

Although it is sometimes used synonymously with *oral*, *verbal* literally means "in words" and can refer to what is spoken or written. To avoid possible confusion, do not use *verbal* if you can use *written* or *oral*.

> CHANGE   He made a *verbal* agreement to complete the work.
> TO   He made a *written* agreement to complete the work.
> OR   He made an *oral* agreement to complete the work.

> CHANGE   Avoid *verbal* orders.
> TO   Avoid *oral* orders.

When you must refer to something both written and spoken, you should use both *written* and *oral* to make your meaning clear.

> EXAMPLE   He demanded either a *written* or an *oral* agreement before he would continue the project.

organization

Organization is achieved, first, by developing your topic in a way that will enable your **reader** to understand your message and then by **outlining** your material on the basis of that **method of development.**

A logical method of development satisfies the reader's need for a controlling shape and structure for your subject. For example, if you were writing a proposal to your manager, trying to get him or her to fund a project you wanted to undertake, you would probably use an **order-of-importance method of development** topic. You would start with the most important reason that the project should be funded and end with the least important reason. On the other hand, if you were a Federal Aviation Administration agent reporting on an airplane crash, you would probably use a **cause-and-effect method of development** starting with the cause of the crash and leading up to the crash (or starting with the crash and tracing back to the cause).

An appropriate method of development is the writer's tool for keeping things under control and the reader's means of following the writer's development of a theme. Many different methods of development are available to the writer. This book includes those that are likely to be used by technical people: **sequential, chronological, increasing-order-of-importance, decreasing-order-of-importance, division-and-classification, comparison, spatial, specific-to-general,**

general-to-specific, and cause-and-effect. As the writer, you must choose the method of development that best suits your subject, your reader, and your **purpose.**

Outlining provides structure to your writing by ensuring that it has a beginning, a middle, and an end. It gives proportion to your writing by making sure that one step flows smoothly to the next without omitting anything important, and it enables you to emphasize your key points by placing them in the positions of greatest importance. Using an outline makes larger and more difficult subjects easier to handle by breaking them into manageable parts. Finally, by forcing you to organize your subject and structure your thinking in the outline stage, creating a good outline allows you to concentrate exclusively on writing when you begin the rough draft.

organizational charts

An organizational chart shows how the various components of an organization are related to one another. It is useful when you want to give your **readers** an overview of an organization or to show them the lines of authority within it.

The **title** of each organizational component (office, section, division) is placed in a separate box. These boxes are then linked to a central authority. (See Figure 1.) If your readers need the information, include the name of the person occupying the position identified in each box. (See Figure 2.)

O

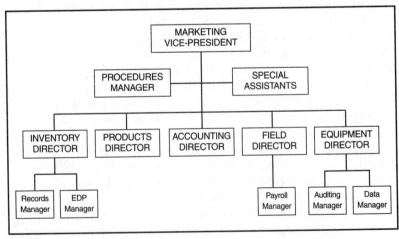

FIGURE 1 Organizational Chart Showing Positions

```
                ┌─────────────────────────┐
                │     FIELD DIRECTOR       │
                │      J. Ann Carlson      │
                └─────────────────────────┘
```

FIGURE 2   Organizational Box with Name of Person and Position

As with all **illustrations,** place the organizational chart as close as possible to the text that refers to it.

orient/orientate

Orientate is merely a long variant of *orient,* meaning "locate in relation to something else." The shorter form should be used because it is simpler.

> CHANGE   Let me *orientate* your group on our operation.
> TO   Let me *orient* your group on our operation.

outlining

Organizing your information before you write has two important advantages. First, it forces you to reexamine the information you plan to include to be sure that you have sufficient facts and details to satisfy your **reader's** needs and achieve the **purpose** of your writing. Second, it forces you to order the information logically, so that your reader understands it as clearly as you do.

Not all writing benefits from a full-scale outline, of course. For relatively short items, such as memos and letters, you may need only to jot down a few notes to make sure that you haven't left out any important information and that you have arranged the information in a logical order. These notes then guide you as you write the draft.

Longer documents generally require more elaborate planning, such as a formal outline. Any outline is tentative and represents your best thinking at that point. The following information introduces you to the conventions used to create simple and complex outlines and provides techniques for verifying that your outline is soundly constructed.

An outline consists of **phrases** arranged according to the logical development of your subject. To create an outline, begin by dividing your **topic** into its major sections. Then identify all items by a number or a letter, according to the following system.

    I. Major section
      A. First-level subsection
         1. Second-level subsection
           a. Third-level subsection
              1) Fourth-level subsection

Outlining permits you to recognize at a glance the relative importance of divisions within your subject. Your subject will seldom require this many divisions, but the system allows for a highly detailed outline if one is necessary. **Word-processing** programs have simplified the mechanics of creating outlines and you may want to take advantage of those features.

When you are ready to write, you should know your topic well enough to be able to identify its major sections. Begin your outline by writing them down. Then consider them carefully to make sure they divide the topic logically. For example, assume that you are writing an article about computers for a company magazine. You might use the following major sections.

    I. Development of computers
   II. Present applications
  III. Future benefits

Once you have established your major sections, look for minor divisions within each section.

    I. Development of computers
      A. History of computing aids
      B. Development of the computer to a practical size
      C. New industry created around the more compact computer
   II. Present applications
      A. Business applications of the computer
      B. Social, scientific, and technical applications of the computer
  III. Future benefits
      A. Potential impact of the computer on business and industry
      B. Potential social impact of the computer

Now you are ready to insert any notes that you may have compiled under the appropriate major and minor divisions.

    I. Development of computers
      A. History of computing aids
         1. Abacus as first computing tool
         2. First electronic computer patented in 1944

B. Development of the computer to a practical size
  1. Early computers very large, impractical for commercial use
  2. First electronic computer weighed 30 tons, took 1,500 square feet of floor space
  3. Introduction of transistor in 1958 made computer commercially practical
C. New industry created around the more compact computer
  1. Transition from research phase to commercial application phase
  2. Corporate applications
II. Present applications
  A. Business applications of the computer
    1. Route long-distance telephone calls
    2. Monitor airline reservations
    3. Keep records and aid project management
    4. Communicate mail electronically
    5. Set type
    6. Word processing
  B. Social, scientific, and technical applications of the computer
    1. Prepare weather forecasts
    2. Direct city traffic
    3. Maintain data banks on crime and criminals
    4. Monitor the condition of patients in hospitals
    5. Compare chemical characteristics of drugs
    6. Navigate ships and planes
    7. Monitor the performance of automobile engines
III. Future benefits
  A. Potential impact of the computer on business and industry
    1. More efficient day-to-day operations
    2. Greater productivity
    3. More leisure time for employees, enhancing the recreation industry
  B. Potential social impact of the computer
    1. Make education and government more effective
    2. Help find cures for diseases, translate languages, land jetliners without human aid

Treat **illustrations** as an integral part of your outline, noting approximately where each should appear throughout the outline. At each place, either make a rough pencil sketch of the visual or write "illustration of . . ." and enclose each suggestion in a box. Like other information in an outline, these boxes and sketches can be moved, amended, or deleted, as required. Planning your graphics requirements from the beginning stages of your outline ensures their harmonious integration throughout all versions of the draft to the finished document.

When you have finished, you have a complete outline. Even though the outline looks final at this point, you still must check for a number of things. For example, make sure that corresponding divisions are equally important (that major divisions are equal to one another and minor divisions are equal to one another).

NOT   II. Present applications
        A. Economic applications of the computer
        B. Social, scientific, and technical applications of the computer
        C. Future benefits

BUT   II. Present applications
        A. Economic applications of the computer
        B. Social, scientific, and technical applications of the computer
     III. Future benefits

Make sure that all division heads at any one level are in parallel form (see **parallel structure**). If you begin one with an **adjective** phrase, for example, make sure that all heads at that level begin with adjective phrases.

NOT   A. Potential economic impact of the computer
PARALLEL   B. The computer's potential social impact

PARALLEL   A. Potential economic impact of the computer
        B. Potential social impact of the computer

Finally, scan your outline for completeness, determining whether you need additional information. If you find that your **research** is not really complete, return to your sources and dig out the missing data.

The outline samples shown to this point use a combination of numbers and letters to differentiate the various levels of information. But you could also use a decimal numbering system, such as the following, for your outline.

    1   FIRST-LEVEL SECTION
    1.1   Second-level section
    1.2   Second-level section
    1.2.1   Third-level section
    1.2.2   Third-level section
    1.2.2.1 Fourth-level section
    1.2.2.2 Fourth-level section

1.3   Second-level section
2   FIRST-LEVEL SECTION

This system should not go beyond the fourth level, because the numbers get too cumbersome beyond that point. In many documents, the decimal numbering system is carried over from the outline to the final version of the document for ease of cross-referencing sections.

Remember that the outline is only a means to an end, not an end in itself. Don't view it as being cast in concrete. Outlines are preliminary by their nature. If you suddenly see a better way to organize your material while you are writing the draft, depart from your outline and follow the better approach. The main purpose of the outline is to bring order and shape to your writing *before* you begin to write the draft.

outside of

In the **phrase** *outside of*, the word *of* is redundant.

     CHANGE   Place the rack *outside of* the incubator.
        TO   Place the rack *outside* the incubator.

In addition, do not use *outside of* to mean "aside from" or "except for."

     CHANGE   *Outside of* his frequent absences, Jim has a good work record.
        TO   *Except for* his frequent absences, Jim has a good work record.

O

over with

In the expression *over with*, the word *with* is redundant; moreover, the word *completed* often better expresses the thought.

     CHANGE   You may enter the test chamber now that the experiment is *over with*.
        TO   You may enter the test chamber now that the experiment is *over*.
        OR   You may enter the test chamber now that the experiment has been *completed*.

P

pace

Pace is the speed at which the writer presents ideas to the **reader.** Your goal should be to achieve a pace that fits both your reader and your subject; at times you may need a fast pace, at other times a slow pace. The more knowledgeable the reader is, the faster your pace can be—but be careful not to lose control of it. In the first passage of the following example, facts are piled on top of each other at a rapid pace. In the second passage, the same facts are spread out over two sentences and presented in a more easily assimilated manner, even though the length is no greater; in addition, a different and more desirable **emphasis** is achieved.

CHANGE The generator is powered by a 90-horsepower engine, is designed to operate under normal conditions of temperature and humidity, produces 110 volts at 60 Hertz, is designed for use under emergency conditions, and may be phased with other units of the same type to produce additional power when needed.

TO The generator, which is powered by a 90-horsepower engine, produces 110 volts at 60 Hertz under normal conditions of temperature and humidity. Designed especially for use under emergency conditions, this generator may be phased with other units of the same type to produce additional power when needed.

paragraphs

A *paragraph* is a group of **sentences** that support and develop a single idea; it may be thought of as an essay in miniature, for its function is to expand the core idea stated in its topic sentence. A paragraph may use a particular **method of development** to expand this idea. The following paragraph uses the **general-to-specific method of development.** Its topic sentence is italicized.

> *The cost of training new employees is high.* In addition to the cost of classroom facilities and instructors, an organization must pay employees a salary to sit in the classroom while they are learning. We have determined that for the company to break even on professional employees, they must stay on the job for which they have been trained for at least one year.

The paragraph performs three functions: (1) it develops the unit of thought stated in the topic sentence; (2) it provides a logical break in the material; and (3) it creates a physical break on the page, which in turn provides visual assistance to the **reader.**

Topic Sentence

A topic sentence states the controlling idea of a paragraph; the rest of the paragraph supports and develops that statement with carefully related details. The topic sentence may appear anywhere in the paragraph, which permits the writer to achieve **emphasis** and variety in **style.** A topic statement may also be more than one sentence if necessary.

The topic sentence is often the first sentence because it states the subject the paragraph is to develop. The topic sentence is effective in this position because the reader knows immediately what the paragraph is about.

P

> EXAMPLE *The arithmetic of searching for oil is stark.* For all his scientific methods of detection, the only way the oil driller can actually know for sure that there is oil in the ground is to drill a well. The average cost of drilling an oil well is over $300,000, and drilling a single well may cost over $8,000,000! And once the well is drilled, the odds against its containing any oil at all are 8 to 1! Even after a field has been discovered, one out of every four holes drilled in developing the field is a dry hole because of the uncertainty of defining the limits of the producing formation. The oil driller can never know what Mark Twain once called "the calm confidence of a Christian with four aces in his hand."

On rare occasions, the topic sentence logically falls in the middle of a paragraph, as the topic sentence in italics in the following paragraph demonstrates.

EXAMPLE It is perhaps natural that psychologists should awaken only slowly to the possibility that behavioral processes may be directly observed, or that they should only gradually put the older statistical and theoretical techniques in their proper perspective. But it is time to insist that science does not progress by carefully designed steps called "experiments," each of which has a well-defined beginning and end. *Science is a continuous and often a disorderly and accidental process.* We shall not do the young psychologist any favor if we agree to reconstruct our practices to fit the pattern demanded by current scientific methodology. What the statistician means by the design of experiments is design that yields the kind of data to which his techniques are applicable. He does not mean the behavior of the scientist in his laboratory devising research for his own immediate and possibly inscrutable purposes.

Although the topic sentence is usually most effective early in the paragraph, a paragraph can lead up to the topic sentence; this is sometimes done to achieve emphasis. When a topic sentence concludes a paragraph, it can also serve as a summary or **conclusion,** based on the details that were designed to lead up to it.

EXAMPLE Energy does far more than simply make our daily lives more comfortable and convenient. Suppose you wanted to stop— and reverse—the economic progress of this nation. What would be the surest and quickest way to do it? Find a way to cut off the nation's oil resources! Industrial plants would shut down; public utilities would stand idle; all forms of transportation would halt. The economy would plummet into the abyss of national economic ruin. *Our economy, in short, is energy-based.*

Because multiple paragraphs are sometimes used to develop different aspects of an idea, not all paragraphs have topic sentences. In this situation, **transition** between paragraphs helps the reader know that the same idea is being developed through several paragraphs.

Topic sentence for all three paragraphs
To conserve valuable memory space, a large portion of the software package remains on disc; only the most frequently used portion resides in internal memory all of the time. The disc-resident software is organized into small modules that are called into memory as needed to perform specific functions.

Transition
The memory-resident portion of the operating system maintains strict control of processing. It consists of routines, subroutines, lists, and tables that are used to perform common program

functions, such as processing input/output operations, calling other software routines from disc as needed, and processing errors.

Transition

The disc-resident portion of the operating system contains routines that are used less frequently in system operation, such as the peripheral-related software routines that are useful for correcting errors encountered on the various units, and the log and display routines that record unusual operating conditions in the system log. The disc-resident portion of the operating system also contains Monitor, the software program that supervises the loading of utility routines and the user's programs.

In this example, the reason for breaking the development of the idea expressed in the topic sentence into three paragraphs is to help the reader assimilate the fact that the main idea has two separate parts.

Paragraph Length

Paragraph length should be tailored to aid the reader's understanding of ideas. A series of short, undeveloped paragraphs can indicate poor **organization** of material, in which case you should look for a larger idea to which the ideas in the short paragraphs relate and then make the larger idea the topic sentence for a single paragraph. A series of short paragraphs can also sacrifice unity by breaking a single idea into several pieces. A series of long paragraphs, on the other hand, can fail to provide the reader with manageable subdivisions of thought. A good rule of thumb is that a paragraph should be just long enough to deal adequately with the subject of its topic sentence. A new paragraph should begin whenever the subject changes significantly. Occasionally, a one-sentence paragraph is acceptable if it is used as a **transition** between larger paragraphs or in letters and **memorandums** in which one-sentence openings and closings are sometimes appropriate.

Writing Paragraphs

Careful paragraphing reflects the writer's accurate thinking and logical organization. Clear and orderly paragraphs help the reader follow the writer's thoughts more easily.

Outlining is the best guide to paragraphing. It is easy to group ideas into appropriate paragraphs when you follow a good working outline. Notice how the following outline plots the course of subsequent paragraphs:

Outline

    I. Advantages of Chicago as location for new plant
      A. Transport facilities
        1. Rail
        2. Air
        3. Truck
        4. Sea (except in winter)
      B. Labor supply
        1. Engineering and scientific personnel
          a. Many similar companies in the area
          b. Several major universities
        2. Technical and manufacturing personnel
          a. Existing programs in community colleges
          b. Possible special programs designed for us

Resulting Paragraphs

Probably the greatest advantage of Chicago as a location for our new plant is its excellent transport facilities. The city is served by three major railroads. Both domestic and international air cargo service is available at O'Hare International Airport. Chicago is a major hub of the trucking industry, and most of the nation's large freight carriers have terminals there. Finally, except in the winter months, when the Great Lakes are frozen, Chicago is a seaport, accessible through the St. Lawrence Seaway.

A second advantage of Chicago is that it offers an abundant labor force. An ample supply of engineering and scientific personnel is assured not only by the presence of many companies engaged in activities similar to ours but also by the presence of several major universities in the metropolitan area. Similarly, technicians and manufacturing personnel are in abundant supply. The seven colleges in the Chicago City College system, as well as half a dozen other two-year colleges in the outlying areas, produce graduates with associate degrees in a wide variety of technical specialties appropriate to our needs. Moreover, three of the outlying colleges have expressed an interest in establishing special courses attuned specifically to our requirements.

Consider not only the nature of the material you are developing but also the appearance of your page. An unbroken page looks forbidding.

Paragraph Coherence and Unity

A good paragraph has unity, **coherence,** and adequate development. Unity is singleness of purpose, based on a topic sentence that states the core idea of the paragraph. When every sentence in the paragraph contributes to developing the core idea, the paragraph has unity. Coherence is holding to one **point of view,** one attitude, one **tense;** it is the joining of sentences into a logical pattern. Coherence is advanced by the careful choice of transitional words so that ideas are tied together as they are developed.

Topic sentence	*Any company which operates internationally today faces a host of difficulties.* Inflation is worldwide. Most countries are strug-
Transition	gling with other economic problems *as well. In addition,* there are many monetary uncertainties and growing economic na-
Transition	tionalism directed against multinational companies. *Yet* there is ample business available in most developed countries if you have the right products, services, and marketing organization. To maintain the growth we have achieved overseas, we recently restructured our international operations into four major
Transition	trading areas. *This step* will improve the services and support that the Corporation can provide to its subsidiaries around the world. *At the same time* it establishes firm management
Transition	control, ensuring consistent policies around the world. *So* you might say the problems of doing business abroad will be more difficult this year but we are better organized to meet those problems.

Simple enumeration (*first, second, then, next,* and so on) can also provide effective transition within paragraphs. Notice how the italicized words and **phrases** give coherence to the following paragraph:

Most adjustable office chairs have nylon hub tubes that hold metal spindle rods. To ensure trouble-free operation, lubricate these spindle rods occasionally. *First,* loosen the set screw in the adjustable bell. *Then* lift the chair from the base so that the entire spindle rod is accessible. *Next,* apply the lubricant to the spindle rod and the nylon washer, using the lubricant sparingly to prevent dripping. *When you have finished,* replace the chair and tighten the set screw.

Sometimes a paragraph is used solely for transition, as in the following example:

. . . that marred the progress of the company.

Two other setbacks to the company's fortunes that year contributed to its present shaky condition: the loss of many skilled workers through the early retirement program and the intensification of the devastating rate of inflation.

The early retirement program . . .

parallel structure

Parallel structure requires that sentence elements that are alike in function be alike in construction as well, as in the following example (in which similar actions are stated in similar **phrases**):

> **EXAMPLE**   We need a supplementary work force to *handle* peak-hour activity, *free* full-time employees from routine duties, *relieve* operators during lunch breaks, and *replace* vacationing employees.

Parallel structure achieves an economy of words, clarifies meaning, expresses the equality of its ideas, and pleases the **reader** aesthetically. This technique assists readers because they are able to anticipate the meaning of a sentence element on the basis of its parallel construction. When they recognize the similarity of word order (or construction), readers know that the relationship between the new sentence element and the subject is the same as the relationship between the last sentence element and the subject. Because of this they can go from one idea to another more quickly and confidently.

Parallel structure can be achieved with words, phrases, or clauses.

> **EXAMPLES**   The computer instruction contains *fetch, initiate,* and *execute* stages. (parallel words)
>
> The computer instruction contains *a fetch stage, an initiate stage,* and *an execute stage.* (parallel phrases)
>
> *The computer instruction contains a fetch stage, it contains an initiate stage,* and *it contains an execute stage.* (parallel clauses)

Parallelism is most frequently accomplished by the use of phrases.

> **EXAMPLES**   I was convinced of their competence *by their conduct, by their reputation,* and *by their survival* in a competitive business. (prepositional phrases)
>
> *Filling the gas tank, testing the windshield wipers,* and *checking tire pressure* are essential to preparing for a long trip. (gerund phrases)
>
> From childhood the artist had made it a habit *to observe people, to store up the impressions they made upon him,* and *to draw conclusions about human beings from them.* (infinitive phrases)

Correlative **conjunctions** (*either . . . or, neither . . . nor, not only . . . but also*) should always be followed by parallel structure. Both members of these pairs should be followed immediately by the same grammatical form: two words, two similar phrases, or two similar **clauses.**

EXAMPLES Viruses carry either *DNA* or *RNA,* never both. (words)

Clearly, neither *serologic tests* nor *virus isolation studies* alone would have been adequate. (phrases)

Either *we must increase our operational efficiency* or *we must decrease our production goals.* (clauses)

To make a parallel construction clear and effective, it is often best to repeat a **preposition,** an article, a **pronoun,** a subordinating **conjunction,** a helping **verb,** or the mark of an infinitive.

EXAMPLES The Babylonians had *a* rudimentary geometry and *a* rudimentary astronomy. (article)

My father and *my* teacher agreed that I was not really trying. (pronoun)

To run and be elected is better than *to* run and be defeated. (mark of the infinitive)

The driver *must* be careful to check the gauge and *must* move quickly when the light comes on. (helping verb)

New teams were being established *in* New York and *in* Hawaii. (preposition)

The history of factories shows both *the* benefits and *the* limits of standardization. (article)

Parallel structure is especially important in creating your **outline,** your **table of contents,** and your **headings** because it enables your reader to know the relative value of each item in your table of contents and each head in the body of your document.

P

Faulty Parallelism

Faulty parallelism results when joined elements are intended to serve equal grammatical functions but do not have equal grammatical form. Avoid this kind of partial parallelism. Make certain that each element in a series is similar in form and structure to all others in the same series.

CHANGE Before mowing the lawn, check the following items: the dipstick for proper oil level, the gas tank for fuel, the spark plug wire attachment, and that no foreign objects are under or near the mower.

TO Before mowing the lawn, check the following items: the dipstick *for proper oil level,* the gas tank *for fuel,* the spark plug wire *for attachment,* and the lawn *for foreign objects.*

In work-related writing, **lists** often cause problems with parallel structure. When you use a list that consists of phrases or clauses, each phrase or clause in the list should begin with the same **part of speech.**

CHANGE   The following recommendations were made regarding the Cost Containment Committee's position statement:

1. *Stress* that this statement is for all departments.
2. *Start* the statement with "If the company continues to grow, the following steps should be taken."
3. *The statement* should emphasize that it applies both to department managers and staff.
4. *Such strong words* as *obligation, owe,* and *must* should be replaced with words that are less harsh.

TO   The following recommendations were made regarding the Cost Containment Committee's position statement:

1. *Stress* that this statement is for all departments.
2. *Start* the statement with "If the company continues to grow, the following steps must be taken."
3. *Emphasize* that it applies both to department managers and staff.
4. *Replace* such strong words as *obligation, owe,* and *must* with words that are less harsh.

In the first list the items are not grammatically parallel in structure: items 1 and 2 begin with verbs, but items 3 and 4 do not. Notice how much more smoothly the corrected version reads, with all items beginning with verbs.

P

Faulty parallelism sometimes occurs because a writer tries to compare items that are not comparable.

CHANGE   The company offers special training to help nonexempt employees move into professional and technical careers like data processing, bookkeeping, *customer engineers,* and *sales trainees.* (Notice that occupations—data processing and bookkeeping—are being compared to people—customer engineers and trainees.)

TO   The company offers special training to help nonexempt employees move into professional and technical careers like data processing, bookkeeping, *customer service,* and *sales.*

CHANGE   We are not only *responsible to our stockholders* but also *to our customers.* (different grammatical forms: an **adjective** with a modifying prepositional phrase, *responsible to our stockholders;* and a prepositional phrase, *to our customers*)

TO   We are responsible not only *to our stockholders* but also *to our customers.* (parallel grammatical forms: prepositional phrases)

OR   We are not only *responsible to our stockholders* but also *responsible to our customers.* (parallel grammatical forms: verb phrases)

paraphrasing

When you *paraphrase* a written passage, you rewrite it to state the essential ideas in your own words. Because you do not quote your source word for word when paraphrasing, it is unnecessary to enclose the paraphrased material in **quotation marks.** However, paraphrased material should be footnoted because the ideas are taken from someone else whether or not the words are identical.

Original Material

> One of the major visual cues used by pilots in maintaining precision ground reference during low-level flight is that of object blur. We are acquainted with the object-blur phenomenon experienced when driving an automobile. Objects in the foreground appear to be rushing toward us while objects in the background appear to recede slightly. There is a point in the observer's line of sight, however, at which objects appear to stand still for a moment, before once again rushing toward him with increasing angular velocity. The distance from the observer to this point where objects appear stationary is sometimes referred to as the "blur threshold" range.
>
> —Wesley E. Woodson and Donald W. Conover, *Human Engineering Guide for Equipment Design*

Paraphrased

> *Object blur refers to the phenomenon by which observers in a moving vehicle look out and see the foreground objects appear to rush at them, while background objects appear to recede. But objects at some point appear temporarily stationary. The distance separating observers from this point is sometimes called the "blur threshold" range.*

Note that the paraphrased version includes only the essential information from the original passage. Strive to put the original ideas into your words without distorting them. (See also **notetaking, plagiarism,** and **quotations.**)

parentheses

Parentheses ( ) are used to enclose words, **phrases,** or **sentences.** The material within parentheses can add clarity to a statement without altering its meaning. Parentheses de-emphasize (or play down) an inserted element. Parenthetical information may not be essential to a sentence, but it may be interesting or helpful to some **readers.**

> EXAMPLE   Aluminum is extracted from its ore (called bauxite) in three stages.

Parenthetical material applies to the word or phrase immediately preceding it.

> **EXAMPLE**   The development of International Business Machines (IBM) is a uniquely American success story.

Parentheses may be used to enclose figures or letters that indicate sequence. Enclose the figure or letter with two parentheses rather than using only one parenthesis.

> **EXAMPLE**   The following sections deal with (1) preparation, (2) research, (3) organization, (4) writing, and (5) revision.

Parenthetical material does not affect the **punctuation** of a sentence. If a parenthesis closes a sentence, the ending punctuation should appear after the parenthesis. Also, a **comma** following a parenthetical word, phrase, or **clause** appears outside the closing parenthesis.

> **EXAMPLE**   These oxygen-rich chemicals, as for instance potassium permanganate ($KmnO_4$) and potassium chromate ($KCrO_4$), were oxidizing agents (they added oxygen to a substance).

However, when a complete sentence within parentheses stands independently, the ending punctuation goes inside the final parenthesis.

> **EXAMPLE**   The new marketing approach appears to be a success; most of our regional managers report sales increases of 15 to 30 percent. (The only important exceptions are the Denver and Houston offices.) Therefore, we plan to continue . . .

P

Do not follow a spelled-out number with a figure in parentheses representing the same number.

> **CHANGE**   Send five (5) copies of the report.
> **TO**   Send five copies of the report.

In some footnote forms, parentheses enclose the publisher, place of publication, and date of publication.

> **EXAMPLE**   [1]J. Demarco, *Nuclear Reactor Theory* (Reading, MA: Addison, 1991) 9.

Use **brackets** to set off a parenthetical item that is already within parentheses.

> **EXAMPLE**   We should be sure to give Emanuel Foose (and his brother Emilio [1812–1882] as well) credit for his part in founding the institute.

Do not overuse parentheses. And guard against using parentheses when other marks of punctuation are more appropriate.

participial phrases (see phrases)

participles (see verbs)

parts of speech

Parts of speech is a term used to describe the class of words to which a particular word belongs, according to its function in the sentence; that is, each function in a sentence (naming, asserting, describing, joining, acting, modifying, exclaiming) is performed by a word belonging to a certain part of speech.

If a word's function is to *name* something, it is a **noun** or **pronoun.** If a word's function is to make an *assertion* about something, it is a **verb.** If its function is to *describe* or *modify* something, the word is an **adjective** or an **adverb.** If its function is to *join* or *link* one element of the sentence to another, it is a **conjunction** or a **preposition.** If its function is to *express an exclamation,* it is an **interjection.** (See also **functional shift.**)

party

In legal language, *party* refers to an individual, group, or organization.

> **EXAMPLE**   The injured *party* brought suit against my client.

The term is inappropriate to general writing; when an individual is being referred to, use *person.*

> **CHANGE**   The *party* whose file you requested is here now.
> **TO**   The *person* whose file you requested is here now.

Party is, of course, appropriate when it refers to a group.

> **EXAMPLE**   Arrangements were made for the members of our *party* to have lunch after the tour.

passive voice (see voice)

per

Per is a common business term that means "by means of," "through," or "on account of," and in these senses it is appropriate.

EXAMPLES   *per* annum, *per* capita, *per* diem, *per* head

When used to mean "according to" (*per* your request, *per* your order), the expression is business **jargon** and should be avoided, as should the phrase *as per.*

CHANGE   *Per your request,* I am enclosing the production report.
TO   *As you requested,* I am enclosing the production report.

CHANGE   *As per our discussion,* I will send revised instructions.
TO   *As we agreed,* I will send revised instructions.

per cent/percent/percentage

Percent, which is replacing the two-word *per cent,* is used instead of the **symbol** (%) except in **tables.**

EXAMPLE   Only 25 *percent* of the members attended the meeting.

Percentage, which is never used with numbers, indicates a general size.

EXAMPLE   Only a small *percentage* of the managers attended the meeting.

periods

P

A *period* (also called a full stop or end stop) usually indicates the end of a declarative **sentence.** Periods also link (when used as leaders) and indicate omissions (when used as **ellipses**).

Although the primary function of periods is to end declarative sentences, periods also end imperative sentences that are not emphatic enough for an **exclamation mark.**

EXAMPLE   Send me any information you may have on the subject.

Periods may also end questions that are really polite requests and questions to which an affirmative response is assumed.

EXAMPLE   Will you please send me the specifications.

Periods end minor sentences when the meaning is clear from the context. These sentences are common in advertising. (See also **sentence faults.**)

EXAMPLE   When you insert your key into the ignition of your new luxury automobile, your seat adjusts to your body. *Instantly.*

In Quotations

Do not use a period after a declarative sentence that is quoted in the context of another sentence.

CHANGE   "There is every chance of success." she stated.
TO   "There is every chance of success," she stated.

A period is conventionally placed inside **quotation marks.**

EXAMPLES   He liked to think of himself as a "tycoon."
He stated clearly, "My vote is yes."

With Parentheses

A sentence that ends in a **parenthesis** requires a period *after* the parenthesis.

EXAMPLE   The institute was founded by Harry Denman (1902–1972).

If a whole sentence (beginning with an initial **capital letter**) is in parentheses, the period (or any other end mark) should be placed inside the final parenthesis.

EXAMPLE   The project director listed the problems facing her staff. (This was the third time she had complained to the board.)

Conventional Uses of Periods

Use periods after initials in names.

EXAMPLES   W. T. Grant, J. P. Morgan

Use periods as decimal points with **numbers.**

EXAMPLES   109.2, $540.26, 6.9

Use periods to indicate **abbreviations.**

EXAMPLES   Ms., Dr., Inc.

When a sentence ends with an abbreviation that ends with a period, do not add another period.

EXAMPLE   Please meet me at 3:30 p.m.

Use periods following the numbers in numbered **lists.**

EXAMPLE   1.
2.
3.

P

As Leaders

When spaced periods are used in a **table** to connect one item to another, they are called leaders. The purpose of leaders is to help the reader align the data.

150 lbs . . . . 1.7 psi
175 lbs . . . . 2.8 psi
200 lbs . . . . 3.9 psi

The most common use of leaders is in **tables of contents.**

Period Faults

The incorrect use of a period is sometimes referred to as a *period fault.* When a period is inserted prematurely, the result is a **sentence fragment.**

CHANGE   After a long day at the office in which we finished the report. We left hurriedly for home.

TO   After a long day at the office in which we finished the report, we left hurriedly for home.

When a period is left out, the result is a "fused," or run-on, sentence. Be careful never to leave out necessary periods.

CHANGE   Bill was late for ten days in a row Ms. Sturgess had to fire him.

TO   Bill was late for ten days in a row. Ms. Sturgess had to fire him.

(See also **ellipses** and **sentence faults.**)

person

Person refers to the form of a personal **pronoun** that indicates whether the pronoun represents the speaker, the person spoken to, or the person (or thing) spoken about. A pronoun representing the speaker is in the first person.

EXAMPLE   *I* could not find the answer in the manual.

A pronoun representing the person or persons spoken to is in the second person.

EXAMPLE   *You* are going to be a good supervisor.

A pronoun representing the person or persons spoken about is in the third person.

EXAMPLE   *They* received the news quietly.

Identifying pronouns by person helps the writer avoid illogical shifts from one person to another. A common error is to shift from the third person to the second person.

CHANGE   *Managers* should spend the morning hours on work requiring mental effort, for *your* mind is freshest in the morning.

TO   *Managers* should spend the morning hours on work requiring mental effort, for *their* minds are freshest in the morning.

OR   *You* should spend the morning hours on work requiring mental effort, for *your* mind is freshest in the morning.

The following table shows first, second, and third person pronouns.

Person	Singular	Plural
First	I, me, my	we, ours, us
Second	you, your	you, your
Third	he, him, his	they, them, their
	she, her, hers	
	it, its	

(See also **number, case,** and **one.**)

personal/personnel

Personal is an **adjective** meaning "of or pertaining to an individual person."

EXAMPLE   He left work early because of a *personal* problem.

Personnel is a **noun** meaning a "group of people engaged in a common job."

EXAMPLE   All *personnel* should pick up their paychecks on Thursday.

Be careful not to use *personnel* when the word you need is *persons* or *people*.

CHANGE   The remaining two *personnel* will be moved next Thursday.
TO   The remaining two *persons* will be moved next Thursday.

personal pronouns (see **pronouns**)

persons/people

When we use *persons,* we are usually referring to individual people thought of separately.

EXAMPLE   We need to find three qualified *persons* to fill the vacant positions.

When we say *people,* we are identifying a large or anonymous group.

> **EXAMPLE**   Many *people* have never even heard of our product.

persuasion

Persuasion attempts to convince the **reader** to adopt the writer's point of view. The range of persuasive writing required on the job varies widely. Technical writers may use persuasive techniques when writing **memorandums,** pleading for safer working conditions in a shop or factory, justifying the expense of a new program, or writing a **proposal** for a multimillion-dollar contract.

Statement of position	For maximum chain life, the chain should be properly tensioned. Maintaining the proper tension has always been difficult with solid-nosed bars because the chain develops a great deal of heat from friction as it slides over the nose of the bar. The heat causes the chain to expand and hence become too slack. If the chain is properly tensioned when cold, it will be too slack when running. If the chain is tensioned after it warms up, then stopping the saw long enough to refuel it might result in the chain shrinking to the bar so tight that it will not turn.
Supporting facts	The first attempt at solving these problems was the roller-nosed bar. This bar cuts down on the friction to a considerable extent, but is somewhat fragile, and since the chain is unsupported during part of its travel, it is not good practice to cut with the tip of the bar. The sprocket-nosed bar seems to have solved these problems. Here a thin sheet-metal sprocket, roughly the thickness of the drive links, is all but hidden inside the nose of the bar. As the chain approaches the end of the bar, the sprocket lifts the drive link and therefore keeps the saw from running on the rails of the bar is it travels around the nose. This cuts friction while supporting the chain well enough so that the nose
Acknowledgment of disadvantage	of the bar can be used for cutting. It is true that a saw with a sprocket-nosed bar will kick back harder than other saws because there is so little friction between the chain and the bar, and the sprocket nose may be somewhat more fragile, particularly if it is pinched in a cut. On the other hand, its advantages often outweigh its disadvantages. It gives faster cutting, better chain life, and better chain tension, which in itself is a safety factor. Most sprockets and bearings for bars can be replaced by a dealer, although if the nose portion of the bar is bent, the bar is ruined. Some sprocket-nosed bars are available with the entire tip of the bar made as a section that may be easily replaced by the user.

—R. P. Sarna, *Chain Saw Manual*

FIGURE 1   Example of Persuasive Writing

In persuasive writing, the way that you present your ideas is as important as the ideas themselves. You must support your appeal with **logic,** facts, statistics, and examples. You must also acknowledge any real or possible conflicting opinions. Only then can you demonstrate and argue for the merit of your point of view. But remember your audience and take into account your readers' feelings. Always maintain a positive **tone.** And be careful not to wander from your main point—avoid ambiguity and never make trivial, irrelevant, or false claims. But if a reader may be skeptical of your main idea, you may need to build up to it with strong, specific points or arguments that support it. (This *inductive* method is illustrated in **specific-to-general method of development.**)

The passage presented in Figure 1, on page 434, is taken from a guide on the selection and proper use of chainsaws. The intended readers, loggers, use chainsaws daily in their work. The author explains the necessity of maintaining the proper tension on the saw's chain. In explaining some of the technical problems that had to be overcome so that the proper tension could be maintained under all conditions, the author convincingly makes the case that chainsaws of a particular design are the most effective.

Notice that the author acknowledges that this design is not entirely free of potentially harmful side effects: a saw with this bar will kick back harder than others will, and the bar itself is more fragile. By acknowledging negative details or opposing views, a writer also gains credibility through the readers' impression of him or her as a person. (See also **methods of development** and **forms of discourse.**)

P

phenomenon/phenomena

A *phenomenon* is an observable thing, fact, or occurrence. Its plural form is *phenomena.*

> EXAMPLES The natural *phenomenon* of earth tremors *is* a problem we must anticipate in designing the California installation.
>
> The *phenomena* associated with atomic fission *were* only recently understood.

photographs

Photographs are the best way to show the surface appearance of an object or to record an event or the development of a phenomenon over a period of time. Not all representations, however, call for photographs. Photographs cannot depict the internal workings of a mechanism or below-the-surface details of objects or structures. Such details are better represented in **drawings** or **schematic diagrams.**

Highlighting Photographic Subjects

Stand close enough to the object so that it fills your picture frame. To get precise and clear photographs, choose camera angles carefully. A camera will photograph only what it is aimed at; accordingly, select the important details and the camera angles that will record them. To show relative size, place a familiar object—such as a ruler, a book, or a person—near the object being photographed.

Using Color

The preparation and printing of color photographs are complex tasks performed by graphics and printing experts. If you are planning to use color photographs in your publication, discuss with these experts the type and quality of photographs required. Generally, they prefer color transparencies (slides) and color negatives (negatives of color prints) for reproduction.

The advantages of using color illustrations are obvious: color is a good way of communicating crucial information. In medical, chemical, geological, and botanical publications, to name but a few, readers often need to know exactly what an object or phenomenon looks like. In these circumstances, color reproduction is the only legitimate option available.

Be aware, however, that color reproduction is significantly more expensive than black and white. Color can also be tricky to reproduce accurately without loss of contrast and vividness. For this reason, the original must be sharply focused and rich in contrasts.

Tips for Using Photographs

Like all illustrative materials, photographs must be handled carefully. When preparing photographs for printing, observe the following guidelines:

1. Mount photographs on white bond paper with spray adhesive (available at art supply stores) and allow ample margins.
2. If the photograph is the same size as the paper, type the caption, figure and page numbers, and any other important information on labels and fasten them to the photograph with spray adhesive. The

FIGURE 1   Example of a Photograph

labels that identify key features in the photograph are referred to as *callouts*. (Photographs are given figure numbers in sequence with **illustrations** in a publication; see Figure 1.)

3. Position the figure number and caption so that the reader can view them and the photograph from the same orientation.

4. Do not draw crop marks (lines showing where the photo should be trimmed for reproduction) directly across a photograph. Draw them at the very edge of the photograph.

5. Do not write on a photograph, front or back. Tape a tissue-paper overlay over the face of the photograph, and then write very lightly on

the overlay with a soft-lead pencil. Never write on the overlay with a ball-point pen.

6. Do not use paper clips or staples directly on photographs.
7. Do not fold or crease photographs.
8. If only a color photograph or slide is available for black-and-white reproduction, have a photographer produce a black-and-white glossy copy or slide for printing. Otherwise, the printed image will not have an accurate tone.

phrases

Below the level of the sentence, there are two ways to combine words into groups: by forming **clauses,** which combine subjects and **verbs,** and by forming *phrases,* which are based on **nouns,** nonfinite verb forms, or verb combinations without subjects. A phrase is the most basic meaningful group of words; unlike a clause, a phrase cannot make a full statement, because it does not contain a subject and a verb.

EXAMPLE   He encouraged his staff (clause) *by his calm confidence.* (phrase)

A phrase may function as an **adjective,** an **adverb,** a noun, or a verb.

EXAMPLES   The subjects *on the agenda* were all discussed. (adjective)
We discussed the project *with great enthusiasm.* (adverb)
Working hard is her way of life. (noun)
The chief engineer *should have been notified.* (verb)

Even though phrases function as adjectives, adverbs, nouns, or verbs, they are normally named for the kind of word around which they are constructed—**preposition,** participle, infinitive, gerund, verb, or noun. A phrase that begins with a preposition is a *prepositional phrase;* a phrase that begins with a noun is a *noun phrase,* and so on.

Prepositional Phrases

A preposition is a word that shows relationship and combines with a noun or **pronoun** (its **object**) to form a modifying phrase. A prepositional phrase, then, consists of a preposition plus its object and the object's modifiers.

EXAMPLE   *After the meeting,* the district managers adjourned *to the executive dining room.*

Prepositional phrases, because they normally modify nouns or verbs, usually function as adjectives or adverbs.

A prepositional phrase may function as an adverb of motion.

> EXAMPLE Turn the dial four degrees *to the left.*

A prepositional phrase may function as an adverb of manner.

> EXAMPLE Answer customers' questions *in a courteous fashion.*

A prepositional phrase may function as an adverb of place.

> EXAMPLE We ate lunch *in the company cafeteria.*

When functioning as adverbs, prepositional phrases may appear in different places.

> EXAMPLES *In residential and farm wiring,* one of the wires must always be grounded.
>
> One of the wires must always be grounded *in residential and farm wiring.*

A prepositional phrase may function as an adjective. When functioning as adjectives, prepositional phrases follow the nouns they modify.

> EXAMPLE Garbage *with a high paper content* has been turned into protein-rich animal food.

Be careful when using prepositional phrases, because separating a prepositional phrase from the noun it modifies can cause **ambiguity.**

> CHANGE *The man* standing by the drinking fountain *in the gray suit* is our president.
>
> TO *The man in the gray suit* who is standing by the drinking fountain is our president.

(See also **misplaced modifiers.**)

Participial Phrases

A *participle* is any form of a verb that is used as an adjective. A participial phrase consists of a participle plus its object and its modifiers.

> EXAMPLES The division *having the largest sales increase* wins the trophy.
> *Finding the problem resolved,* he went to the next item.
> *Having begun,* we felt that we had to see the project through.

The relationship between a participial phrase and the rest of the sentence must be clear to the **reader.** For this reason, every sentence containing a participial phrase must have a noun or pronoun that the participial phrase modifies; if it does not, the result is a dangling participial phrase that is misplaced in the sentence and so appears to modify the wrong noun or pronoun. Both of these problems are discussed below.

Dangling Participial Phrases. A dangling participial phrase occurs when the noun or pronoun that the participial phrase is meant to modify is not stated but only implied in the sentence. (See also **dangling modifiers.**)

> CHANGE   *Being unhappy with the job,* his efficiency suffered. (His efficiency was not unhappy with the job; what the participial phrase really modifies—*he*—is not stated but merely implied.)
> TO   *Being unhappy with the job,* he grew less efficient. (Now what that participial phrase modifies—*he*—is explicitly stated.)

Misplaced Participial Phrases. A participial phrase is misplaced when it is too far from the noun or pronoun it is meant to modify and so appears to modify something else. This is an error that can sometimes make the writer look ridiculous indeed. (See also **modifiers.**)

> CHANGE   *Rolling around in the bottom of the vibration test chamber,* I found the missing bearings.
> TO   I found the missing bearings *rolling around in the bottom of the vibration test chamber.*

> CHANGE   We saw a large warehouse *driving down the highway.*
> TO   *Driving down the highway,* we saw a large warehouse.

P

Infinitive Phrases

An infinitive is the bare form of a verb *(go, run, talk)* without the restrictions imposed by **person** and **number;** an infinitive is generally preceded by the word *to* (which is usually a preposition but in this use is called the sign, or mark, of the infinitive). An *infinitive phrase* consists of the word *to* plus an infinitive and any objects or modifiers.

> EXAMPLE   *To succeed in this field,* you must be willing *to assume responsibility.*

Do not confuse a prepositional phrase beginning with *to* with an infinitive phrase. In the infinitive phrase, the *to* is followed by a **verb;** in the prepositional phrase, *to* is followed by a **noun** or **pronoun.**

> EXAMPLES   We went *to the building site.* (prepositional phrase)
>
> Our firm tries *to provide a comprehensive training program.* (infinitive phrase)

The implied subject of an introductory infinitive phrase should be the same as the subject of the sentence. If it is not, the phrase is a dangling modifier. In the following example, the implied subject of the infinitive is *you*, or *one*, not *practice*.

CHANGE   *To learn shorthand,* practice is needed.
TO   *To learn shorthand,* you must practice.
OR   *To learn shorthand,* one must practice.

Gerund Phrases

A *gerund phrase*, which consists of a gerund plus any objects or modifiers, always functions as a noun.

EXAMPLES   *Preparing an annual report* is a difficult task. (subject)
She liked *running the department.* (direct object)

Verb Phrases

A verb phrase consists of a main verb and its helping verb.

EXAMPLES   He *is* (helping verb) *working* (main verb) hard this summer.

Company officials discovered that a computer *was emitting* more data than it *had been asked for.* After investigation, police suggested that an unauthorized program *had been run* through the computer at the coded command of another computer belonging to a different company. The police obtained a warrant to search for electronic impulses in the memory of the suspect machine. The case (involving trade secrets) and its outcome *should make* legal history.

Words can appear between the helping verb and the main verb of a verb phrase.

EXAMPLE   He *is* always *working.*

The main verb is always the last verb in a verb phrase.

EXAMPLES   You *will file* your tax return on time if you begin early.
You *will have filed* your tax return on time if you begin early.
You *are* not *filing* your tax return too early if you begin now.

Questions often begin with a verb phrase.

EXAMPLE   *Will* he *audit* their account soon?

The **adverb** *not* may be appended to a helping verb in a verb phrase.

EXAMPLES   He *cannot work* today.
He *did not work* today.

Noun Phrases

A noun phrase consists of a noun and its modifiers.

EXAMPLES   *Many large companies* use computers.
Have *the two new employees* fill out *these forms.*

plagiarism

To use someone else's exact words without **quotation marks** and appropriate credit, or to use the unique ideas of someone else without acknowledgment, is known as plagiarism. In publishing, plagiarism is illegal; in other circumstances it is, at the least, unethical. (For detailed guidance on quoting correctly, see **quotations.**)

You may quote or **paraphrase** the words and ideas of another if you document your source. (See **documenting sources.**) Although you need not enclose paraphrased material in quotation marks, you must document the source. Paraphrased ideas are taken from someone else whether or not the words are identical. Paraphrasing a passage without citing the source is permissible only when the information paraphrased is common knowledge in a field. (Common knowledge refers to historical, scientific, geographical, technical, and other types of information on a topic readily available in handbooks, manuals, atlases, and other references.) If you intend to publish or reproduce and distribute material in which you have included quotations from published works, you may have to obtain written permission to do so from the **copyright** holder.

plurals (see **nouns, pronouns,** and **verbs**)

point of view

Point of view indicates the writer's relation to the information presented, as reflected in the use of **person.** The writer usually expresses point of view in first-, second-, or third-person personal **pronouns.** The use of the first person indicates that the narrator is a participant or observer ("This happened to me," "I saw that"). The second and third person indicate that the narrator is writing about other people or something impersonal or is giving directions, instructions, or advice ("This happened to her, to them, to it"; "Enter the data after pressing the ENTER key once"). The writer may also avoid using personal pronouns, thus adopting an impersonal point of view.

EXAMPLE   *It is regrettable* that your shipment of March 12 is not acceptable.

Consider the same sentence written from the personal, first-person point of view.

EXAMPLE   *I* regret that we cannot accept your shipment of March 12.

Although the meaning of both sentences is the same, the sentence with the personal point of view indicates that two people are involved in the communication. Years ago technical people preferred the impersonal point of view because they thought it made writing sound more objective and professional. Unfortunately, however, the impersonal point of view often prevents clear and direct communication and can cause misunderstanding. Today, good technical writers adopt a more personal point of view whenever it is appropriate.

Writers should not avoid *I* by using *one* when they are really talking about themselves. They do not increase objectivity but merely make the statement impersonal.

CHANGE   *One* can only conclude that the absorption rate is too fast.
TO   *I* can only conclude that the absorption rate is too fast.

Writers should never use *the writer* to replace *I* in a mistaken attempt to sound formal or dignified.

CHANGE   *The writer* believes that this project will be completed by the end of June.
TO   *I* believe that this project will be completed by the end of June.

However, the personal point of view should not be used when an impersonal point of view would be more appropriate or more effective.

CHANGE   *I* am inclined to think that each manager should attend the final committee meeting to hear the committee's recommendations.
TO   *Each manager* should attend the final committee meeting to hear the committee's recommendations.

The preceding examples make clear that when the impersonal point of view is used, the focus shifts from the writer to the person or thing being discussed. For the subject matter to receive more emphasis than either the writer or the reader does, an impersonal point of view should be used.

EXAMPLE   *The evidence* suggests that the absorption rate is too fast.

In a letter on company stationery, use of the pronoun *we* may be interpreted as reflecting company policy, whereas *I* clearly reflects personal opinion. Which pronoun to use should be decided according to whether the matter discussed in the letter is a corporate or an individual concern.

> **EXAMPLES** *I* appreciate your suggestion regarding our need for more community activities.
>
> *We* appreciate your suggestion regarding our need for more community activities.

The pronoun *we* may commit an organization to what the writer says. If the following statement accurately reflects an organization's opinion, with its implication that the proposal being singled out will be accepted, it is justified.

> **EXAMPLE** *We* believe that your proposal was the best submitted.

If, however, the writer can speak only personally instead of for the organization, the statement should be hedged so that the reader understands that it represents the writer's opinion rather than the organization's official position.

> **EXAMPLE** I *personally* believe that your proposal was the best submitted, but the policy committee will make the final decision.

Whether the writer adopts a personal or an impersonal point of view should depend on the objective and the reader. For example, in a business letter to an associate, the writer will most likely adopt a personal point of view. But in a report to a large group, the writer will probably choose to emphasize the subject by adopting an impersonal point of view.

P

positive writing

Presenting positive information as though it were negative is a trap that technical writers fall into quite easily because of the complexity of the information they must write about. It is a practice that confuses the **reader,** however, and one that should be avoided.

> **CHANGE** If the error does *not* involve data transmission, the special function will *not* be used.
> **TO** The special function is used only if the error involves data transmission.

In the first sentence, the reader must reverse two negatives to understand the exception that is being stated; the second sentence presents an exception to a rule in a straightforward manner.

On the other hand, negative facts or conclusions should be stated negatively; stating a negative fact or conclusion positively can mislead the reader.

> **CHANGE**   For the first quarter of this year, employee exposure to airborne lead has been maintained to within 10 percent of acceptable state health standards.
>
> **TO**   For the first quarter of this year, employee exposure to airborne lead continues to be 10 percent above acceptable state health standards.

Even if what you are saying is negative, do not use any more negative words than are necessary or words that are more negative than necessary.

> **CHANGE**   We are withholding your shipment until we receive your payment.
>
> **TO**   We will forward your shipment as soon as we receive your payment.

possessive case

A **noun** or **pronoun** is in the possessive case when it represents a person or thing owning or possessing something.

> **EXAMPLE**   A recent scientific analysis of *New York City's* atmosphere concluded that New Yorkers on the street took into *their* lungs the equivalent in toxic materials of 38 cigarettes per person daily.

Singular nouns show the possessive by adding an **apostrophe** and *s*.

> **EXAMPLE**   company/company's

Nouns that form their plurals by adding an *s* show the possessive by placing an apostrophe after the *s* that forms the plural.

> **EXAMPLES**   a managers' meeting/the technicians' handbooks

The following **list** shows the relationships among singular, plural, and possessive nouns that form their plurals by adding s or changing *y* to *ies*.

singular	company	employee
singular possessive	company's	employee's
plural	companies	employees
plural possessive	companies'	employees'

Nouns that do not add *s* to form their plurals add an **apostrophe** and an *s* to show possession in both the plural and the singular forms.

singular	child	man
singular possessive	child's	man's
plural	children	men
plural possessive	children's	men's

Singular nouns that end in *s* form the possessive either by adding only an apostrophe or by adding both an apostrophe and an *s*.

EXAMPLES   a *waitress'* uniform/an *actress'* career
a *waitress's* uniform/an *actress's* career

Singular nouns of one syllable always form the possessive by adding both an apostrophe and an *s*.

EXAMPLE   The *boss's* desk was cluttered.

Although exceptions are relatively common, it is a good rule of thumb to use an apostrophe and an *s* with nouns referring to persons and living things and to use an *of* phrase for possessive nouns referring to inanimate objects.

EXAMPLES   The *chairperson's* address was well received.
The keys *of the keyboard* were sticking.

If this rule leads to awkwardness or wordiness, however, be flexible.

P

EXAMPLES   The *company's* plants are doing well.
The *plane's* landing gear failed.

Several indefinite pronouns (*any, each, few, most, none,* and *some*) form the possessive only in *of* phrases (*of any, of some*). Others, however, use an apostrophe (*anyone's*).

The established **idiom** calls for the possessive case in many stock phrases.

EXAMPLES   a *day's* journey, a *day's* work, a *moment's* notice, at his *wit's* end, the *law's* delay

In a few cases, the idiom even calls for a double possessive using both the *of* and the *'s* forms.

EXAMPLE   That colleague *of George's* was at the conference.

When several words make up a single term, add *'s* to the last word only.

EXAMPLES   The *chairman of the board's* statement was brief.

The *Department of Energy's* new budget shows increased revenues of $192 million for uranium enrichment.

With coordinate nouns, the last noun takes the possessive form to show joint possession.

EXAMPLE   *Michelson and Morely's* famous experiment on the velocity of light was made in 1887.

To show individual possession with coordinate nouns, each noun should take the possessive form.

EXAMPLE   The difference between *Thomasson's* and *Silson's* test results were statistically insignificant.

To form the possessive of a compound word, add *'s*.

EXAMPLES   the *vice-president's* car, the *pipeline's* diameter, the *antibody's* reaction.

When a noun ends in multiple consecutive *s* sounds, form the possessive by adding only an apostrophe.

EXAMPLES   *Jesus'* disciples, *Moses'* journey

Do not use an apostrophe with possessive pronouns.

EXAMPLES   *yours, its, his, ours, whose, theirs*

It's is a **contraction** of *it is,* not the possessive form of *it*.

In the names of places and institutions, the apostrophe is often omitted.

P

EXAMPLES   Harpers Ferry, Writers Book Club

The use of possessive pronouns does not normally cause problems except with gerunds and indefinite pronouns. Several indefinite pronouns *(all, any, each, few, most, none,* and *some)* require *of* phrases to form the possessive case.

EXAMPLE   Both dies were stored in the warehouse, but rust had ruined the surface *of each.*

Others, however, use an apostrophe.

EXAMPLE   *Everyone's* contribution is welcome.

Only the possessive form of a pronoun should be used with a gerund.

EXAMPLES   The safety officer insisted in *my* wearing a respirator.
Our monitoring was not affected by changing weather conditions.

practicable/practical

Practicable means that something is possible or feasible. *Practical* means that something is both possible and useful.

> **EXAMPLE**   The program is *practical,* but considering the company's recent financial problems, is it *practicable?*

Practical, not *practicable,* is used to describe a person, implying common sense, a commitment to what works, rather than to theory.

predicates (see **sentence construction**)

prefixes

A prefix is a letter, or letters, placed in front of a root word that changes its meaning, often causing the new word to mean the opposite of the root word. When a prefix ends with a vowel and the root word begins with the same vowel, the prefix may be separated from the root word with a **hyphen** (re-enter, co-operate, re-elect); the second vowel may be marked with a dieresis (reënter, coöperate, reëlect), although this practice is rare now; or the word may be written with neither hyphen nor dieresis (reenter, cooperate, reelect). The hyphen and the dieresis are visual aids that help the **reader** recognize that the two vowels are pronounced differently.

Except between identical vowels, the hyphen rarely appears between a prefix and its root word. At times, however, a hyphen is necessary for clarity of meaning; for example, *reform* means "correct" or "improve," and *re-form* means "change the shape of."

preparation

The *preparation* stage of the writing process is analogous to focusing a camera before taking a picture. By determining who your **reader** is, what your **purpose** is, and what your **scope** of coverage should be, you are bringing your whole writing effort into focus before beginning to write. As a result, your **topic** will be much more clearly focused for your reader.

Purpose

What do you want your readers to know, believe, or be able to do when they have finished reading your document? When you can answer this question, you will have determined the objective of your writing. A good test of whether you have formulated your objective adequately is to state it in a single sentence. This statement may also function as the thesis statement in your outline.

Readers

Learn certain key facts about your readers, such as their educational level, technical knowledge, and needs relative to your subject. Their level of technical knowledge, for example, should determine whether you need to cover the fundamentals of your subject and which terms you must define.

Scope

If you know the purpose of your writing project and your readers' needs, you will know the type and amount of detail you must include in your writing to accomplish your objective and meet your readers' needs. This is your required scope of coverage.

prepositional phrases (see **phrases**)

prepositions

A *preposition* is a word that links a **noun** or **pronoun** (its **object**) to another sentence element, by expressing such relationships as direction (*to, into, across, toward*), location (*at, in, on, under, over, beside, among, by, between, through*), time (*before, after, during, until, since*), or figurative location (*for, against, with*). Although only about seventy prepositions exist in the English language, they occur frequently. Together, the preposition, its object, and the object's **modifiers** form a prepositional **phrase,** which acts as a modifier.

The object of a preposition, the word or phrase following the preposition, is always in the objective **case.** This situation gives rise to a problem in such constructions as "between you and *me,*" a phrase that is frequently and incorrectly written as "between you and *I.*" *Me* is the objective form of the pronoun, and *I* is the subjective form.

Many words that function as prepositions also function as **adverbs.** If the word takes an object and functions as a connective, it is a preposition; if it has no object and functions as a modifier, it is an adverb.

EXAMPLES   The manager sat *behind* the desk *in* his office. (prepositions)

The customer lagged *behind;* then she came *in* and sat *down.* (adverbs)

Avoiding Errors with Prepositions

Do not use redundant prepositions, such as "off *of,*" "in back *of,*" and "inside *of.*"

CHANGE   The client arrived *at about* four o'clock.
TO   The client arrived *at* four o'clock. (to be exact)
OR   The client arrived *about* four o'clock. (to be approximate)

Do not omit necessary prepositions.

CHANGE   He was oblivious and not distracted by the view from his office window.
TO   He was oblivious *to* and not distracted *by* the view from his office window.

Avoid unnecessarily adding the preposition *up* to **verbs.**

CHANGE   Call *up* and see if he is in his office.
TO   Call and see if he is in his office.

Using a Preposition at the End of a Sentence

If a preposition falls naturally at the end of a sentence, leave it there.

EXAMPLE   I don't remember which file I put it *in.*

Be aware, however, that a preposition at the end of a sentence can be an indication that the sentence is awkwardly constructed.

CHANGE   The branch office is where he was *at.*
TO   He was at the branch office.

Common Uses of Prepositions

Certain verbs, adverbs, and **adjectives** are used with certain prepositions. For example, we say "interested *in,*" "aware *of,*" "devoted *to,*" "equated *with,*" "abstain *from,*" "adhere *to,*" "conform *to,*" "capable *of,*" "comply *with,*" "object *to,*" "find fault *with,*" "inconsistent *with,*" "independent *of,*" "infer *from,*" and "interfere *with.*" (See also **idioms.**)

Using Prepositions in Titles

When a preposition appears in a title, it is capitalized if it is the first word in the title or if it has four letters or more. (See also **capital letters.**)

EXAMPLES   *Composition for the Business and Technical World* was reviewed recently in the *Journal of Technical Communications.*

The book *Managing Like Mad* should be taken seriously in spite of its title.

principal/principle

Principal, meaning an "amount of money on which interest is earned or paid" or a "chief official in a school or court proceeding," is sometimes confused with *principle,* meaning a "basic truth or belief."

EXAMPLES   The bank will pay 6.5 percent per month on the *principal.*
She is a person of unwavering *principles.*
He sent a letter to the *principal* of the high school.

Principal is also an adjective, meaning "main" or "primary."

EXAMPLE   My *principal* objection is that it will be too expensive.

process explanation

Many kinds of technical writing explain a process, an operation, or a procedure. The explanation involves putting in the appropriate order the steps that a specific mechanism or system uses to accomplish a certain result. The process itself might range from the legal steps necessary to form a corporation, to the steps necessary to develop a roll of film. At the least, you should divide the process into distinct steps and present them in their normal order.

In your opening **paragraph,** you might tell your **reader** why it is important to become familiar with the process you are explaining. Before you explain the steps necessary to form a corporation, for example, you could cite the tax savings that incorporation would permit. To provide your reader with a framework for the details that will follow, you might present a brief overview of the process. Finally, you might describe how the process works in relation to a larger whole. In explaining the air brake system of a large dump truck, you might note that the braking system is one part of the vehicle's air system, which also controls the throttle and transmission-shifting mechanisms.

A process explanation can be long or short, depending on how much detail is necessary. The following description of the way in which a camera controls light to expose a photographic film fits into one paragraph:

EXAMPLE   The camera is the basic tool for recording light images. It is simply a box from which all light is excluded except that which passes through a small opening at the front. Cameras are equipped with various devices for controlling the light rays as they enter this opening. At the press of a button, a mechanical blade or curtain, called a shutter, opens and closes automatically. During the fraction of a second that the shutter is open, the light reflected from the subject toward which the camera is aimed passes into the camera through a piece of optical glass called the lens. The lens focuses, or projects, the light rays onto the wall at the back of the camera. These light reflections are captured on a sheet of film attached to the back wall.

Many process explanations require more details and will, of course, be longer than a paragraph. The example presented in Figure 1 discusses several methods of surface mining for coal. The writer begins with an overview of the elements common to all the processes, defines the terms important to the explanation, and then describes each separately. Transitional words and **phrases** serve to achieve **unity** within **paragraphs,** and **headings** mark the **transition** from one process to the next. **Illustrations** often help convey your message.

P

Surface Mining of Coal

Overview   The process of removing the earth, rock, and other strata (called *overburden*) to uncover an underlying mineral deposit is generally referred to as surface mining. Strip mining is a specific kind of surface mining in which all the overburden is removed in strips, one cut at a time. Three types of strip-mining methods are used to mine coal: *area, contour,* and *mountain-top removal.* Which method is used depends upon the topography of the area to be mined.

Area Strip Mining

Detailed explanation   Area strip mining is used in regions of flat to gently rolling terrain, like those found in the Midwest and West. Depending on applicable reclamation laws, the topsoil may be removed from the area to be mined, stored, and later reapplied as surface material during reclamation of the mined land. Following removal of the topsoil, a trench is cut through the overburden to expose the upper surface of the coal to be mined. The length of the cut generally corresponds to the length of the property or of the deposit. The overburden from the first cut is placed on unmined land adjacent to the cut. After the first cut is completed, the coal is removed and a second cut is made parallel to the first. The overburden (now referred to as spoil) from each of the succeeding cuts is deposited

FIGURE 1 Explaining a Process

in the adjacent pit from which the coal was just removed. The final cut leaves an open trench equal in depth to the thickness of the overburden plus the coal bed, bounded on one side by the last spoil pile and on the other side by the undisturbed soil. The final cut may be as far as a mile from the first cut. The overburden from all the cuts, unless graded and leveled, resembles the ridges of a giant washboard.

Contour Strip Mining

In areas of rolling or very steep terrain, such as in the eastern United States, contour strip mining is used. In this method, the overburden is removed from the mineral seam in a pattern that follows the contour line around the hillside. The overburden is then deposited on the downslope side of the cut until the depth of the overburden becomes too great for economical recovery of the coal. This method leaves a bench, or shelf, on the side of the hill, bordered on the inside by a highwall (30 to 100 feet high) and on the other side by a high ridge of spoil.

A method of mining that is often used in conjunction with contour mining is *auger mining*. This method is employed when the overburden becomes too thick and renders contour mining uneconomical and when extraction by underground mining would be too costly or unsafe. In auger mining an instrument bores holes horizontally into the coal seam. The coal can then be removed like the shavings produced by a drill bit. The exposed coal seam in the highwall is left with a continuous series of bore holes from which the coal was removed.

Mountain-Top Removal

In areas of rolling or steep terrain an adaptation of area mining to conventional contour mining is used; it is called the mountain-top removal method. With this method entire mountain tops are removed down to the coal seam by a series of parallel cuts. This method is economical when the coal lies near the tops of mountains, ridges, or knobs. If there is excess overburden that cannot be stored on the mined land, it may be transported elsewhere.

Margin note, left: Detailed explanation
Margin note, left: Detailed explanation

FIGURE 1   Explaining a Process *(continued)*

progress and activity reports

Progress reports and *activity reports* differ: progress reports are submitted by a vendor to a client company and activity reports are submitted by an employee to a superior.

Progress Reports

A *progress report* provides information about a project—its current status, whether it is on schedule, whether it is within the budget, and so on. Progress reports are often submitted by a contracting company to a client company. They are issued at regular intervals throughout the life of a project and state what has been done in a specified interval and what has yet to be done before the project can be completed. The progress report is used primarily with projects that involve many steps over a period of weeks, months, or even years. Progress reports help keep projects running smoothly by allowing management to assign workers, adjust schedules, allocate budgets, or schedule supplies and equipment, as necessary.

All reports in a series of progress reports should have the same format. Since progress reports are normally sent outside the company, they are often written in letter **format.**

The **introduction** to the first progress report should identify the project, any methods and necessary materials, and the date by which the project is to be completed. Subsequent reports should then summarize the progress since the first report.

The body of the progress report should describe the project's present status, including such details as schedules and costs.

The report should end with **conclusions** and recommendations about changes in the schedule, materials, techniques, and so on. The conclusion may also include a statement of the work done and an estimate of future progress.

The example shown in Figure 1 is the first progress report submitted by an electrical contractor (REMCON) to a client (the Arena Committee).

Activity Reports

Within an organization, professional employees often submit reports on the progress and current status of all ongoing projects assigned to them by their managers. The managers combine these *activity reports* (also called *status reports*) into larger activity reports that they, in turn, submit to their managers. Activity reports help keep all levels of management aware of the progress and status of projects within an organization.

An activity report includes information about the status of all current projects, including any problems, the action currently being taken to resolve any problems, and the writer's plans for the coming month. A manager's activity report should also indicate the current number of employees.

Because the activity report is issued periodically (often monthly) and contains material familiar to its readers, it normally needs no introduction or conclusion. The format of an activity report may vary from company to company, or even among different parts of the same company. The following sections are typical and would be adequate for most situations:

─────────── **REMCON ELECTRIC** ───────────
5099 Seventh Street, St. Paul, Minnesota 55101 (612) 555-1212

August 20, 19—

Arena Committee
Minnesota Sports
708 N. Case St.
St. Paul, MN 55101

Subject: Arena Rewiring Progress Report

Intro-
duction
identifies
project and
completion
date

This report, as agreed to in our contract, covers the progress on the rewiring program at the Sports Arena from May 5 to August 15 of 19—. Although the costs of certain equipment are higher than our original bid indicated, we expect to complete the project by December 23, without going over cost; the speed with which the project is being completed will save labor costs.

Work Completed
On August 15, we finished installing the circuit-breaker panels and meters of Level I service outlets and of all subfloor rewiring. Lighting fixture replacement, Level II service outlets, and the upgrading of stage lighting equipment are in the preliminary stages (meaning that the wiring has been completed but installation of the fixtures has not yet begun).

Body
details
schedules
and costs

Costs
Equipment used up to this point has cost $10,800, and labor has cost $31,500 (including some subcontracted plumbing). My estimate for the rest of the equipment, based on discussions with your lighting consultant, is $11,500; additional labor costs should not exceed $25,000.

Work Schedule
I have scheduled the upgrading of stage lighting equipment from August 16 to October 5, the completion of Level II service outlets from October 6 to November 12, and the replacement of lighting fixtures from November 15 to December 17.

Conclusion
indicates
pertinent
changes

Conclusion
Although my original estimate on equipment ($20,000) has been exceeded by $2,300, my original labor estimate ($60,000) has been reduced by $3,500; so I will easily stay within the limits of my original bid. In addition, I see no difficulty in having the arena finished for the Christmas program on December 23.

Sincerely,

John Remcon
John Remcon
President and Owner

ics

FIGURE 1 Progress Report

Current Projects
This section lists every project assigned to the employee or manager and summarizes its current status.

Current Problems
This section details any problems confronting the employee or manager and explains the steps being taken to resolve them.

Plans for the Next Period
This section states what the writer expects to achieve on each project during the next month or reporting period.

Current Staffing Level
This section, included managers and project leaders, lists the number of subordinates assigned to the writer and correlates that number with the staffing level considered necessary for the projects assigned.

The activity report shown in Figure 2, using a **memorandum** format, was submitted by a manager of software development (Wayne Tribinski) who supervises 11 employees. The reader of the report (Kathryn Hunter), Tribinski's manager, is the director of engineering.

INTEROFFICE MEMORANDUM

Date: June 5, 19—
To: Kathryn Hunter, Director of Engineering
From: Wayne Tribinksi, Manager, Applications Programs *W.T.*

Subject: Activity Report for May, 19—

We are dealing with the following projects and problems, as of May 31.

Projects

Status of current projects

1. The Problem-Tracking System now contains both software and hardware problem-tracking capabilities. The system upgrade took place over the weekend of May 11 and 12 and was placed online on the 13th.
2. For the *Software Training Mailing Campaign,* we anticipate producing a set of labels for mailing software-training information to customers by June 10.
3. The *Search Project* is on hold until the PL/I training has been completed, probably by the end of June.
4. The project to provide a data base for the *Information Management System* has been expanded in scope to provide a data base for all training activities. We are in the process of rescheduling the project to take the new scope into account.
5. The *Metering Reports* project is part of a larger project called "Reporting Upgrade." We have completed the Final Project Requirements and sent it out for review. The Resource Requirements estimate has also been completed, and Phase Three is scheduled for completion by June 17.

Problems

Existing problems

The *Information Management System* has been delayed. The original schedule was based on the assumption that a systems analyst who was familiar with the system would work on this project. Instead, the project was assigned to a newly hired systems analyst who was inexperienced and required much more learning time than expected.

Bill Michaels, whose activity report is attached, is correcting a problem in the *CNG Software.* This correction may take a week.

The *Beta Project* was delayed for approximately one week because of two problems: interfacing and link handling. The interfacing problem was resolved rather easily. The link-handling problem, however, was more severe, and Debra Mann has gone to the customer's site in France to resolve it.

Plans for Next Month

Plans for the coming months

- Complete the Software Training Mailing Campaign.
- Resume the Search Project.
- Restart the project to provide a data base on information management with a schedule that reflects its new scope.
- Complete the Phase Three project.
- Write a report to justify the addition of two software engineers to my department.
- Congratulate publicly the recipients of Meritorious Achievement Awards: Bill Thomasson and Nancy O'Rourke.

Current Staffing Level
Current staff: 11
Open requisitions: 0

ja
Enclosure

FIGURE 2   Activity Report

pronoun reference

Avoid vague and uncertain references between a **pronoun** and its antecedent.

CHANGE   Studs and thick treads make snow tires effective. *They* are implanted with an air gun.

TO   Studs and thick treads make snow tires effective. *The studs* are implanted with an air gun.

CHANGE   We made the sale and delivered the product. *It* was a big one.

TO   We made the sale, *which* was a big one, and delivered the product.

The **noun** to which a pronoun refers must be clear. Three problems are encountered in regard to pronoun references.

1. One is an ambiguous reference, or one that can be interpreted in more than one way.

   CHANGE   Jim worked with Tom on the report, but *he* wrote most of it. (Who wrote most of it, Jim or Tom?)

   TO   Jim worked with Tom on the report, but *Tom* wrote most of it.

2. A general (or broad) reference, or one that has no real antecedent, is another problem that often occurs when the word *this* is used by itself.

   CHANGE   He deals with personnel problems in his work. *This* helps him in his personal life.

   TO   He deals with personnel problems in his work. *This experience* helps him in his personal life.

3. A hidden reference, or one that has only an implied antecedent, is the third problem.

   CHANGE   A high lipid, low carbohyrdate diet is "ketogenic" because it favors *their* formation.

   TO   A high lipd, low carbohyrdate diet is "ketogenic" because it favors the formation of *ketone bodies.*

For the sake of **coherence,** pronouns should be placed as close as possible to their antecedents. The danger of creating an ambiguous reference increases with distance. Don't force your **reader** to go back too far to find out what a pronoun stands for.

CHANGE   The *house* in the meadow at the base of the mountain range was resplendent in *its* coat of new paint.

TO   The *house*, resplendent in *its* coat of new paint, was nestled in a meadow at the base of the mountain range.

P

Do not repeat an antecedent in **parentheses** following the pronoun. If you feel that you must identify the pronoun's antecedent in this way, you need to rewrite the sentence.

> CHANGE   The senior partner first met Bob Evans when he *(Evans)* was a trainee.
>
> TO   *Bob Evans* was a trainee when the senior partner first met him.

When referring to a noun that includes both sexes (student, teacher, clerk, everyone), the pronouns *he* and *his* have traditionally been used.

> EXAMPLE   Each employee is to have *his* annual X-ray taken by Friday.

However, it is now generally accepted that this conventional use of the masculine personal pronoun implies sexual bias. If there is danger of offending, it is often best to avoid the problem by substituting an article or changing the statement from singular to plural. (See also **he/she.**)

> EXAMPLES   Each employee is to have *the* annual X-ray taken by Friday.
>
> All employees are to have *their* annual X-rays taken by Friday.

pronouns

A *pronoun* is a word that is used as a substitute for a **noun** (the noun for which a pronoun substitutes is called its *antecedent*). Using pronouns in place of nouns relieves the monotony of repeating the same noun over and over. Pronouns fall into several different categories: personal, demonstrative, relative, interrogative, indefinite, reflexive, intensive, and reciprocal. (See also **pronoun reference.**)

Personal Pronouns

Personal pronouns refer to the person or persons speaking *(I, me, my, mine; we, us, our, ours)*, the person or persons spoken to *(you, your, yours)*, or the person or thing spoken of *(he, him, his; she, her, hers; it, its; they, them, their, theirs)*. (See also **person.**)

> EXAMPLES   *I* wish *you* had told *me* that *she* was coming with *us.*
>
> If *their* figures are correct, *ours* must be in error.

Don't attempt to avoid using the personal pronoun *I* when it is natural. The use of unnatural devices to avoid using *I* is more likely to call attention to the writer than would use of the pronoun *I*.

CHANGE	*The writer* wishes to point out that the proposed solution has several weaknesses.
TO	*I* wish to point out that the proposed solution has several weaknesses.
OR	The proposed solution has several weaknesses.

Also, avoid substituting the reflexive pronouns *myself* for *I* or *me,* or *ourselves* for *we* or *us.*

CHANGE	Joe and *myself* worked all day on it.
TO	Joe and *I* worked all day on it.
CHANGE	He gave it to my assistant and *myself.*
TO	He gave it to my assistant and *me.*
CHANGE	This decision cannot be made by managers like *myself.*
TO	This decision cannot be made by managers like *me.*
CHANGE	Nobody can solve this problem but *ourselves.*
TO	Nobody can solve this problem but *us.*

The use of *we* for general reference ("*we* are living in a time of high inflation") is acceptable. But using the editorial *we* to avoid *I* is pompous and stiff. It may also assume, presumptuously, that "we" all agree.

CHANGE	*We* must state unequivocally that *we* disapprove of such practices.
TO	*I* must state unequivocally that *I* disapprove of such practices.

Bear in mind that the use of *we* can be legally construed to commit the writer's company as well as the writer personally. If you are speaking for yourself—even in your role as an employee—it is better to use *I.*

One is also acceptable for general reference. Be aware, however, that it is impersonal and somewhat stiff and that overdoing it makes your writing awkward.

CHANGE	*One* must be careful about *one's* work habits lest *one* become sloppy.
TO	*People* must be careful about *their* work habits lest *they* become sloppy.

(See also **one** and **point of view.**)

Demonstrative Pronouns

The antecedent of a *demonstrative pronoun (this, that, these, those)* must be either the last noun of the preceding sentence—not the idea of the sentence—or clearly identified within the same sentence, as in the following examples:

EXAMPLES   *This* is the *specification* as we wrote it.
These are the *statistics* we needed to begin the project.
That was a *proposal* for drilling a water well.
Those were *salesmen*.

Do not use the pronouns *this* and *that* to make broad or ambiguous references to a whole sentence or to an abstract thought. (See also **pronoun reference.**) One way to avoid the problem is to insert the noun immediately after the pronoun.

CHANGE   The inadequate quality-control procedure has resulted in an equipment failure. *This* is our most serious problem at present.

TO   The inadequate quality-control procedure has resulted in an equipment failure. *This failure* is our most serious problem at present.

Relative Pronouns

Relative pronouns (the most common are *who, whom, which, what,* and *that*) perform a dual function: They substitute for a noun and link the dependent **clause** in which they appear to a main clause. In each of the following examples, two words are italicized: the relative pronoun and the word for which it is substituting:

EXAMPLES   The specifications were sent to the *manager, who* immediately approved them.

The project was assigned to *Ed Jones, whom* everyone respects for his skill as a designer.

The customer's main *objection, which* can probably be overcome, is to the high cost of installation.

The *memo* from Martha Goldstein *that* you received yesterday sets forth the most important points.

Sometimes a relative pronoun does not stand for another specific noun but does act as a noun in serving as the subject of a clause. In the following example, *what* acts as a subject in the dependent clause while also linking that clause to the main clause:

EXAMPLE   We invited the committee to come to see *what* was being done about the poor lighting.

The relative pronoun *who (whom)* normally refers to a person; *that* and *which* normally refer to a thing; and *that* can refer to either. The pronoun *that* is normally used with restrictive clauses and the pronoun *which* with nonrestrictive clauses. (See also **restrictive** and **nonrestrictive elements.**)

EXAMPLES   The annual report, *which was distributed yesterday,* shows that sales increased 20 percent last year. (nonrestrictive)

Companies *that adopt the plan* nearly always show profit increases. (restrictive)

Relative pronouns can often be omitted.

EXAMPLE   I received the handcrafted ornament (that) I ordered.

Do not omit the word *that,* however, when it is used as a **conjunction** and its omission could cause the **reader** to misread a sentence.

CHANGE   This time delay ensures the operator has time to load the card hopper.

TO   This time delay ensures *that* the operator has time to load the card hopper.

(See also **who/whom.**)

Interrogative Pronouns

Interrogative pronouns (who or *whom, what,* and *which)* ask questions. They differ from relative pronouns in two ways: They are used only to ask questions, and they may introduce independent interrogative sentences, whereas relative pronouns connect or show relationship and introduce only dependent clauses.

Notice that, unlike relative pronouns, interrogative pronouns normally lack expressed antecedents; the antecedent is actually in the expected answer.

EXAMPLES   *Who* said that?
What is the trouble?
Which of these is best?

Indefinite Pronouns

Indefinite pronouns specify a class or group of persons or things rather than a particular person or thing. *All, any, another, anyone, anything, both, each, either, everybody, few, many, most, much, neither, nobody, none, several, some,* and *such* are indefinite pronouns.

EXAMPLE   Not *everyone* liked the new procedures; *some* even refused to follow them.

Most indefinite pronouns require singular verbs.

EXAMPLES   If *either* of the vice-presidents *is* late, we will delay the meeting.
Neither of them *was* available for comment.
Each of the writers *has* a unique style.
Everyone in our department *has* completed the form.

Some indefinite pronouns require plural verbs.

EXAMPLES   *Many are* called, but *few are* chosen.
Several of them *are* aware of the problem.

A few indefinite pronouns may take either plural or singular verbs, depending on whether the nouns they stand for are plural or singular.

EXAMPLES   *Most* of the employees *are* pleased, but *some are* not.
Most of the oil *is* imported, but *some is* domestic.

Reflexive Pronouns

A *reflexive pronoun*, which always ends with the **suffix** *-self* or *-selves*, indicates that the subject of the sentence acts upon itself.

EXAMPLE   The lathe operator cut *himself.*

The reflexive form of the pronoun may also be used as an intensive to give **emphasis** to its antecedent. The reflexive pronouns are *myself, yourself, himself, herself, itself, oneself, ourselves, yourselves,* and *themselves.*

EXAMPLE   I collected the data *myself.*

A common mistake is substituting a reflexive pronoun for *I, me,* or *us.*

CHANGE   John and *myself* completed the report.
TO   John and *I* completed the report.

CHANGE   He gave it to my assistant and *myself.*
TO   He gave it to my assistant and *me.*

CHANGE   Nobody can solve the problem but *ourselves.*
TO   Nobody can solve the problem but *us.*

Intensive Pronouns

Intensive pronouns are identical in form with the reflexive pronouns, but they perform a different function: to give emphasis to their antecedents.

EXAMPLE   I *myself* asked the same question.

Reciprocal Pronouns

Reciprocal pronouns (one another and *each other)* indicate the relationship of one item to another. *Each other* is commonly used when referring to two persons or things and *one another* when referring to more than two.

EXAMPLES   They work well with *each other.*
The crew members work well with *one another.*

(See also **case, gender,** and **number.**)

proofreaders' marks

Publishers have established **symbols,** called *proofreaders' marks,* which writers and editors use to communicate with printers in the production of publications. A familiarity with these symbols makes it easy for you to communicate your changes to others. (See Figure 1 on page 464.)

proofreading

The introduction of the computer spell checker has been at the same time a boon and a bane to *proofreading.* It has been a boon because it does much of the work for us, but the downside is that it tends to make us overconfident, and we get a little sloppy in our proofreading. If a typographical error results in a legitimate English word, the spell checker will not flag it—so we must still proofread. Also, spell checkers are not foolproof—they can and do fail to detect misspelled words. *So don't fail to proofread your work.* Proofread printed copy; do not proofread only on the computer screen.

Proofreading should be done in several stages: read through the material three times, looking for specific things each time, and then read it a final time for content.

During the first time through, ask yourself, "Does it look right?" Look for:

- aesthetic page placement of material
- acceptable **format**
- correct **spelling** of names and places
- accuracy of **numbers**

Mark in Margin	Instruction	Mark on Manuscript	Corrected Type
℮	Delete	the ~~lawyer's~~ bible	the bible
lawyer's	Insert	The bible	The lawyer's bible
(stet)	Let stand	the ~~lawyer's~~ bible	the lawyer's bible
(cap)	Capitalize	the bible	the Bible
(lc)	Make lowercase	the Law	the law
(ital)	Italicize	the lawyer's bible	the *lawyer's* bible
(tr)	Transpose	the bible/lawyer's	the lawyer's bible
⌣	Close space	the Bi ble	the Bible
(sp)	Spell out	②bibles	two bibles
#	Insert space	The/Bible	The Bible
¶	Start paragraph	¶The lawyer's . . .	The lawyer's . . .
(run in)	No paragraph	marks. Below is a	. . . marks. Below is a . . .
(sc)	Set in small capitals	the bible	The BIBLE
(rom)	Set in roman type	The(bible)	The bible
(bf)	Set in boldface	The bible	The **bible**
(lf)	Set in lightface	The(**bible**)	The bible
⊙	Insert period	The lawyers have their own bible	The lawyers have their own bible.
ˆ	Insert comma	However we cannot . . .	However, we cannot . . .
⁼/⁼	Insert hyphens	half and half	half-and-half
⊙	Insert colon	We need the following	We need the following:
;	Insert semicolon	Use the law don't . . .	Use the law; don't . . .
ˇ	Insert apostrophe	Johns law book	John's law book
❡/❡	Insert quotation marks	The law is law.	The "law" is law.
(/)/	Insert parentheses	John's law book	John's (law) book
[/]/	Insert brackets	John 1920-1962 went . . .	John [1920-1962] went . . .
$\frac{1}{N}$	Insert en dash	1920 1962	1920–1962
$\frac{1}{M}$	Insert em dash	Our goal victory	Our goal—victory
ˇ	Insert superior type	$3^2 = 9$	$3^2 = 9$
ˆ	Insert inferior type	HSO_4	H_2SO_4
ˇ	Insert asterisk	The law	The law*
†	Insert dagger	The law	The law†
‡	Insert double dagger	The bible	The bible‡
§	Insert section symbol	Research	§Research

FIGURE 1 Proofreader's Marks

During the second time, ask, "Am I following the rules?" Check for:

- typographical errors
- **capitalization**
- **punctuation**
- **spelling**
- **grammar**

During the third time, ask, "Is it complete?" Look for:

- omissions
- deletions
- doubly typed words

When you read the material for content, make certain that everything is there, that it is correct, and that it is in the right place.

proper nouns (see nouns)

proposals

A *proposal* is a document written to persuade someone to follow a plan or course of action. It may be internal to an organization or be sent outside the organization to a potential client.

Internal proposals usually recommend a change or improvement within an organization. The recommendation might be to expand the cafeteria service, to combine multiple manufacturing operations, or to introduce new working procedures. A capital appropriations proposal, for example, is an internal proposal submitted for management approval to spend large sums of money. An internal proposal is normally prepared by an employee or department and then sent to a higher-ranking person in the organization for approval.

External proposals include both sales proposals and government proposals. Both represent a company's offer to provide goods or services to a potential buyer within a certain amount of time and at a specified cost. The purpose of the proposal is to present a product or service in the best possible light and to explain why a buyer should choose it over the competitors. Remember that the survival of some companies depends on their ability to produce effective proposals.

Since proposals offer plans to fill a need, your **readers** will evaluate your plan according to how well your written presentation answers their questions about what you are proposing to do, how you plan to do it, when you plan to do it, and how much it is going to cost.

P

Make certain that your proposal is written at your reader's level of knowledge. If you have more than one reader—and proposals often require more than one level of approval—take into account all your readers. For example, if your immediate reader is an expert on your subject but the next higher level of management (which must also approve the proposal) is not, provide an **executive summary** written in language that is as nontechnical as possible. You might also include a **glossary** of technical terms used in the body of the proposal, or an **appendix** that explains the technical information in nontechnical language. On the other hand, if your immediate reader is not an expert but the person at the next level is, write the proposal with the non-expert in mind and include an appendix that contains the technical details.

Proposals usually consist of an **introduction,** a body, and a **conclusion.** The introduction should summarize the problem you are proposing to solve and your solution. It may also indicate the benefits that your reader will receive from your solution and its total cost. The body should explain in detail (1) how the job will be done, (2) what methods will be used to do it (and if applicable, the materials to be used and any other pertinent information), (3) when work will begin, (4) when the job will be completed, and (5) a cost breakdown for the entire job. The conclusion should emphasize the benefits for the reader and should urge him or her to take action. Your conclusion should have an encouraging, confident, and reasonably assertive tone.

There are basically three kinds of proposals: government proposals, internal proposals, and sales proposals.

P

Government Proposals

Government proposals are usually prepared as a result of either an *Invitation for Bids* or a *Request for Proposals* that has been issued by a government agency.

Invitation for Bids. An Invitation for Bids is inflexible; the rules are rigid and the terms are not open to negotiation. Any proposal prepared in response to an Invitation for Bids must adhere strictly to its terms.

The Invitation for Bids clearly defines the quantity and type of item that a government agency intends to purchase. It is prepared at the request of the agency that will use the item, such as the Department of Defense. An advertisement indicating that the government intends to purchase the item is published in an official government publication, such as the *Commerce Business Daily.* This publication is required reading for government sales-oriented organizations. The goods or services

Advantages of the Proposed 5,000 H.P. Design

Statement of problem

The Navy Department has indicated to Burdlorn Manufacturing Company a need for main diesel propulsion engines to be used in conjunction with gas turbines on a program with the code name of CODAG. We have been advised that the power requirement for this engine is 5,000 horsepower, plus 10 percent overload. This engine must be contained within a space 24 feet long, 11 feet high, and 7 feet wide and must not exceed 60,000 pounds in weight. In addition, at a 5,000 horsepower operating level, it is not expected that the engine will require replacement or renewal of wearing parts or other major components more frequently than every 5,000 hours. The proposed design changes will result in a very light-weight engine (12 pounds/BHP) but one with durability characteristics that will not be equaled by any engine throughout the world in this compact high-horsepower output.

Brief statement of solution

5,000 Horsepower Engine

Company's technical capability

Burdlorn Manufacturing Company has completed shop tests on NOBS 72393, addenda 12 and 13. During these tests, the Navy engine serial GR-1047-0836 was operated at loads up to 4,950 brake horsepower (just 1 percent short of the 5,000 horsepower goal). As a result of these tests, we are satisfied that the 16-cylinder, 11″ bore, 12″ stroke V-engine, with a rating of 5,000 horsepower at 1,000 rpm, is well within practical limits. On the basis of an intensive review of the service history of the four engines installed on the LST-1176 and all of the research and development work at the Burdlorn factory, coupled with a substantial amount of analytical design work, Burdlorn Manufacturing Company recommends that certain redesigns be effected, based on the above information.

As a result of the recent tests at loads up to 4,950 horsepower that have been run in our factory, Burdlorn Manufacturing Company has successfully dealt with the thermo-dynamic problems of air/fuel requirements, air flow, heat transfer, and other related problems. In proposing design criteria, which are conservative to insure the desired reliability, we have applied sufficient analysis to the individual components to make specific recommendations.

FIGURE 1   Government Proposal *(continued on pages 468–472)*

to be procured are defined in the Invitation for Bids by references to performance standards called **specifications.** If, for instance, the item to be purchased is a truck, the important characteristics of that truck (height, weight, speed, carrying capacity, and so on) will be listed in the machine specification. More than one specification may apply to a single purchase; that is, one specification may apply to the item, another to the manuals, another to welding procedures, and so forth. Bidders must be prepared to prove that their product will meet all requirements of all specifications.

<table>
</table>

Advantages to be outlined

In order to summarize findings that we feel the Navy needs in order to evaluate our proposal, we will outline in numerical sequence the improvements and advantages based on our total experience with the 11″ bore, 12″ stroke, 16-cylinder V-engine.

1. Frame. Burdlorn Manufacturing Company recommends that the present 17 inches center distance from one cylinder to the adjacent cylinder be increased by 3 inches to provide 20 inches center distance from cylinder to cylinder in a given bank. The additional 3 inches per cylinder will mean an overall increase in engine length of 24 inches. This will mean an engine length of approximately 22 feet with the engine-driven auxiliary equipment. This will be well within the 24-foot maximum covered by the Navy specifications. The additional 3 inches will provide the following advantages.

   a. The principal load-carrying member in the frame, which is the athwartship plate, will be increased in thickness from ½ inch to ¾ inch.
   b. Additional clearance is provided for welding, and machining, particularly on the intermediate deck.
   c. Additional length is provided for the main and crankpin journals. This will increase the bearing areas for the main and crankpin bearings and thus gain the advantage of reduced bearing loads.

Minor problem and proposed solution

The results of our experimental stress analysis have indicated a substantial bending movement in the top deck of the present design. Accordingly, Burdlorn Manufacturing Company proposes to increase the thickness of the top deck. The present design uses forgings. We are proposing to use controlled quality steel castings of substantially heavier construction to reduce the stress levels.

Burdlorn Manufacturing Company has done a substantial amount of stress analysis on a similar frame on our Model FS-13½″ × 16½″ V-engine. On the basis of this experience, we recommend a minor redesign of the main bearing saddles to create a more even stress distribution in transmitting the load through the athwartship plates to the main bearing saddles. These main bearing saddles would be made from steel casts such as our present 13½″ × 16½″ engines currently use.

FIGURE 1 Government Proposal (continued)

A specification is restrictive, binding the bidder to produce an item that meets the exact requirements of the specification. For example, if a truck is the specified item, a company may not bid to furnish a cargo helicopter, even though the helicopter might do the job more efficiently than the truck. An Invitation for Bids requires the bidder to furnish a specific product that meets the specification parameters, and nothing else will be accepted or considered. An alternative suggestion will render the bid "nonresponsive," and it will be rejected.

<table>
<tr><td>Introduction to illustration</td><td>To provide for reduction in stress levels in the sidesheets of the engine, we propose increasing the plate thickness from ½ inch to ⅝ inch. In addition, we have found through extensive testing of the Navy engine and our 13½″ × 16½″ V-engine and also by exhaustive photo-elastic stress analysis techniques that the present "hourglass" design results in a stress concentration factor at the "waist" of the frame. Accordingly, we propose a modification in this shape that will reduce the stress concentration factor due to this shape effect (see Figure 9).</td></tr>
</table>

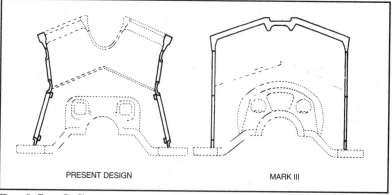

PRESENT DESIGN   MARK III

Figure 9. Frame Profiles

<table>
<tr><td>Another problem and proposed solution with illustration</td><td>One problem experienced repeatedly on the LST-1176 was associated with the bosses for the handhole covers. Stress analysis work has shown a high stress concentration factor around these bosses because of unfused weld roots and also the increased localized stiffness caused by these bosses. Accordingly, we have designed handhole covers held in place by clamping action, which do not impose additional stress in the sidesheet due to the torquing down of the bolts against the bosses. We therefore gain the advantage of this direct means of eliminating a superimposed stress of 5,000 psi (see Figure 10).</td></tr>
</table>

P

FIGURE 1   Government Proposal *(continued)*

Bearing in mind that the product will be tested, measured, and evaluated to see that it does, in fact, meet the requirements of the specification, price is the main criterion that enters into the selection of the vendor. However, an organization with superior engineering skills can often supply a product that meets the specification at a price well below that of a less sophisticated competitor. When a high degree of technological competence, unusual facilities, or other valid requirements make it necessary, the procuring agency may require the bidder

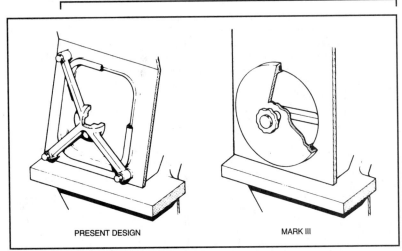

PRESENT DESIGN

MARK III

Figure 10. Side Cover Details

An additional problem experienced on earlier frames was cracking in the air manifold. The present design provides for the air header to be welded in as an integral stress member of the frame. A number of problems are associated with this type of design, some of which are stress concentration factors in the welding together of the top deck, the air header, and the athwartship plate. A second problem associated with this type of construction is the thermogradient experienced because the air manifold temperature is regulated to approximately 100°F., whereas other heat sinks at higher temperature levels create thermal stresses in the frame that are difficult to manage. Accordingly, Burdlorn proposes a separate and independent air header that will be bolted to the frame in such a manner as not to contribute to the overall stresses or stress concentration factors of the frame weldment.

2. Valves. During the 1,000 rpm test of the engine under Navy contract NOBS 72393, a wear rate of approximately .050 inch per 1,000

FIGURE 1 Government Proposal *(continued)*

to possess certain minimum qualifications in order to be considered; a paper clip manufacturer employing ten people, for instance, could not qualify to bid on a procurement of military aircraft.

Request for Proposals. In contrast to the Invitation for Bids, the Request for Proposals is flexible. It is open to negotiation, and it does not necessarily specify exactly what goods or services are required. Often, a Request for Proposals will define a problem and allow those who respond to it to suggest possible solutions. As an example, if a pro-

hours was experienced on the inlet valves. The wear rate on the exhaust valves was approximately .010 inch per 1,000 hours. A total wear of .100 inch would be considered acceptable for this size valve and insert. Accordingly, it will be appreciated that a wear life of more than 5,000 hours could be expected on the exhaust valve, but a much shorter life was experienced on the inlet valve. A careful analysis of all available information has revealed only two differences between the intake valve and the exhaust valve that would contribute to this difference in wear rates. These two differences are as follows:

a. The exhaust cam is driven through a hollow shaft that is 4 inches O.D. by 2½ inches I.D. The inlet cam is driven by a solid shaft that runs within the hollow exhaust camshaft and is 2½ inches in diameter. It will be recognized that the relative stiffness of these two shafts in torsion is in the ratio of 5½:1 as revealed by our calculations (see Figure 11).

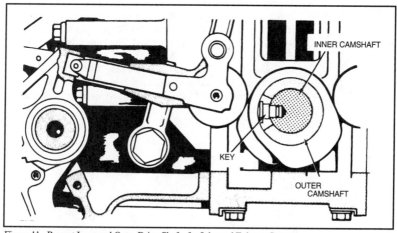

Figure 11. Present Inner and Outer Drive Shafts for Inlet and Exhaust Cams

FIGURE 1 Government Proposal *(continued)*

curing agency wanted to develop a way to make foot soldiers more mobile (capable of covering difficult terrain at high speed), a Request for Proposals would normally be the means used to find the best method and select the most qualified vendor. In some instances, Requests for Proposals are presented in two or more stages, the first being development of a concept, the second being a "prototype" machine or device, and the third being the manufacture of the device selected. The procedure for preparing a proposal in response to a Request for Proposals is as follows:

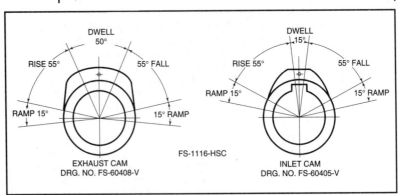

Figure 12. Comparison of Inlet and Exhaust Cam Profiles

b. The lift profiles of the intake and exhaust cams are identical; however, there is a greater dwell on the exhaust cam than there is on the inlet cam. The inlet cam has 15 degrees dwell, whereas the exhaust cam has 50 degrees dwell. The longer dwell on the exhaust cam, which cannot be incorporated in the inlet cam due to timing requirements, provides additional time in which the vibratory amplitudes in the valve train dampen or attentuate to a greater extent than the 15 degrees dwell provides on the inlet cam (see Figure 12). . . .

FIGURE 1 Government Proposal *(continued)*

The procuring agency defines the problem and publishes it as a Request for Proposals in one or more business journals. Companies interested in government business scan the appropriate publications daily. Upon finding a project of interest, the sales department of such a company obtains all available information from the procuring agency. It then presents the data to management for a decision on whether the company is interested in the project. If the decision is positive, the corporate technical staff is assigned the task of developing a "concept" to accomplish the task posed by the Request for Proposals.

The technical staff normally considers several alternatives, selecting one that combines feasibility and price. The staff's concept is presented to management for a decision on whether the company wishes to present a proposal to the requesting government agency.

Assuming that the decision is to proceed, preparation of a proposal is the next step. At a minimum, the proposal should provide a clear-cut statement of the problem and the proposed solution. It should include data to show that the company is financially and technologically sound (this may require a résumé of the qualifications of the people in charge and an **organizational chart** to show the chain of command). A discussion of the company's manufacturing capabilities and any other advantages is also advisable. However, in dealing with government agencies, bear in mind that unnecessarily elaborate and costly proposals may be construed as a lack of cost consciousness.

In many instances, the final cost to the government is determined by negotiation after the best overall concept has been selected. However, if your concept is one that provides a simple and inexpensive solution to the problem, cost is an advantage that you should point out. If your concept is expensive, explain the benefits of advanced engineering, speed, increased capacity, and so forth.

The excerpt presented in Figure 1 on pages 467–472 comes from a proposal that was prepared in response to a Request for Proposals. The rest of the proposal continues to explain the design of the engine's cylinder heads, bearings, crankshaft, pistons, cylinder liners, connecting rod, and reversing mechanism.

P

Internal Proposal

The purpose of an *internal proposal* is to suggest a change or an improvement within an organization. For example, one might be written to propose initiating a new management reporting system, expanding the cafeteria service, or changing the policy on parking privileges. An internal proposal, usually in a **memo** format, is prepared by a person or a department and is sent to a higher-ranking person who has the authority to accept or reject the proposal.

In the **opening** of a proposal, you must establish that a problem exists that needs a solution. If the person to judge the proposal is not convinced that there is a problem, your proposal will have no chance of success. Notice how the sample proposal on page 474 states its problem directly.

The body of a proposal should offer a practical solution to the problem. In building a case for a solution, be as specific as possible.

To: Ronald Wepner, General Manager
From: Nevil Broadmoor, Data Center Manager *N. B.*

Date: June 14, 19—

Subject: Expansion of the Data Center

Problem

Because the data center is now operating at its maximum capacity, we have found ourselves having to decline new business. By expanding our equipment, staff, and facility, we could attract new customers while offering our present customers additional and faster services. I have outlined a plan to expand the data center that would allow us to increase our business—and thus our net income—about $300,000 a year.

Requirements

Solution

Equipment. To provide the computer time required to process new applications, we need to reduce our present time requirements. By upgrading our processor to a V8565M, upgrading our Century software to VRX software, and adding four 658 disc units, we would reduce the time required for our present applications by as much as 50 percent. The addition of six online terminals and a multiplexer would reduce our program maintenance time as much as 60 percent, which would allow our staff to convert applications to take advantage of the new software.

Staff. To convert present applications to VRX software and to develop and support new applications, we need to increase our programming staff by three people (one senior programmer and two junior programmers).

Facility. To house the additional staff, we need to expand our office area and to purchase additional office furniture for the six online terminals we would place in the programmers' offices. The V8565M requires less space than the present processor, however, so we could

P

FIGURE 2 Internal Proposal

When it is appropriate, include (1) a breakdown of costs; (2) information about equipment, material, and personnel requirements; and (3) a schedule for completing the task. Such information can help your readers to think about the proposal and thus may stimulate them to act. Consider how the examples on pages 475 and 476 do it.

The conclusion should be brief but must tie everything together. It is a good idea to restate recommendations here. Be careful, of course, to conclude in a spirit of cooperation, offering to set up a meeting, rearrange the computer room to provide space for the multiplexer and additional disc units.

Benefits

On the basis of the business available in this area, we could expect to increase our net income by $300,000 a year. We would be able to attract new customers—as well as new business from our present customers—by offering additional and faster services. The increase in our operating budget would actually be small because of the reduction in time requirements provided by the new equipment.

Costs

Equipment. The purchase of new equipment—reduced by the trade-in allowance on our old processor—would be $383,000.

Cost
breakdown

Processor	$250,000
Integrated Disc Controller	20,000
Disc Units (four @ $20,000 each)	80,000
Integrated Multiplexer	10,000
Online Terminals (six @ $3,000 each)	18,000
VRX Software	5,000

Budget. The annual increase in our operating budget would be $86,800.

Salaries	$75,000
Benefits	6,800
Office Supplies	1,000
Utilities	1,000
Equipment Maintenance	3,000

Facility. The cost of expanding our facility would be $33,000.

Office Area	$20,000
Computer Room	10,000
Furniture	3,000

FIGURE 2   Internal Proposal *(continued)*

supply further facts and figures, or providing any other assistance that might solve the problem, as is done in the example on page 476.

A proposal to commit large sums of money is a common internal proposal. Although such internal proposals have various names, *capital appropriations proposal* is a common and descriptive name for them.

The introduction to such a proposal should provide any background information your reader might need in order to make the decision in question. The introduction should also briefly describe any feasibility

---

Schedule

To achieve the expected net income increase, we would need to meet the following schedule in the expansion.

    July 1—Order the required equipment.
    September 1—Begin recruiting efforts.
    September 15—Start expanding the office area.
    October 1—Order the required office furniture.
    November 1—Start soliciting new business.
    November 5—Finish expanding office area.
    November 5—Fill programming positions
    November 10—Start new applications design.
    November 10—Start conversion effort.
    December 1—Install new equipment.
    December 15—Implement the first new customer application.

Conclusion

I feel that this expansion plan would enable us to meet the requirements of the businesses in this area. I have looked at the alternative equipment and believe that the V8565M offers better performance than the other hardware available in this price range. The design of the V8565M would allow us to grow with a minimum of effort as the business in this area grows. I would like to meet with you Friday to discuss this plan and will be happy to explain any details or to gather any additional information that you may desire. Please let me know if we can get together then to discuss this proposed expansion of the data center.

P

FIGURE 2   Internal Proposal *(continued)*

---

study you may have conducted, the conclusions the study led you to reach, and the decision you made on the basis of those conclusions.

The body of the proposal should identify all the equipment or property you are asking to acquire and should state the economic justification for the expenditure involved. If you need to purchase equipment, be sure to list all of it so that later you don't have to prepare another proposal to purchase the rest of it. Include an economic justification of either the calculated return on investment or the internal rate of return. Your proposal should include all of the following items that are applicable to your particular request:

1. A brief description of the item to be purchased.
2. The alternative choices you considered before deciding on the one you are recommending.
3. The specific products or projects to be supported by the item.
4. The volume of usage that the item would receive over a specified period of time, such as the next two years.
5. If the company already has some of the item, an explanation of why you require an additional purchase.
6. The life expectancy of the item and of the product or project it is to support.
7. Any future products or projects that the item could be used to support.
8. Whether a leasing arrangement is available and more economical than purchasing the item.
9. The impact on the company if the item is not acquired.
10. The latest feasible date of acquisition, including the impact that a ninety-day delay would have.
11. The calculated return on investment, or the internal rate of return.

With your capital appropriations request, you should also submit a cover **memorandum.** The memo, which should be no more than one page long, should give a brief description of the item and the reason it is needed. Figure 2 on pages 474–476 is such a capital appropriations proposal.

P

Sales Proposal

The *sales proposal,* one of the major marketing tools in use in business and industry today, is a company's offer to provide specific goods or services to a potential buyer within a specified period of time and for a specified price. The primary purpose of a sales proposal is to demonstrate that the prospective customer's purchase of the seller's products or services will solve a problem, improve operations, or offer benefits. Indeed, the sales proposal should sell solutions to problems rather than equipment or services.

Sales proposals vary greatly in size and relative sophistication. Some may be a page or two long and be written by one person; others may be many pages long and be written by several people; still others may take hundreds of pages and be written by a team of professional proposal writers. A short sales proposal might bid for the construction of a single home; a sales proposal of moderate length might bid for the installation of a network of computer systems; and a very long sales proposal might be used to bid for the construction of a multibillion-dollar aircraft carrier.

Your first task in writing a sales proposal is to find out exactly what your prospective customer needs. The information in a proposal should be gathered in a survey of the potential customer's business and needs. You must then determine whether your organization can satisfy the customer's needs. Before preparing a sales proposal, you should try to find out who your principal competitors are. Then compare your company's strengths with those of the competing firms, determine your advantages over your competitors, and emphasize them in your proposal. For example, a small software company bidding for an Air Force contract at a local base would be familiar with its competitors. If the proposal writer believes that his or her company has better-qualified personnel than do its competitors, he or she might include the **résumés** of the key people who would be involved in the project, as a way of emphasizing that advantage.

Solicited and Unsolicited Proposals. Sales proposals may be either solicited or unsolicited. The latter are not so unusual as they may sound: companies often operate for years with a problem they have never recognized (unnecessarily high maintenance costs, for example, or poor inventory-control methods). If you learn of such a company, you might prepare an unsolicited proposal if you were convinced that the potential customer could realize substantial benefits by adopting your solution to the problem. You would, of course, need to persuade the customer of the excellence of your idea and of his or her need for what you are proposing.

To ensure that you don't waste time and effort, you might precede an unsolicited proposal with an inquiry to determine any potential interest. If you receive a positive response, you might then need to conduct a detailed study of the prospective customer's needs and requirements in order to determine whether you can be of help and, if so, exactly how. You would then prepare your proposal on the basis of your study.

An unsolicited proposal should clearly identify the potential customer's problem, but at the same time the proposal should be careful not to overstate the problem. Furthermore, the proposal must convince the customer that the problem needs to be solved. One way to do this is to emphasize the benefits the customer will realize from the solution being proposed.

The other type of sales proposal—the solicited sales proposal—is a response to a request for goods or services. Procuring organizations that would like competing companies to bid for a job commonly issue a *Request for Proposals.* The Request for Proposals is the means used by many companies and government agencies to find the best method of doing a job and the most qualified company to do it. A Request for Proposals may be rigid in its specifications governing how the proposal should be organized and what it should contain, but it is normally quite flexible about the approaches that the bidding firms may propose. Normally, the Request for Proposals simply defines the basic work that the firm needs.

The procuring organization generally publishes its Request for Proposals in one or more journals, in addition to sending it to certain companies that have good reputations for doing the kind of work needed. Some companies and government agencies even hold a conference for the competing firms at which they provide all pertinent information about the job being bid for.

Managers interested in responding to Requests for Proposals regularly scan the appropriate publications. Upon finding a project of interest, an executive in the sales department of such a company obtains all available information from the procuring company or agency. The data is then presented to management for a decision on whether the company is interested in the project. If the decision is positive, the technical staff is assigned to develop an approach to the work described in the Request for Proposals. The technical staff normally considers several alternatives, selecting one that combines feasibility and a price that offers a profit. The staff's concept is presented to higher management for a decision on whether the company wishes to present a proposal to the requesting organization. Assuming that the decision is to proceed, preparation of the proposal is the next step.

When you respond to a Request for Proposals, pay close attention to any specifications in the request governing the preparation of the proposal, and follow them carefully.

Writing a Simple Sales Proposal.   Even a short and uncomplicated sales proposal should be carefully planned. The introduction should state the purpose and scope of the proposal. It should indicate the dates on which you propose to begin and complete work on the project, any special benefits of your proposed approach, and the total cost of the project. The introduction could also refer to any previous association your company may have had with the potential customer, assuming that the association was positive and mutually beneficial.

PROPOSAL
TO LANDSCAPE THE NEW CORPORATE HEADQUARTERS
OF THE
WATFORD VALVE CORPORATION

Submitted to: Ms. Tricia Olivera, Vice-President
Submitted by: Jerwalted Nursery, Inc.
Date Submitted: February 1, 19—

Introduction states purpose and scope of proposal, indicates when project can be started and completed.

Jerwalted Nursery, Inc., proposes to landscape the new corporate headquarters of the Watford Valve Corporation, on 1600 Swason Avenue, at a total cost of $8,000. The lot to be landscaped is approximately 600 feet wide and 700 feet deep. Landscaping will begin no later than April 30, 19— and will be completed by May 31.

The following trees and plants will be planted, in the quantities and sizes given and at the prices specified.

Body lists products to be provided, cost per item.

    4 maple trees (not less than 7 ft.) @ $40 each—$160
   41 birch trees (not less than 7 ft.) @ $65 each—$2,665
    2 spruce trees (not less than 7 ft.) @ $105 each—$210
   20 juniper plants (not less than 18 in.) @ $15 each—$300
   60 hedges (not less than 18 in.) @ $7 each—$420
  200 potted plants (various kinds) @ $2 each—$400

                          Total Cost of Plants = $4,155
                                     Labor = $3,845
                                Total Cost = $8,000

Conclusion specifies time limit of proposal, expresses confidence, and looks forward to working with prospective customer.

All trees and plants will be guaranteed against defect or disease for a period of 90 days, the warranty period to begin June 1, 19—.

The prices quoted in this proposal will be valid until June 30, 19—.

Thank you for the opportunity to submit this proposal. Jerwalted Nursery, Inc., has been in the landscaping and nursery business in the St. Louis area for thirty years, and our landscaping has won several awards and commendations, including a citation from the National Association of Architects. We are eager to put our skills and knowledge to work for you, and we are confident that you will be pleased with our work. If we can provide any additional information or assistance, please call.

FIGURE 3   Short Sales Proposal

The body of a sales proposal should itemize the products and services you are offering. It should include, if applicable, a discussion of the procedures you would use to perform the work and any materials to be used. It should also present a time schedule indicating when each stage of the project would be completed. Finally, the body should include a breakdown of the costs of the project.

The conclusion should express your appreciation for the opportunity to submit the proposal and your confidence in your company's ability to do the job. You might add that you look forward to establishing good working relations with the customer and that you would be glad to provide any additional information that might be needed. Your conclusion could also review any advantages your company may have over its competitors. It should specify the time period during which your proposal can still be considered a valid offer. If any supplemental materials, such as blueprints or price sheets, accompany the proposal, include a list of them at the end of the proposal. Figure 3 shows a typical short sales proposal.

Writing a Long Sales Proposal. Wherever possible in a large sales proposal, use eye-catching graphics instead of sentences and paragraphs. Lists and flowcharts are particularly effective. Customize your proposal whenever possible.

EXAMPLE   A PROPOSAL PREPARED EXPRESSLY FOR
Mr. William A. Kurtz
Vice-President
The Cambridge Company
Tucson, Arizona

Also, whenever possible, include testimonial letters from other users, stressing specific benefits. Be sure to include the appropriate advertising literature, a copy of your company's annual report, and the contract.

A large sales proposal should contain the following elements:

• Transmittal letter
• Title page
• Executive summary
• Description of method currently being used
• Description of proposed method
• Comparison of present and proposed methods
• Equipment recommendations
• Cost analysis
• Delivery schedule
• Summary of advantages
• Statement of responsibilities
• Description of vendor
• Advertising literature
• Contract

P

The **cover letter** expresses your appreciation for the opportunity to submit your proposal and for any assistance you may have received in studying the customer's requirements. The letter should acknowledge any previous association with the customer, assuming that it was a

---

THE WATERS CORPORATION
17 North Waterloo Blvd., Tampa, Florida 33607

September 1, 19—

Mr. John Yeung, General Manager
Cookson's Retail Stores, Inc.
Savannah, Georgia 31499

Dear Mr. Yeung:

Express your appreciation for the opportunity to make the proposal

The Waters Corporation appreciates the opportunity to respond to Cookson's Request for Proposals, dated July 26, 19—. We would like to thank Mr. Becklight, Director of your Management Information Systems Department, for his invaluable contributions to the study of your operations that we conducted before preparing our proposal.

Summarize the proposal's recommendations

Our proposal describes a Waters Interactive Terminal/Retail Processor System designed to meet Cookson's network and processing needs. It will provide all of your required capabilities, from the point-of-sale operational requirements at the store terminals to the host processor. The system uses the proven Retail III modular software, with its point-of-sale applications, and the superior Interactive Terminal with its advanced capabilities and design. This system is easily installed without massive customer reprogramming.

Make a soft sales pitch

The Waters Interactive Terminal/Retail Processor System, which is compatible with much of Cookson's present equipment, not only will answer your present requirements but also will provide the flexibility to add new features and products in the future. The system's unique hardware modularity, efficient microprocessor design, and flexible programming capability will greatly reduce the risk of obsolescence.

Thank you for the opportunity to present this proposal. We are confident that you will select our system, and we look forward to working with you.

Sincerely yours,

Janet A. Curtain

Janet A. Curtain
Account Manager
General Merchandise Systems

JAC/mo

FIGURE 4   Cover Letter for a Proposal

positive experience. Then it should summarize the recommendations offered in the proposal and express your confidence that they will satisfy the customer's needs. Figure 4 shows the cover letter for the proposal illustrated in Figures 5 through 13, a proposal that the Waters Corporation of Tampa provide a computer system for the Cookson chain of retail stores.

An **executive summary** follows the transmittal letter. Written to the executive who will ultimately accept or reject the proposal, it

---

EXECUTIVE SUMMARY

The Waters 319 Interactive Terminal/615 Retail Processor System will give your management the tools necessary to manage people and equipment more profitably with procedures that will yield more cost-effective business controls for Cookson's.

The equipment and applications proposed for Cookson's will respond to your current requirements and allow for a logical expansion in the future.

The features and hardware in the system were determined from data acquired through the comprehensive survey we conducted at your stores in February of this year. The total of 71 Interactive Terminals proposed to service your four stores is based on the number of terminals currently in use and on the average number of transactions processed during normal and peak periods. The planned remodeling of all four stores was also considered, and the suggested terminal placement has been incorporated into the working floor plan. The proposed equipment configuration and software applications have been simulated to determine system performance based on the volumes and anticipated growth rates of the Cookson stores.

The information from the survey was also used in the cost justification, which was checked and verified by your controller, Mr. Deitering. The cost effectiveness of the Waters Interactive Terminal/Retail Processor System is apparent. Expected savings, such as the projected 46-percent reduction in sales audit expenses, are realistic projections based on Waters's experience with other installations of this type.

FIGURE 5 Executive Summary

should summarize in nontechnical language how you plan to approach the work. Figure 5 shows the executive summary of the Waters Corporation proposal.

If your proposal offers products as well as services, it should include a *general description* of the products, as in Figure 6.

GENERAL SYSTEM DESCRIPTION

The point-of-sale system that Waters is proposing for Cookson's includes two primary Waters products. These are the 319 Interactive Terminal and the 615 Retail Processor.

Waters 319 Interactive Terminal

The primary component in the proposed retail system is the Interactive Terminal. It contains a full microprocessor, which gives it the flexibility that Cookson's has been seeking.

The 319 Interactive Terminal gives you freedom in sequencing a transaction. You are not limited to a preset list of available steps or transactions. The terminal program can be adapted to provide unique transaction sets, each designed with a logical sequence of entry and processing to accomplish required tasks. In addition to sales transactions recorded on the selling floor, specialized transactions, such as theater tickets sales and payments, can be designed for your customer service area.

The 319 Interactive Terminal also functions as a credit authorization device, either by using its own floor limits or by transmitting a credit inquiry to the 615 Retail Processor for authorization.

Data-collection formats have been simplified so that transaction editing and formatting are easier. Mr. Sier has already been given the documentation on these formats and has outlined all data-processing efforts that will be necessary to adapt the data to your current systems. These projections have been considered in the cost justification.

Waters 615 Retail Processor

The Waters 615 Retail Processor is a minicomputer system designed to support the Waters family of retail terminals. The processor will reside in the computer room in your data center in Buffalo. Operators already on your staff will be trained to initiate and monitor its activities.

The 615 will collect data transmitted from the retail terminals, process credit, check authorization inquiries, maintain files to be accessed by the retail terminals, accumulate totals, maintain a message-routing network, and control the printing of various reports. The functions and level of control performed at the processor depend on the peripherals and software selected.

Software

The Retail III software used with the system has been thoroughly tested and is operational in many Waters customer installations.

The software provides the complete processing of the transaction, from the interaction with the operator on the sales floor through the data capture on cassette or disk in stores and your data center.

Retail III provides a menu of modular applications for your selection. Parameters adapt each of them to your hardware environment and operation requirements. The selection of hardware will be closely related to the selection of the software applications.

FIGURE 6   General Description of Products

After the executive summary and the *general description,* explain exactly how you plan to do what you are proposing. Because this section will be read by specialists who can understand and evaluate your plan, you can feel free to use technical language and discuss complicated con-

PAYROLL APPLICATION

Current Procedure

Your current system of reporting time requires each hourly employee to sign a time sheet; the time sheet is reviewed by the department manager and sent to the Payroll Department on Friday evening. Since the week ends on Saturday, the employee must show the scheduled hours for Saturday and not the actual hours; therefore, the department manager must adjust the reported hours on the time sheet for employees who do not report on the scheduled Saturday or who do not work the number of hours scheduled.

The Payroll Department employs a supervisor and three full-time clerks. To meet deadlines caused by an unbalanced work flow, an additional part-time clerk is used for 20 to 30 hours per week. The average wage for this clerk is $7.58 per hour.

Advantage of Waters's System

The 319 Interactive Terminal can be programmed for entry of payroll data for each employee on Monday mornings by department managers, with the data reflecting actual hours worked. This system would eliminate the need for manual batching, controlling, and keypunching. The Payroll Department estimates conservatively that this work consumes 40 hours per week.

Hours per week	40
Average wage	× 7.58
Weekly payroll cost	$303.20
Annual savings	$15,766.40

Elimination of the manual tasks of tabulating, batching, and controlling can save .25 units. Improved work flow resulting from timely data in the system without keypunch processing will allow more efficient use of clerical hours. This would reduce payroll by the .50 units currently required to meet weekly check disbursement.

Eliminate manual tasks	.25
Improve work flow	.75
40-hour unit reduction	1.00
Hours per week	40
Average wage	7.58
Savings per week	$319.20
Annual savings	$15,984.00

TOTAL SAVINGS: $31,750.40

FIGURE 7 Detailed Solution

cepts. Figure 7 shows one part of the detailed solution appearing in the Waters Corporation proposal, which included several other applications in addition to the payroll application. Notice that this discussion, like that found in an unsolicited sales proposal, begins with a state-

COST ANALYSIS

This section of our proposal provides detailed cost information for the Waters 319 Interactive Terminal and the 615 Waters Retail Processor. It then extends these major elements by the quantities required at each of your four locations.

319 Interactive Terminal

	Price	Maint. (1 yr)
Terminal	$2,895	$167
Journal Printer	425	38
Receipt Printer	425	38
Forms Printer	525	38
Software	220	═══
TOTALS	$4,490	$281

615 Retail Processor

	Price	Maint. (1 yr)
Processor	$57,115	$5,787
CRT I/O Writer	2,000	324
Matrix Printer	4,245	568
Software	12,480	═══
TOTALS	$75,840	$6,679

The following breakdown itemizes the cost per store.

Store No. 1

Description	Qty.	Price	Maint.
Terminals	16	$68,400	$4,496
Digital Cassette	1	1,300	147
Thermal Printer	1	2,490	332
Software	16	3,520	═══
TOTALS		$75,710	$4,975

FIGURE 8   Cost Analysis

ment of the customer's problem, follows with a statement of the solution, and concludes with a statement of the benefits to the customer.

Essential to any sales proposal are a *cost analysis* and a *delivery schedule.* The cost analysis itemizes the estimated cost of all the

Store No. 2

Description	Qty.	Price	Maint.
Terminals	20	$85,400	$5,620
Digital Cassette	1	1,300	147
Thermal Printer	1	2,490	332
Software	20	4,400	---
TOTALS		$93,590	$6,099

Store No. 3

Description	Qty.	Price	Maint.
Terminals	17	$72,590	$4,777
Digital Cassette	1	1,300	147
Thermal Printer	1	2,490	332
Software	20	3,740	---
TOTALS		$80,120	$5,256

Store No. 4

Description	Qty.	Price	Maint.
Terminals	18	$76,860	$5,058
Digital Cassette	1	1,300	147
Thermal Printer	1	2,490	332
Software	18	3,960	---
TOTALS		$84,610	$5,537

Data Center at Buffalo

Description	Qty.	Price	Maint.
Processor	1	$57,115	$5,787
CRT I/O Writer	1	2,000	324
Matrix Printer	1	4,245	568
Software	1	12,480	---
TOTALS		$75,840	$6,679

The following summarizes all costs

Location	Hardware	Maint.	Software
Store No. 1	$72,190	$4,975	$3,520
Store No. 2	89,190	6,099	4,400
Store No. 3	76,380	5,256	3,740
Store No. 4	80,650	5,537	3,960
Data Center	63,360	6,679	12,480
Subtotals	$381,770	$28,546	$28,100

TOTAL $438,416

P

FIGURE 8   Cost Analysis *(continued)*

products and services that you are offering; the delivery schedule commits you to a specific timetable for providing those products and services. Figure 8 shows the cost analysis, and Figure 9 shows the delivery schedule of the Waters Corp. proposal.

DELIVERY SCHEDULE

Waters is normally able to deliver 319 Interactive Terminals and 615 Retail
Processors within 120 days from the date of the contract. This can vary
depending on the rate and size of incoming orders.

All the software recommended in this proposal is available for immediate
delivery. We no not anticipate any difficulty in meeting your tentative delivery
schedule.

P

FIGURE 9 Delivery Schedule

Also essential if your recommendations include modifying the customer's physical facilities is a *site preparation description* that details the
required modifications. Figure 10 shows the site preparation section of
the Waters proposal.

SITE PREPARATION

Waters will work closely with Cookson's to ensure that each site is properly prepared prior to system installation. You will receive a copy of Waters's installation and wiring procedures manual, which lists the physical dimensions, service clearance, and weight of the system components in addition to the power, logic, communication-cable, and environmental requirements. Cookson's is responsible for all building alterations and electrical facility changes, including the purchase and installation of communication cables, connecting blocks, and receptacles.

Wiring

For the purpose of future site considerations, Waters's in-house wiring specifications for the system call for two twisted pair wires and twenty-two shielded gauges. The length of communications wires must not exceed 2,500 feet.

As a guide for the power supply, we suggest that Cookson's consider the following:

1. The branch circuit (limited to 20 amps) should service no equipment other than 319 Interactive Terminals.
2. Each 20-amp branch circuit should support a maximum of three Interactive Terminals.
3. Each branch circuit must have three equal-sized conductors—one hot leg, one neutral, and one insulated isolated ground.
4. Hubbell IG 5362 duplex outlets or the equivalent should be used to supply power to each terminal.
5. Computer-room wiring will have to be upgraded to support the 615 Retail Processor.

P

FIGURE 10 Site Preparation Section

If the products you are proposing require training the customer's personnel, your proposal should specify the *required training* and its cost. Figure 11 shows the training section of the Waters proposal.

TRAINING

To ensure a successful installation, Waters offers the following training course for your operators.

Interactive Terminal/Retail Processor Operations

Course number: 8256
Length: three days
Tuition: $500.00 per person

This course provides the student with the skills, knowledge, and practice required to operate an Interactive Terminal/Retail Processor System. On-line, clustered, and stand-alone environments are covered.

We recommend that students have department store background and that they have some knowledge of the system configuration with which they will be working.

P

FIGURE 11 Training Section

To prevent misunderstandings about what you and your customer's responsibilities will be, you should draw up a *statement of responsibilities,* as shown in Figure 12. It usually appears toward the end of the proposal. Also toward the end of the proposal is a

---

RESPONSIBILITIES

On the basis of its years of experience in installing information-processing systems, Waters believes that a successful installation requires a clear understanding of certain responsibilities.

Generally, it is Waters's responsibility to provide its users with needed assistance during the installation so that live processing can begin as soon thereafter as is practical.

Waters's Responsibilities

- Provide operations documentation for each application that you acquire from Waters.
- Provide forms and other supplies as ordered.
- Provide specifications and technical guidance for proper site planning and installation.
- Provide advisory assistance in the conversion from your present system to the new system.

Customer's Responsibilities

- Identify an installation coordinator and system operator.
- Provide supervisors and clerical personnel to perform conversion to the system.
- Establish reasonable time schedules for implementation.
- Ensure that the physical site requirements are met.
- Provide competent personnel to be trained as operators and ensure that other employees are trained as necessary.
- Assume the responsibility for implementing and operating the system.

FIGURE 12 Statement of Responsibilities

description of the vendor (Figure 13), which should give a factual description of your company, its background, and its present position in the industry. After this, many proposals add what is known as an *institutional sales pitch* (Figure 14). Up to this point, the pro-

---

DESCRIPTION OF VENDOR

The Waters Corporation develops, manufactures, markets, installs, and services total business information-processing systems for selected markets. These markets are primarily in the retail, financial, commercial, industrial, health-care, education, and government sectors.

The Waters total system concept encompasses one of the broadest hardware and software product lines in the industry. Waters computers range from small business systems to powerful general-purpose processors. Waters computers are supported by a complete spectrum of terminals, peripherals, data-communication networks, and an extensive library of software products. Supplemental services and products include data centers, field service, systems engineering, and educational centers.

The Waters Corporation was founded in 1934 and presently has approximately 26,500 employees. The Waters headquarters is located at 17 North Waterloo Boulevard, Tampa, Florida, with district offices throughout the United States and Canada.

FIGURE 13 Description of Vendor

posal has attempted to sell specific goods and services. The sales pitch, striking a somewhat different chord, is designed to sell the company and its general capability in the field. Less factual than the vendor description, the sales pitch promotes the company and concludes the proposal on an upbeat note.

WHY WATERS?

Corporate Commitment to the Retail Industry

Waters's commitment to the retail industry is stronger than ever. We are continually striving to provide leadership in the design and implementation of new retail systems and applications that will ensure our users of a logical growth pattern.

Research and Development

Over the years, Waters has spent increasingly large sums on research and development efforts to ensure the availability of products and systems for the future. In 19—, our research and development expenditure for advanced systems design and technological innovations reached the $70 million level.

Leading Point-of-Sale Vendor

Waters is a leading point-of-sale vendor, having installed over 150,000 units. The knowledge and experience that Waters has gained over the years from these installations ensure well-coordinated and effective systems implementations.

P

FIGURE 14 Institutional Sales Pitch

proved/proven

Both *proved* and *proven* are acceptable past participles of *prove,* although *proved* is currently in wider use.

EXAMPLES They had *proved* more obstinate than expected.
They had *proven* more obstinate than expected.

Proven is more commonly used as an **adjective.**

> EXAMPLE   She was hired because of her *proven* competence as a manager.

pseudo/quasi

As a **prefix,** *pseudo,* meaning "false or counterfeit," is joined to a word without a **hyphen** unless the word begins with a **capital letter.**

> EXAMPLES   *pseudo*science (false science)
> *pseudo*-Americanism (pretended Americanism)

Pseudo is sometimes confused with *quasi,* meaning "somewhat" or "partial." Unlike *semi, quasi* does not mean half. Quasi is usually hyphenated in combinations.

> EXAMPLE   *quasi*-scientific literature

(See also **bi-/semi-.**)

punctuation

Punctuation is a system of **symbols** that helps the **reader** understand the structural relationship within (and the intention of) a **sentence.** Marks of punctuation may link, separate, enclose, indicate omissions, terminate, and classify sentences. Most of the thirteen punctuation marks can perform more than one function. The use of punctuation is determined by grammatical conventions and the writer's intention—in fact, punctuation often substitutes for the writer's facial expressions. Misuse of punctuation can cause your reader to misunderstand your meaning. Detailed information on each mark of punctuation is given in its own entry. The following are the thirteen marks of punctuation:

apostrophe	'
brackets	[ ]
colon	:
comma	,
dash	—
exclamation mark	!
hyphen	-
parentheses	( )
period	.
question mark	?

quotation marks	" "
semicolon	;
slash	/

purpose

What do you want your **readers** to know or be able to do when they have read your finished writing project? When you have answered this question, you have determined the *purpose,* or *objective,* of your writing project. Too often, however, beginning writers state their purpose in broad terms that are of no practical value to their readers. Such a purpose as "to explain the Model 6000 Accounting System" is too general to be of any real help. But "to explain how to operate a Model 6000 Accounting System" is a specific purpose that will help keep the writer on the right track.

The writer's purpose is rarely simply to "explain" something, although on occasion it may be. You must ask yourself, "Why do I need to explain it?"

A writer for a company magazine who has been assigned to write an article about cardiopulmonary resuscitation, in answer to the question *what,* could state the purpose as "to show the importance of CPR." In answer to the question *why,* the writer might state, "so employees will sign up for evening classes."

If you answer these two questions and put your answers in writing as your stated objective, your job will be made easier and you will be confident of ultimately reaching your goal. As a test of whether you have adequately formulated your objective, try to state it in a single sentence. If you find that you cannot, continue to formulate your objective until you can.

Even a specific objective is of no value, however, unless you keep it in mind as you work. Guard against losing sight of your objective as you become involved with the other steps of the writing process.

P

Q

question marks

The *question mark* (?) has the following uses:

Use a question mark to end a sentence that is a direct question.

EXAMPLE   Where did you put the specifications?

Use a question mark to end any statement with an interrogative meaning (a statement that is declarative in form but asks a question).

EXAMPLE   The report is finished?

Use a question mark to end an interrogative **clause** within a declarative **sentence.**

EXAMPLE   It was not until July (or was it August?) that we submitted the report.

Retain the question mark in a title that is being cited, even though the sentence in which it appears has not ended.

EXAMPLE   *Should Engineers Be Writers?* is the title of her book.

When used with **quotations,** the question mark indicates whether the writer who is doing the quoting or the person being quoted is asking the question. When the writer doing the quoting asks the question, the question mark is outside the **quotation marks.**

EXAMPLE   Did she say, "I don't think the project should continue"?

If, on the other hand, the quotation itself is a question, the question mark goes inside the quotation marks.

EXAMPLE   She asked, "When will we go?"

If the writer doing the quoting and the person being quoted both ask questions, use a single question mark inside the quotation marks.

> EXAMPLE   Did she ask, "Will you go in my place?"

Question marks may follow a series of separate items within an interrogative sentence.

> EXAMPLE   Do you remember the date of the contract? its terms? whether you signed it?

A question mark should never be used at the end of an indirect question.

> CHANGE   He asked me whether sales had increased this year?
> TO   He asked me whether sales had increased this year.

When a directive or command is phrased as a question, a question mark is usually not used. However, a request (to a customer or a superior, for instance) almost always requires a question mark.

> EXAMPLES   Will you make sure that the machinery is operational by August 15.
>
> Will you please telephone me collect if your entire shipment does not arrive by June 10?

questionnaires

A *questionnaire*—a series of questions on a particular **topic,** sent out to a number of people—is a sort of **interview** on paper (**email** and online techniques are also used). It is used, of course, to gather information for a **report** or a presentation. A questionnaire allows you to sample many more people than personal **interviews** can. It enables you to obtain responses from people in different parts of the country. Even people who live near you may be easier to reach by mail than in person. Those responding to a questionnaire do not face the constant pressure posed by someone jotting down their every word—a fact that can produce more thoughtful answers from questionnaire respondents. And the questionnaire reduces the possibility that the interviewer might influence an answer by tone of voice or facial expression. Finally, the cost of a questionnaire is lower than the cost of numerous personal interviews.

Questionnaires have drawbacks too. People who have strong opinions on a subject are more likely to respond to a questionnaire than those who do not, which can skew the results. An interviewer can follow up on an answer with a pertinent question; at best, a questionnaire can be designed to let one question lead logically to another. Furthermore, mailing a batch of questionnaires and waiting for replies take considerably longer than conducting a personal interview.

A questionnaire must be properly designed. Your goal should be to obtain as much information as possible from your respondents with as little effort on their part as possible. The first rule to follow is to keep the questionnaire brief. The longer the questionnaire is, the less likely the recipient will be to complete and return it. Next, the questions should be easy to understand. A confusing question will yield confusing results, whereas a carefully worded question will be easy to answer. Ideally, questions should be answerable with a yes or no.

> Would you be willing to work a four-day work week, ten hours a day, with every Friday off?
>
> Yes _____
>
> No _____
>
> No opinion _____

When it is not possible to phrase questions in such a straightforward way, offer an appropriate range of answers. Questions must be phrased neutrally; their wording must not lead respondents to a particular answer.

CHANGE How many hours of overtime would you be willing to work each week?

> 4 hours _____   10 hours _____
>
> 6 hours _____   More than 10 hours _____
>
> 8 hours _____   No overtime _____

TO Would you be willing to work overtime?

> Yes _____
>
> No _____
>
> If yes, how many hours of overtime would you be willing to work each week?
>
> 4 hours _____   10 hours _____
>
> 6 hours _____   12 hours _____
>
> 8 hours _____   More than 12 hours _____

MEMORANDUM

May 18, 19—

To: All Company Employees

From: Nelson Barrett, Director of Personnel 𝓃𝐵

Subject: Review of Flexible Working Hours Program

Please complete this questionnaire regarding Luxwear Corporation's trial program of flexible working hours. Your answers will help us to decide whether the program should continue. Return the questionnaire to the Personnel Department by May 28.

If you want to discuss any item in more detail, call Tania Peters in Personnel at extension 8812.

1. What kind of position do you hold?

   _____   _____
   supervisory   nonsupervisory

2. Indicate your exact starting time under flexitime. _____

3. Where do you live?
   Talbot County _____   Greene County _____
   Montgomery County _____   Other, specify _____

4. How do you usually travel to work?
   Drive alone _____   Walk _____
   Car pool _____   Bus _____
   Train _____   Motorcycle _____
   Bicycle _____   Other, specify _____

5. Has flexitime affected your commuting time? If so, please indicate the approximate number of minutes.

   _____   _____   _____
   increase   decrease   no change

6. If you drive, has flexitime affected the amount of time it takes to find a parking space?

   _____   _____   _____
   increase   decrease   no change

FIGURE 1 Sample Questionnaire *(continued on page 500)*

When preparing your questions, remember that you must eventually tabulate the answers; therefore, try to formulate questions whose answers can be readily computed. The easiest questions to tabulate are those for which the recipient does not have to compose an answer. Any questions that require a comment for an answer take time to

7. Do you think that flexitime has affected your productivity?

_____   _____   _____
increase   decrease   no change

8. Have you had difficulty getting in touch with colleagues whose work schedules are different from yours?
Yes _____   No _____

9. Have you had trouble scheduling meetings?
Yes _____   No _____

10. Has flexitime affected the way you feel about your job?

_____   _____   _____
feel better   feel worse   no change

11. How important is it for you to have flexibility in your working hours?
Very _____   Not very_____
Somewhat _____   Not at all _____

12. If you have children, has flexitime made it easier or more difficult for you to obtain babysitting or day-care services?

_____   _____   _____
easier   more difficult   no change

13. Do you recommend that the flexitime program be made permanent?
Yes _____   No _____

14. Do you have suggestions for any changes in the program? If so, please specify.

_____
_____

mo

FIGURE 1 Sample Questionnaire *(continued)*

think about and write and thus lessen your chances of obtaining a response. They are also difficult to interpret. Questionnaires should include a section for additional comments where recipients may clarify their overall attitude toward the subject. If the information will help interpret the answers, include questions about the recipient's age, education, occupation, and so on. Include your name, your address, the purpose of the questionnaire, and the date by which an answer is needed.

A questionnaire sent by mail must be accompanied by a letter explaining who you are, the purpose of the questionnaire, how it will be used, and the date by which you would like to receive a reply. If the information provided will be kept confidential or if the recipient's identity will not be disclosed, say so in the letter.

The sample questionnaire (see Figure 1) was sent to employees in a large organization who had participated in a six-month program of flexible working hours. Under the program, employees worked a forty-hour, five-day week, with flexible starting and quitting times. Employees could start work between 7 and 9 a.m. and leave between 3:30 and 6:30 p.m., provided that they worked a total of eight hours each day and took a one-half-hour lunch period midway through the day.

Select the recipients for your questionnaire carefully. If you want to survey the opinions of all the employees in a small laboratory, simply send each worker a questionnaire. To survey the members of a professional society, mail questionnaires to everyone on the membership list. But to survey the opinions of large groups in the general population—for example, all medical technologists working in private laboratories—is not so easy. Since you cannot include everybody in your survey, you have to choose a representative cross-section.

quid pro quo

Quid pro quo, which is Latin for "one thing for another," can suggest mutual cooperation or "tit for tat" in a relationship between two groups or individuals. The term may be appropriate to business and legal contexts if you are sure your **reader** understands its meaning.

> **EXAMPLE**   It would be unwise to grant their request without insisting on some *quid pro quo.*

quotation marks

Quotation marks (" ") are used to enclose direct **repetition** of spoken or written words. They should not be used to emphasize. There are a variety of guidelines for using quotation marks.

Enclose in quotation marks anything that is quoted word for word (direct quotation) from speech.

> **EXAMPLE**   She said clearly, "I want the progress report by three o'clock."

Do not enclose indirect **quotations**—usually introduced by *that*—in quotation marks. Indirect quotations are **paraphrases** of a speaker's words or ideas.

> EXAMPLE   She said that she wanted the progress report by three o'clock.

Handle quotations from written material the same way: place direct quotations, but not indirect quotations, within quotation marks.

> EXAMPLES   The report stated, "The potential in Florida for our franchise is as great as in California."
>
> The report indicated that the potential for our franchise is as great in Florida as in California.

If you use quotation marks to indicate that you are quoting, you may not make any changes in the quoted material unless you clearly indicate what you have done. (See **quotations** and **brackets.**)

Quotations longer than four typed lines (at least fifty characters per line) are normally indented (*all* the lines) five spaces from the left margin, single-spaced, and *not* enclosed in quotation marks.

Unless it is indented as just described, a quotation of more than one **paragraph** is given quotation marks at the beginning of each new paragraph, but at the end of only the last paragraph.

Use single quotation marks (on a keyboard use the **apostrophe** key) to enclose a quotation that appears within a quotation.

> EXAMPLE   John said, "Jane told me that she was going to 'hang tough' until the deadline is past."

Use quotation marks to set off special words or technical terms only to point out that the term is used in context for a unique or special purpose (used, that is, in the sense of *the so-called*).

> EXAMPLE   Typical of deductive analyses in real life are accident investigations: What chain of events caused the sinking of an "unsinkable" ship such as the *Titanic* on its maiden voyage?

However, slang, colloquial expressions, and attempts at humor, although infrequent in technical writing, seldom rate being set off by quotation marks.

> CHANGE   Our first six months in the new office amounted to little more than a "shakedown cruise" for what lay ahead.
>
> TO   Our first six months in the new office amounted to little more than a shakedown cruise for what lay ahead.

Use quotation marks to enclose titles of reports, short stories, articles, essays, radio and television programs, short musical works, paintings and other works of art.

    EXAMPLE   Did you see the article, "No-Fault Insurance and Your Motorcycle" in last Sunday's *Journal?*

Titles of books and periodicals are underlined (to be typeset in **italics**).

    EXAMPLE   Articles in the *Business Education Forum* and *Scientific American* quoted the same passage.

Some titles, by convention, are neither set off by quotation marks nor underlined, although they are capitalized.

    EXAMPLES   Technical Writing (college course title), the Bible, the Constitution, Lincoln's Gettysburg Address, the Montgomery Ward Catalog.

Commas and **periods** always go inside closing quotation marks.

    EXAMPLE   "Reading *Space Technology* gives me the insider's view," he says, adding "it's like having all the top officials sitting in my office for a bull session."

Semicolons and **colons** always go outside the closing quotation marks.

    EXAMPLES   He said, "I will pay the full amount"; this certainly surprised us. The following are her favorite "sports": eating and sleeping.

All other **punctuation** follows the logic of the context: if the punctuation is a part of the material quoted, it goes inside the quotation marks; if the punctuation is not part of the material quoted, it goes outside the quotation marks.

    Quotation marks may be used as ditto marks, instead of repeating a line of words or numbers directly beneath an identical set. In formal writing, this use is confined to **tables** and **lists**.

    EXAMPLE   A is at a point equally distant from L and M.
                     B " " "   "   "   "   "   S and T.
                     C " " "   "   "   "   "   R and Q.

quotations

When you have borrowed words, facts, or ideas of any kind from someone else's work, acknowledge your debt by giving your source credit in a footnote. Otherwise, you will be guilty of **plagiarism.** Also be sure that you have represented the original material honestly and accurately. (See also **documenting sources.**)

Direct Quotations

Direct word-for-word quotations are enclosed in **quotation marks.** They are usually, although not always, separated from the rest of the sentence in which they occur by either a **comma** or a **colon.**

> EXAMPLES   The noted economist says, "If monopolies could be made to behave as if they were perfectly competitive, we would be able to enjoy the benefits both of large-scale efficiency and of the perfectly working price mechanism."
>
> The noted economist pointed out: "There are three options available when technical conditions make a monopoly the natural outcome of competitive market forces: private monopoly, public monopoly, or public regulation."

When a quotation is divided, the material that interrupts the quotation is set off, before and after, by commas, and quotation marks are used around each part of the quotation.

> EXAMPLE   "Regulation," he said in a recent interview, "cannot supply the dynamic stimulus that in other industries is supplied by competition."

At the end of a quoted passage, commas and **periods** go inside the quotation marks, and colons and **semicolons** go outside the quotation marks.

Q

Indirect Quotations

Indirect quotations, which are essentially **paraphrases** and are usually introduced by *that,* are not set off from the rest of the **sentence** by **quotation marks.**

> DIRECT   He said in a recent interview, "Regulation cannot supply the dynamic stimulus that in other industries is supplied by competition." (direct quotation)
>
> INDIRECT   In a recent interview he said that regulation does not stimulate the industry as well as competition does. (indirect quotation)

Deletions or Omissions

Deletions or omissions from quoted material are indicated by three ellipsis dots within a sentence and four **ellipsis** dots at the end of a sentence, the first of the four dots being the period that ends the sentence.

EXAMPLE   "If monopolies could be made to behave . . . we would be able to enjoy the benefits of . . . large-scale efficiency. . . ."

When a quoted passage begins in the middle of a sentence rather than at the beginning, however, ellipsis dots are not necessary; the fact that the first letter of the quoted material is not capitalized tells the **reader** that the quotation begins in midsentence.

CHANGE   He goes on to conclude that ". . . coordination may lessen competition within a region."

TO   He goes on to conclude that "coordination may lessen competition within a region."

Such quotations should be worked into one of your sentences rather than left standing alone. When quoted material is worked into a sentence, be sure that it is related logically, grammatically, and syntactically to the rest of the sentence.

Inserting Material into Quotations

When it is necessary to insert a clarifying comment within quoted material, use **brackets.**

EXAMPLE   "The industry is organized as a relatively large, integrated system serving an extensive [geographic] area, with smaller systems existing as islands within the larger system's sphere of influence."

When quoted material contains an obvious error, or might in some other way be questioned, the expression *sic*, enclosed in brackets, follows the questionable material to indicate that the writer has quoted the material exactly as it appeared in the original. (See also **sic.**)

EXAMPLE   In *Basic Astronomy*, Professor Jones noted that the "earth does not revolve around the son *[sic]* at a constant rate."

Incorporating Quotations into Text

Depending on the length, there are two mechanical methods of incorporating quotations into your text.

Quotations of four or fewer lines are incorporated into your text and enclosed in quotation marks. (Many writers use this method regardless of length. When this method is used with multiple **paragraphs,** quotation marks appear at the beginning of each new paragraph, but at the end of only the last paragraph.)

Material longer than five lines is now usually inset; that is, it is set off from the body of the text by being indented five spaces from the left margin and by triple-spacing above and below the quotation. The quoted passage is single-spaced and not enclosed in quotation marks.

EXAMPLE   Among the symptoms that patients report with osteoarthritis are pain, loss of range of motion, and stiffness. In managing this disease, the aims are to relieve pain and stiffness and restore function. In the control of pain, Donner (237) has pointed to the role of analgesia:

> Although patient pain may be adequately controlled by occasional, or even regular, analgesics, a significant number benefit from the regular use of nonsteroidal anti-inflammatory drugs. Twice-daily dosage interfered less with work than more frequent dosage.

Donner points out that medication regimes for the elderly should be kept simple. The elderly metabolize drugs less completely and more slowly than younger persons. They may also have trouble opening medication containers.

Overquoting

Do not rely too heavily on the use of quotations in the final version of your **report** or paper. If you do, your work will appear merely derivative. The temptation to overquote during the **note-taking** phase of your **research** can be avoided if you concentrate on summarizing what you read. Quote word for word only when your source concisely sums up a great deal of information or points to a trend or development important to your subject. As a rule of thumb, avoid quoting anything over one paragraph.

Q

R

raise/rise

Both *raise* and *rise* mean "move to a higher position." However, *raise* is a transitive verb and always takes an **object** (*raise* crops), whereas *rise* is an intransitive verb and never takes an object (heat rises).

> **EXAMPLES** *Raise* the lever to the ON position.
> Turn off the motor if the internal temperature *rises* above 110°.

re

Re (and its variant form *in re*) is business and legal **jargon** meaning "in reference to" or "in the case of." *Re* is sometimes used in **memorandums.**

> **EXAMPLE** To: Elaine Barr
> From: Edgar Roden
> *Re:*   Revised Office Procedures

Re, however, is now giving way to *subject.*

> **EXAMPLE** To: Elaine Barr
> From: Edgar Roden
> *Subject:*   Revised Office Procedures

readers/audience

The first rule of effective writing is to help the *reader.* The responsibility that technical writers have to their readers is expressed nicely in the Society of Technical Communications's *Code for Communicators,* which calls for technical writers to hold themselves responsible for how well their audience understands their message and to satisfy the readers' need for information rather than the writer's need for self-expression. This code reminds writers to concentrate on writing for the needs of specific readers rather than merely about certain subjects. Too often, writers are blinded by their own familiarity with the technical subject and therefore tend to overlook their readers' lack of knowledge.

As a technical writer, you must usually assume that your readers are less familiar than you are with the subject. You have to be careful, for example, in writing **instructions** for operators of equipment that you designed or in writing a **report** about a technical system for executives whose training is in such nontechnical areas as management, accounting, law, and so on. Such readers are unlikely to have had training in your technical specialty and thus need definitions of technical terms as well as clear explanations of principles that you, as a specialist, take for granted, Even if you write a **trade journal article** for others in the field, you must remember to explain new or special uses of standard terms and principles.

Before starting to write, you must determine who your readers are and then adjust the amount of detail and technical vocabulary to their training and experience. For example, to describe a new technique for repairing a vibration dampener to an experienced auto mechanic, you could use standard auto repair terms and procedures. To describe the same technique in a training manual, however, would require a much fuller explanation of terms and procedures.

As the previous example implies, when you write for many readers (hundreds of auto mechanic trainees, for example), try to visualize a single, typical member of that group and write for that reader. You might also make a list of the characteristics (experience, training, and work habits, for example) of that reader. This technique, used widely by professional writers, enables you to decide what should or should not be explained according to that reader's needs. When you are writing to a group, it is usually best to aim at those with the least training and experience. The extra information and explanation will not insult those who know more; they can simply read or follow the instructions more quickly, perhaps skipping sections with which they are familiar.

When your reading audience includes people with widely varied backgrounds, consider aiming various sections of the document at different sets of readers. Recommendations, **executive summaries,** and **abstracts** can be aimed at busy executives who will be reading to understand the general implications of projects or technical systems. **Appendixes** containing **tables, graphs,** and raw technical data can be aimed at specialists who wish to examine or use such supporting

data. The body of a report or **proposal** should be aimed at those readers with the most serious interest or those who need to make decisions
based on the details in the contents.

Routine **memorandums** and short reports that are written for one
individual reader do not require such elaborate design. Simply remember that person's exact needs as you write.

Letters written to readers outside your organization, however, require
a special tone. In such letters, you must concern yourself not only with
the readers' understanding of your topic but with their reaction to it as
well. You must be as careful in representing your organization as you
are in explaining your subject matter. (See **correspondence** for a discussion of the extra courtesy required in such business letters.)

reason is because

The **phrase** *reason is because* is a colloquial expression to be avoided in
writing. *Because,* which in this phrase only repeats the notion of cause,
should be replaced by *that.*

CHANGE	Sales have increased more than 20 percent. The *reason is because* our sales force has been more aggressive this year.
TO	Sales have increased more than 20 percent. The *reason is that* our sales force has been more aggressive this year.
OR	Sales have increased more than 20 percent this year *because* our sales force has been more aggressive.

reciprocal pronouns (see **pronouns**)

reference books (see **library research**)

R

reference letters

Almost everyone is called upon at some time to provide a recommendation or reference for a colleague, an acquaintance, or an employee.
Reference letters may range from completing an admission form for a
prospective student to composing a detailed description of professional accomplishments and personal characteristics for someone
seeking employment.

In order to write an effective letter of recommendation, you must
be familiar enough with the applicant's abilities or actual performance
to offer an evaluation. Then you must *truthfully,* without embellish-

510 reference letters

IVY COLLEGE
DEPARTMENT OF BUSINESS
WEST LAFAYETTE, IN 47906
(691) 423-1719
(691) 423-2239 (FAX)
IVCO@AOL.EDU (EMAIL)

January 14, 19—

Mr. Phillip Lester
Personnel Director
Thompson Enterprises
201 State Street
Springfield, IL 62705

Dear Mr. Lester:

How long writer has known applicant and the circumstances

As her employer and her former professor, I am happy to have the opportunity to recommend Kerry Hawkins. I've known Kerry for the last four years, first as a student in my class and for the last year as a research assistant.

Outstanding characteristics of applicant

Kerry is an excellent student, with above average grades in our program. On the basis of a GPA of 5.6 (A=6), Kerry was offered a research assistantship to work on a grant under my supervision. In every instance, Kerry completed her library search assignment within the time agreed upon. The material provided in the reports Kerry submitted met the requirements for my work and more. These reports were always well written. While working 15 hours a week on this project, Kerry has maintained a class load of 12 hours per semester.

Recommendation and summary of qualifications

I strongly recommend Kerry for her ability to work independently, to organize her time efficiently, and for her ability to write clearly and articulately. Please let me know if I can be of further service.

Sincerely yours,

Michael Paul
Michael Paul
Professor of Business

MP:mt

R

FIGURE 1 Reference Letter

---

ment, communicate that evaluation to the inquirer. For the reference letter to serve as a valid selection device, you must address specifically the applicant's skills, abilities, knowledge, and personal characteristics in relation to the requested objective. You may also be asked to say what you know about the person's former employment and education or even his or her military discharge and personnel records.

In a reference letter, always respond directly to the inquiry, being careful to address the specific questions asked. For the record, you must identify yourself: name, title or position, employer, and address.

You will also have to state how long you have known the applicant and the circumstances of your acquaintance. You should mention with as much substantiation as possible, one or two outstanding characteristics of the applicant. Organize the details in your letter in order of importance; put the most important details first. Conclude with a statement of recommendation and a brief summary of the applicant's qualifications.

The Privacy Act of 1974 has had a major impact on reference letters. Under the act, applicants have a legal right to examine the materials in an organization's files that concern them—unless they sign a waiver of their right to do so. Organizations seeking to avoid the privacy issue may offer applicants the option of signing a waiver against reading letters of reference written about them. When you are requested to serve as a reference or asked to supply a letter of reference, you should thus inquire about the confidentiality status of your letter—whether the applicant's file will be open or closed. Figure 1 is a typical reference letter.

reflexive pronouns (see **pronouns**)

refusal letters

When you receive a **complaint letter** or an **inquiry letter** to which you must give a negative reply, you may need to write a refusal letter. For example, a professional organization may have asked you to speak at their regional meeting, but they require your firm to pay your expenses. Because your firm has declined the request, you must write a refusal letter turning down the invitation. The refusal letter is difficult to write because it contains bad news; however, you can tactfully and courteously convey the bad news.

In your letter you should lead up to the refusal. To state the bad news in your **opening** would certainly affect your **reader** negatively. The ideal refusal letter says no in such a way that you not only avoid antagonizing your reader but keep his or her goodwill. To achieve such a **purpose,** you must convince your reader of the justness of your refusal *before refusing.* The following pattern is an effective way to deal with this problem:

1. A buffer beginning.
2. A review of the facts.
3. The bad news, based on the facts.
4. A positive and pleasant closing.

The primary purpose of a buffer beginning is to establish a pleasant and positive **tone.** You want to convince your reader that you are a reasonable person. One way is to indicate some form of agreement with, or approval of, the reader or the project. If your refusal is in response to a complaint letter, do not begin by recalling the reader's disappointment ("We regret your dissatisfaction . . ."). Keep your buffer **paragraph** positive and pleasant. You can express appreciation for your reader's time and effort, if appropriate. Stating such appreciation will soften the disappointment and pave the way for future good relations.

> EXAMPLE   Thank you for your cooperation, your many hours of extra work answering our questions, and your patience with us as we struggled to make a very difficult decision. We believe our long involvement with your company certainly indicates our confidence in your products.

Next you should analyze the circumstances of the situation sympathetically. Place yourself in your reader's position and try to see things from his or her **point of view.** Establish clearly the reasons you cannot do what the reader wants—even though you have not yet said you cannot do it. A good explanation of the reasons should so thoroughly justify your refusal that the reader will accept your refusal as a logical conclusion.

> EXAMPLE   The Winton Check Sorter has all the features that your Abbott Check sorter offers and, in fact, offers two additional features that your sorter does not. The more important of the two is a backup feature that retains totals in memory, even if the power fails, so that processing doesn't have to start again from scratch following a power failure. The second additional feature that the Winton Sorter offers is stacked pockets, which take less space than do the linear pockets on your sorter.

Don't belabor the bad news—state your refusal quickly, clearly, and as positively as possible.

> EXAMPLE   After much deliberation, we have decided to purchase Winton Check Sorters because of the extra features they offer.

Close your letter with a friendly remark, whether to assure the reader of your high opinion of his product or merely to wish him success.

> EXAMPLE   We have enjoyed our discussions with your company and feel we have learned a great deal about check processing. Although we did not select your sorter, we were very favorably impressed with your systems, your people, and your company. Perhaps we will work together in the future on another project.

> ### Titus Packaging, Inc.
> #### 2063 Eldorado Dr.
> #### Billings, Montana 59102
>
> April 4, 19—
>
> Ms. Edna Kohls
> Graphic Arts Services, Inc.
> 936 Grand Avenue
> Billings, Montana 59102
>
> Dear Ms. Kohls:
>
> We appreciate your interest in establishing an open account at Titus Packaging, Inc. In the two years since you began operations, your firm has earned an excellent reputation in the business community.
>
> As you know, interest rates have been rising sharply this past year, while sales in general have declined. With the current negative economic climate, and considering the relatively recent establishment of your company, we believe that an open account would not be appropriate at this time.
>
> We will be happy to have you renew your request around the first of next year, when the economic climate is expected to improve and when your company will be even more firmly established. In the meantime, we will be happy to continue our present cash relationship, with a 2-percent discount for payment made in ten days.
>
> Sincerely,
>
> *Conrad C. Atkins*
> Conrad C. Atkins
> Manager, Credit Department
>
> CCA/mo

Marginal notes:

- **Buffer opening**
- **Review of the facts, followed by the bad news**
- **Positive and pleasant closing**

FIGURE 1   Refusal Letter

The example shown in Figure 1 refuses a company credit for establishing an open account.

(See also **resignation letter or memorandum.**)

relative pronouns (see **pronouns**)

repetition

The deliberate use of repetition to build a sustained effort or emphasize a feeling or idea can be quite powerful.

EXAMPLE   Similarly, atoms *come and go* in a molecule, but the molecule *remains;* molecules *come and go* in a cell, but the cell *remains;* cells *come and go* in a body, but the body *remains;* persons *come and go* in an organization, but the organization *remains.*

Repeating key words from a previous sentence or **paragraph** can also be used effectively to achieve **transition.**

EXAMPLE   For many years, *oil* has been a major industrial energy source. However, *oil* supplies are limited, and other sources of energy must be developed.

Be consistent in the word or **phrase** you use to refer to something. In technical writing, it is generally better to repeat a word (so there will be no question in the reader's mind that you mean the same thing) than to use a **synonym** in order to avoid repeating it. Your primary goal in technical writing is effective and precise communication rather than elegance.

CHANGE   Several recent *analyses* support our conclusion. These *studies* cast doubts on the feasibility of long-range forecasting. The *reports,* however, are strictly theoretical.

TO   Several recent theoretical *studies* support our conclusion. These *studies* cast doubts on the feasibility of long-range forecasting. They are, however, strictly theoretical.

Purposeless repetition, however, makes a sentence awkward and hides its key ideas.

CHANGE   He *said that* the customer *said that* the order was canceled.

TO   He *said that* the customer canceled the order.

The harm caused to your writing by careless repetition is not limited to words and phrases; the needless repetition of ideas can be equally damaging.

CHANGE   In this modern world of ours today, the well-informed, knowledgeable executive will be well ahead of the competition.

TO   To succeed, the contemporary executive must be well informed.

(See also **elegant variation.**)

reports

A *report* is an organized presentation of factual information that answers a request by supplying the results of an investigation, a trip, a test, a research project, and the like. All reports can be divided into two broad categories: formal and informal.

Formal reports present the results of projects that may require months of work and involve large sums of money. These projects may be done either for one's own organization or as a contractual requirement for an outside organization. **Formal reports** generally follow a stringent **format** and include some or all of the report elements discussed in formal reports.

Informal reports normally run from a few paragraphs to a few pages and include only the essential elements of a report: introduction, body, conclusions, and recommendations. Because of their brevity, informal reports are customarily written as a letter (if the report is to be sent to someone outside the organization) or as a **memorandum** (if it is to be sent to someone inside the organization).

The **introduction** serves several functions: it announces the subject of the report, states its purpose, and, when appropriate, gives essential background information. The introduction should also summarize any conclusions, findings, or recommendations made in the report. Managers, supervisors, and clients find a concise summary useful because it gives them the essential information at a glance and helps focus their thinking as they read the rest of the report.

The *body* of the report should present a clearly organized account of the report's subject—the results of a test carried out, the status of a construction project, and so on. The amount of detail given depends on the complexity of the subject and on your reader's familiarity with it.

The **conclusion** should summarize your findings and tell the reader what you think their significance may be. In some reports a final section gives recommendations. (Sometimes the conclusions and recommendations sections are combined.) In this section, you make suggestions for a course of action based on the data you have presented.

Many different types of informal reports are described in separate entries in this handbook: **feasibility reports, investigative reports, progress and activity reports, trip reports,** and **trouble reports.** Many of these can be formal or informal, depending on their length and complexity.

research

Research is the process of investigation, the discovery of facts. As part of the writing process, research must be preceded by **preparation** (that is, you must first determine your **reader, purpose,** and **scope**). During the preparation phase, you establish the extent of the coverage needed for your work, and then during the research stage you can logically determine what information you will need and in how much detail. Without this preparation, the research will not be adequately defined or focused.

Researchers frequently distinguish between primary and secondary information, and depending on the goal of the research, the distinction can be important. *Primary information* refers to the "raw" data compiled from observations, surveys, experiments, **questionnaires,** and the like. This information is frequently gathered or located in notebooks, computer printouts, telephone logs, transcripts of speeches and interviews, tape recordings, and unedited film. *Secondary information* refers to primary information that has been analyzed, assessed, evaluated, compiled, or otherwise organized into accessible form. The forms include books, articles, reports, dissertations, operating and procedure manuals, brochures, and so forth. Use the information most appropriate to your research needs, recognizing that some projects will require both types.

Many projects require information from a variety of sources. The most common ones include written materials, interviews with experts, and firsthand observations. However, before investigating any of them, do not overlook the invaluable resource that you may represent. Before going elsewhere, interview yourself to tap into your own knowledge and experience. You may find that you already have enough information to get started. This technique, commonly known as *brainstorming*, may also stimulate your thinking about additional places to seek information.

First, write down as many ideas as you can think of about the subject, as well as the objective and scope of your project, and jot them down in random order *as they occur to you.* (This may be done by a group as well as alone.) Put down what you know and, if possible, where you learned of it. For every idea noted, ask what, when, who, where, how, and why. List the details these additional questions bring to mind. You can write down these ideas on a chalkboard, pad of paper, or note cards. (Note cards are preferable if you are working alone because they can easily be shuffled and rearranged.) Once your memory runs dry, group the items in the most logical order, based on your objective and your reader's needs.

The end result of this process will be a tentative **outline** of your project. The outline will have sketchy or missing sections, thus helping to show where additional research is needed, and it will also help integrate the various details of the additional research.

If you must conduct a formal, structured search of published sources, see **library research.** Scan the written material quickly before reading it in order to get an idea of the kind of information it includes. Then if it seems useful, read it carefully, taking notes of any information that falls within the scope of your research. Be sure to note where you found the information so you can find it again if necessary or compile it for a list of works cited. You may think of additional questions about the topic as you read; jot these down and identify them with a question mark. Some of them may eventually be answered in other research sources; those that remain unanswered can guide you to further research. (For detailed guidance, see **note taking** and **paraphrasing.**)

Technical professionals should review and borrow from in-house documents and other information, revising as necessary for consistency of content, **style,** and **format** with the work into which the information will be incorporated. Passages borrowed from in-house documents, often referred to as "boilerplate," provide an efficient alternative to original research as long as they are accurate, well written, and originated within your company or organization. Although you are free to use and adapt works written within your organization, remember that you must obtain permission to use copyrighted works written elsewhere. (See also **copyright.**)

If after researching these other sources you still need more information, consider talking to an expert. Not only can someone skilled in a field answer many of your questions, but he or she can also suggest other sources of information. (To make the most of such discussions, see **interviewing for information** and **listening.**)

Vast amounts of information are also available in the trade documents of your field, including magazines, **reports,** newsletters, booklets, brochures, conference proceedings, and membership directories. As you seek information from these sources, keep the following guidelines in mind. The more recent the information, the better. **Trade journal articles** are essential sources of "current awareness" information because books take longer to write, publish, and distribute than trade journal articles. Conference proceedings can be an even better source of up-to-date information. They contain papers presented at meetings of trade, industrial, and professional societies about recent research results or work in progress. Much of this information will either not be published elsewhere or will not appear for a year or more in a published journal.

Finally, you can use your own observations or hands-on experience. In fact, this is the only way to obtain certain kinds of information: the behavior of people and animals, certain natural phenomena, mechanical processes, the operation of tools and equipment, and the like. When planning research that involves observation, choose your sites and times carefully. Also, be sure to obtain permission in advance when necessary. Such observations may be illegal, resented, or contrary to an organization's policy. Keep accurate, complete records, indicating date, time of day, and the like. Save interpretations of your observations for future analysis. Be aware that research involving observations may be time-consuming, complicated, and expensive and that your presence may inadvertently influence what you observe.

If you have to write an operating manual for equipment or a procedure for some task, you must first acquire hands-on experience by operting the equipment or performing the task yourself. Afterwards, you will in effect be interviewing yourself based on your experience. To do so, make a rough outline of your experience using the brainstorming technique described in this entry. After writing the draft, have someone unfamiliar with the equipment operate it or someone unfamiliar with the procedure perform it. When he or she has a problem, rewrite the passage until it is clear and easy to follow.

As with observations, obtain permission to operate the equipment or carry out a procedure. Comply with all safety rules, and make them a part of the manual or procedures you write. (See also **questionnaires.**)

resignation letter or memorandum

R

Start your letter or memorandum of resignation on a positive note, regardless of the circumstances under which you are leaving. You might, for example, point out how you have benefited from working for the company. Or you might say something complimentary about how well the company is run. Or you might say something positive about the people with whom you have been associated.

Then explain why you are leaving. Make your explanation objective and factual, and avoid recriminations. Your letter or memorandum will become part of your permanent file with that company, and if it is angry and accusing, it could haunt you in the future: you may need references from the company.

Your letter or memorandum should give enough notice to allow your employer time to find a replacement. It may be no more than

---

INTEROFFICE MEMORANDUM

To: W.R. Johnson, Director of Purchasing
From: J.L. Washburn, Purchasing Agent *JLW*
Date: January 7, 19—
Subject: Resignation from Barnside Appliances,
         effective February 7, 19—

Positive opening

My three years at Barnside Appliances has been an invaluable period of learning and professional development. I arrived as a novice, and I believe that today I am a professional—primarily as a result of the personal attention and tutoring I have received from my superiors and the fine example set both by my superiors and my peers.

Reason for leaving

I believe the time has come for me to move on, however, to a larger company that can give me an opportunity to continue my professional development. Therefore, I have accepted a position with General Electric, where I am scheduled to begin on February 12. Thus, my last day at Barnside will be February 7. However, if you need more time to hire and train my replacement, I can make arrangements to work longer.

Positive closing

Many thanks for the experiences I have gained, and best wishes for the future.

FIGURE 1   Sample Resignation Memo (Resignation to Accept a Better Position)

R

two weeks, or it may be enough time to enable you to train your replacement. Some organizations ask for a notice that is equivalent to the number of weeks of vacation you receive.

The sample memorandum of resignation in Figure 1 is from an employee who is leaving to take a job offering greater opportunities. The memorandum of resignation presented in Figure 2 is from an employee who is leaving under unhappy circumstances. Notice that it opens and closes positively and that the reason for the resignation is stated without apparent anger or bitterness.

INTEROFFICE MEMORANDUM

To: T.W. Haney, Vice-President, Administration
From: L.R. Rupp, Executive Assistant *LRR*
Date: February 12, 19—
Subject: Resignation from Winterhaven, effective
           March 1, 19—

Positive opening

My ten-year stay with the Winterhaven Company has been a very pleasant experience, and I am sure it has been mutually beneficial.

Reason for leaving

Because the recent realignment of my job leaves no career path open to me, however, I have accepted a position with another company that I feel will give me a better future. I am, therefore, submitting my resignation, to be effective on March 1, 19—.

Positive closing

I have enjoyed working for Winterhaven and wish the company success in the future.

FIGURE 2   Sample Resignation Memo (Resignation under Negative Conditions)

respective/respectively

Respective is an **adjective** that means "pertaining to two or more things regarded individually."

EXAMPLE   The committee members prepared their *respective* reports.

Respectively is the **adverb** form of *respective*, meaning "singly, in the order designated."

EXAMPLE   The first, second, and third prizes in the sales contest were awarded to Maria Juarez, Gloria Hinds, and Margot Luce, *respectively*.

Respective and *respectively* are often unnecessary because the meaning of individuality is already clear.

> **CHANGE** The committee members prepared their *respective* reports.
>
> **TO** Each committee member prepared a report.

restrictive and nonrestrictive elements

Modifying **phrases** and **clauses** may be either restrictive or non-restrictive.

A *nonrestrictive* phrase or clause provides additional information about what it modifies, but it does not limit, or restrict, the meaning of what it modifies. Therefore, the nonrestrictive phrase or clause can be removed without changing the essential meaning of the sentence. It is, in effect, a parenthetical element, and so it is set off by **commas** to show its loose relationship with the rest of the sentence.

> **EXAMPLES** This instrument, *called a backscatter gauge,* fires beta particles at an object and counts the particles that bounce back. (nonrestrictive phrase)
>
> The annual report, *which was distributed yesterday,* shows that sales increased 20 percent last year. (nonrestrictive clause)

A *restrictive* phrase or clause limits, or restricts, the meaning of what it modifies. If it were removed, the essential meaning of the sentence would be changed. Because a restrictive phrase or clause is essential to the meaning of the sentence, it is never set off by commas.

> **EXAMPLES** All employees *wishing to donate blood* may take Thursday afternoon off. (restrictive phrase)
>
> Companies *that adopt the plan* nearly always show profit increases. (restrictive clause)

It is important for writers to distinguish between nonrestrictive and restrictive elements. The same sentence can take on two entirely different meanings depending on whether a modifying element is set off by commas (because nonrestrictive) or not set off (because restrictive). The results of a slip by the writer can be not only misleading but also downright embarrassing.

> **CHANGE** I think you will be impressed by our systems engineerss who are thoroughly experienced in projects like yours.
>
> **TO** I think you will be impressed by our systems engeineers, who are thoroughly experienced in projects like yours.

R

The problem with the first sentence is that it suggests that "you may not be so impressed by our other, less experienced, systems engineers."
    Which should be used to introduce nonrestrictive clauses.

> EXAMPLE   After John left the restaurant, *which* is one of the finest in New York, he came directly to my office.

That should be used to introduce restrictive clauses.

> EXAMPLE   Companies *that* diversify usually succeed.

résumés

A *résumé** is an essential element in any **job search** because it is, in effect, a catalog of what you have to offer to prospective employers. Sent out with your **application letter,** it is the basis for their decision to invite you for a job interview. (See **interviewing for a job.**) It tells them who you are, what you know, what you can do, what you have done, and what your job objectives are. It may be the basis for the questions you are asked in the interview, and it can help your prospective employer evaluate the interview. With today's business climate, which requires employees to move from job to job more frequently than in the past, you need to keep your résumé up to date.
    Three steps will help you prepare your résumé:

1. Analyze yourself and your background.
2. Identify those to whom you will submit your résumé (your **readers**).
3. Organize and prepare the résumé.

Analyzing Your Background

The starting point in preparing a paper or electronic résumé is a thorough analysis of your experience, education and training, professional and personal skills, and personal traits.
    List the major points for each category before actually beginning to organize and write your résumé. The following questions can serve as an inventory of what you should consider in each category.

Experience.   Note all your employment—full-time, part-time, vacation jobs, and free-lance work—and analyze each job on the basis of the following list of questions:

**Résumé* uses accent marks (´) to distinguish it from the verb *resume.*

1. What was the job title?
2. What did you do (in reasonable detail)?
3. What experience did you gain that you can apply to another job?
4. Why were you hired for the job?
5. What special skills did you learn on that job?
6. Were you promoted or given a job with more responsibility?
7. Why did you leave it?
8. When did you start and when did you leave the job?
9. Would your former employer give you a reference?
10. What special traits were required of you on the job (initiative, leadership, ability to work with details, ability to work with people, imagination, ability to organize, and so on)?

Education.   For the applicant with little work experience, such as a recent graduate, the education category is of primary importance. For the applicant with extensive work experience, education is still quite important, even though it is now secondary to work experience. List the following information about your education:

1. Colleges attended and the inclusive dates.
2. Degrees and the dates they were awarded.
3. Major and minor subjects.
4. Courses taken or skills acquired that might be important to the job you are applying for.
5. Internships, work-study programs, co-op positions.
6. Cumulative grade averages and academic honors.
7. Extracurricular activities.
8. Scholarships and awards.
9. Special training courses (academic or industrial).

Organizing and Writing the Résumé

After listing all the items in the preceding two categories, you should analyze them in terms of the job you are seeking, evaluating the items, rejecting some of them, and finally selecting those that you will include in your résumé. Base your decision on the following questions:

1. What exact job are you seeking?
2. Who are the prospective employers?
3. What information is most pertinent to that job and those employers?
4. What details should be included and in what order?
5. How can you present your qualifications most effectively to get an interview?

Résumés should be brief—preferably one or two pages, depending on your level of experience—and should include the following information as applicable:

1. Your name, address, phone number, **fax** number, and **email** address.
2. Your immediate and long-range job objectives.
3. Your professional training.
4. Your professional experience, including the firms where you have worked, and your responsibilities.
5. Your specific skills.
6. Pertinent personal information.

Format of the Résumé

A number of different formats can be used, including the electronic résumé discussed later in this entry. Most important is to make sure that your résumé is attractive, well organized, easy to read, and free of errors. A common format uses the following **headings:**

Employment Objective
Education
Employment Experience
Special Skills and Activities
References (optional)

For résumés on paper, underline or capitalize the headings to make them stand out on the page. Whether you list education or experience first depends on which is stronger in your background. If you are a recent graduate, list education first; if you have substantial job experience, list your most recent experience first, your next most recent experience second, and so on.

The Main Heading.   Center your name in all capital letters (and boldface if you're using a word processor) at the top of the page. Follow it with your address, telephone number, fax number, and email address.

Employment Objective.   State both your immediate and long-range employment objectives. However, if an objective is not appropriate for a particular position (a temporary job, for example), you may exclude it.

Education.   List the college or colleges you attended, the dates you attended each one, any degree or degrees received, your major field of study, and any academic honors you earned. First-time applicants frequently give the name of their high school, its location (city, state), and the dates they attended.

Employment Experience. List all your full-time jobs, from the most recent to the earliest that is appropriate. If you have had little full-time work experience, list your part-time and temporary jobs, too. Give the details of your employment, including the job title, dates of employment, and the name and address of your employer. Provide a concise description of your duties only for those jobs whose duties were similar to those of the job you are seeking; otherwise, give only a job title and a brief description. Specify any promotions or pay increases you received. Do not, however, list your present salary. (Salary depends on your experience and your potential value to an employer.) If you have been with one company for a number of years, highlight your accomplishments during that time. List military service as a job. Give the dates you served, your duty specialty, and your rank at discharge. Discuss the duties only if they relate to the job you are applying for.

Some résumés arrange employment experience by function rather than by chronological time. Instead of listing your jobs first, starting with the most recent, you list the functions you've performed ("Management," "Project Development," "Training," "Sales," and the like). The functional arrangement is useful for persons who wish to stress their skills or who have been employed at one job and wish to demonstrate their diversity of experience at that single position. It is also used by those who have gaps in time on their résumé caused by unemployment or illness. Although the functional arrangement can be effective, employment directors know that it may be used to cover weaknesses, and so you should use it carefully and be prepared to explain any gaps.

Special Considerations. Indeed, most career experts say it is important to acknowledge gaps rather than trying to hide them. This is particularly true for women, for example, who are reentering the workforce because they have chosen to devote a full-time period to care for children or dependent adults. Although such work is unpaid, it often provides experience that develops important time-management, problem-solving, organizational, and interpersonal skills. Do not undervalue such work. Following are examples illustrating how you might reflect such experiences in a résumé:

Child Care Provider, 19— to 19—
Furnished full-time care to three preschool children in home environment. Instructed in crafts, beginning scholastic skills,

time management, basics of nutrition, and swimming. Organized activities, managed household, and served as block-
watch captain.

Home Caregiver, 19— to 19—
Provided 60 hours per week in-home care to Alzheimer's
patient. Coordinated medical care, developed exercise programs, completed and processed complex medical forms,
administered medications, organized budget, and managed
home environment.

If you have participated in volunteer work during such a period, list
the experience. Volunteer work often results in the same experience as
does full-time employed work, a fact that your résumé should reflect
as in the following example:

School Association Coordinator, 19— to 19—
Managed special activities of Briarwood High School Parents
Association. Planned and coordinated meetings, scheduled
events, and supervised fund drive operations. Under my direction, the Association raised $10,000 toward refurbishing the
school auditorium.

Special Skills and Activities. This category usually comes near the end
and may replace the Personal Data category. You may include such
skills as knowledge of foreign languages, writing and editing abilities,
specialized knowledge (such as experience with a computer language
or a background in electronics), hobbies, student or community activities, professional or club memberships, and published works. Be
selective: do not overload this category, do not duplicate information
given in other categories, and do not include any items that do not
support your employment objective.

Personal Data. Federal legislation limits the inquiries an employer can
make before hiring an applicant—especially requests for such personal information as age, sex, marital status, race, and religion. Consequently, some job seekers exclude this category because they feel
their personal data could have a negative impact on the employer. You
may also eliminate this category simply because you need the space
for more significant information about your qualifications.

References. You can state on your résumé that "references will be furnished upon request" or include references as part of the résumé (only
if you need to fill space or if including names is traditional to your
profession). Either way, do not list anyone as a reference without first
obtaining his or her permission.

CAROL ANN WALKER
273 East Sixth Street
Bloomington, Indiana 47401
(913) 321-4567
(913) 321-6225 (fax)
email: caw@hbk.edu

EMPLOYMENT OBJECTIVE

Financial research assistant, leading to a management position in corporate finance.

EDUCATION

Indiana University (Bloomington)
Bachelor of Business Administration (Expected June 19—)
Major: Finance   Minor: Computer Science
Dean's List: 3.88 grade point average of possible 4.0.
Senior Honor Society, 19—.

EMPLOYMENT EXPERIENCE

FIRST BANK (Bloomington, Indiana)
Research Assistant Intern, Summer and Fall Quarters, 19—.
Assisted manager of corporate planning and developed computer model for long-range planning.

MARTIN FINANCIAL RESEARCH SERVICES (Bloomington, Indiana)
Editorial Assistant Intern, 19— to 19—.
Provided research assistance to staff and developed a design concept for in-house financial audits.

SPECIAL SKILLS AND ACTIVITIES

Associate Editor, Business School Alumni Newsletter.
Wrote two articles on financial planning with computer models; surveyed business periodicals for potential articles; edited submissions.

President, Women's Transit Program.
Coordinated activities to provide safe nighttime transportation to and from dormitories and campus buildings.

REFERENCES

Available upon request.

R

FIGURE 1   Student Résumé

Style. As you write your résumé, use action **verbs,** and state your ideas concisely. You can avoid the pronoun *I* since the résumé is only about you.

    CHANGE   *I* was promoted to office manager in June 19—.
      TO   Promoted to office manager in June 19—.

CAROL ANN WALKER
1436 W. Schantz Avenue
Dayton, Ohio 45401
(513) 555-1212
Fax: (513) 555-1001
email: caw@hbk.edu

Employment Objective   Senior Research Financial Analyst in corporate offices of growth and research-oriented major manufacturing company.

Employment Experience
(19— to Present)

January 19— to Present   KERFHEIMER CORPORATION (Dayton, Ohio)

Senior Financial Analyst (September 19— to October 19—)
     Report to the Senior Vice-President for Corporate Financial Planning of this $1-billion manufacturer of heavy mining and construction equipment that is purchased by various branches of the federal government and the Department of Defense (DOD). Work with government procurement officers and engineers in developing manufacturing cost estimates for complex and massive machinery, based on prototypes developed in our research lab. Develop program funding requirements and determine best available sources, some of which have been in the $100- to $300-million range. Recipient of the "Financial Planner of the Year" award from the Association of Financial Planners.

Financial Analyst (January 19— to September 19—)
     Reported to Senior Financial Analyst and assisted in the development of funding for major DOD contracts for troop carriers, digging and earth-moving machines, and extension booms. Researched funding options and recommended those with the most favorable rates and terms. Recommended by Senior Financial Analyst as the person to replace him on his retirement.

June 19— to December 19—   FIRST BANK, INC. (Bloomington, IN)

Planning Analyst (September 19— to December 19—)
     Reported to Manager of Corporate Planning and developed computer models for short- and long-range planning that reduced by 65 percent the time required to computer costs.

FIGURE 2   Advanced Résumé: Organized by Job

The following is a list of typical verbs used in résumés:

administered	conducted	designed
analyzed	coordinated	developed
completed	created	directed

CAROL ANN WALKER
Page 2

Education

> Wharton School, University of Pennsylvania (19— to Present)
> > Graduate Seminars Taken
> > > Advanced Corporate Planning and Demographics
> > > Computer Analysis of Corporate Planning
> > > Theory of Corporate Policies
> > > Regulation and Corporate Financial Policy

> University of Wisconsin-Milwaukee (June 19—)
> > Degree: M.B.A.
> > > "Special Executive Curriculum": A special fast-track program
> > > for M.B.A. students who are identified as promising and who
> > > are sponsored by their employer.

> Indiana University (June 19—)
> > Degree: B.B.A. *(magna cum laude)*
> > > Major: Finance
> > > Minor: Computer Science

Special Skills and Activities

> Article published in *Midwest Finance Journal:* "Developing Computer
> Models for Financial Planning" (Vol. 34, No. 2, 19—), pp. 122–136.

> Guest Lecturer at Indiana University (June 29, 19—)

> Senior Member, Association for Corporate Financial Planning

> Member of Audubon Society, Photographer, and Runner

References and a portfolio of computer financial models on request.

R

FIGURE 2   Advanced Résumé: Organized by Job *(continued)*

evaluated	organized	started
increased	oriented	supervised
managed	programmed	trained
operated	recruited	wrote

CAROL ANN WALKER
1436 W. Schantz Avenue
Dayton, Ohio 45401
(513) 555-1212
Fax: (513) 555-1001
email: caw@hbk.com

Employment   Senior Research Financial Analyst in corporate offices of
Objective   growth- and research-oriented major manufacturing company.

Major Accomplishments

Financial Planning

- Provided research on funding options that enabled company to achieve a 23-percent return on investment.
- Developed long-range funding requirements to respond to government and military contracts that totaled over $1 billion.
- Developed computer model for long- and short-range planning, which saved 65 percent in preparation time.

Capital Acquisition

- Developed strategies enabling employers to acquire over $1 billion at 3 percent below market rate.
- Responsible for securing over $100 million through private and government grants for research.
- Developed computer models for capital acquisition that enabled company to increase its long-term debt during several major building expansions.

Research and Analysis

- Developed sophisticated computer models based on cutting-edge research applied to practical problems of corporate finance.
- Served primarily as a researcher for 11 years at two different types of firms.
- Recipient of the "Financial Planner of the Year" award from the Association of Financial Planners, a group composed of both academics and practitioners.
- Published academic article and am taking doctoral-level seminars at the Wharton School, which are described in the education section that follows.

FIGURE 3   Advanced Résumé: Organized by Function

Be truthful in your résumé. If you give false data and are found out, the consequences could be serious. At the very least, you will have seriously damaged your credibility with your employer.

Make your résumé flawless before mailing it by proofing it and verifying the accuracy of the information. Have someone else review it,

CAROL ANN WALKER
Page 2

Education

> Wharton School, University of Pennsylvania (19— to Present)
>> Graduate Seminars Taken
>>> Advanced Corporate Planning and Demographics
>>> Computer Analysis of Corporate Planning
>>> Theory of Corporate Policies
>>> Regulation and Corporate Financial Policy
> University of Wisconsin-Milwaukee (June 19—)
>> Degree: M.B.A.
>>> "Special Executive Curriculum": A special fast-track program for M.B.A. students who are identified as promising and who are sponsored by their employer.
> Indiana University (June 19—)
>> Degree: B.B.A. *(magna cum laude)*
>>> Major: Finance
>>> Minor: Computer Science

Work History

> KERFHEIMER CORPORATION (Dayton, Ohio): January 19— to Present
>> Senior Financial Analyst (19— to 19—)
>> Financial Analyst (19— to 19—)
> FIRST BANK, INC. (Bloomington, IN): June 19— to December 19—
>> Planning Analyst (19— to 19—)
>> Student Intern (19— to 19—)

Special Skills and Activities

> Article published in *Midwest Finance Journal:* "Developing Computer Models for Financial Planning" (Vol. 34, No. 2, 19—), pp. 122–136.
> Guest Lecturer at Indiana University (June 29, 19—)
> Senior Member, Association for Corporate Financial Planning
> Member of Audubon Society, Photographer, and Runner

References and a portfolio of computer financial models on request.

FIGURE 3   Advanced Résumé: Organized by Function *(continued)*

too. And do not pinch pennies when you create the résumé. It, and the accompanying application letter, will be the sole means by which you are known to prospective employers until an interview. Printing your résumé on a high-quality printer produces a more professional-looking résumé than does photocopying.

The sample résumés presented on pages 527–531 are all for the same person. Figure 1 is a student's first résumé, and Figures 2 and 3 show her work experience later in her career (one is organized by job, the other by function). Examine as many résumés as possible, and select the format that best suits you and your goals.

Electronic Résumés

In addition to the traditional paper résumé, a relatively recent development is the use of electronic résumés. As discussed in the entry **job search,** résumés can be displayed and periodically updated at a job-seeker's World Wide Web site. Or a job-seeker can submit a résumé on disc or through **email** to a commercial database service that functions as an electronic employment agency.

Particularly in large companies, electronic résumés are used to facilitate the screening of applicants. These companies usually store electronic résumés in centralized databases. Applicant's résumés are either scanned into computer files from paper documents, or applicants are asked to submit a résumé on a disc along with a paper résumé.

Electronic résumés differ from paper ones in a number of ways. One significant difference is the abundant use of **nouns,** rather than action **verbs** as shown earlier in this entry. Nouns are used as key words (sometimes called "descriptors") because potential employers use such words to screen candidates for specific qualifications and job descriptions. Key words that produce a "hit" in the search process are critical to the success of an electronic résumé. To determine appropriate key words, read job-vacancy postings and note terms used for job classifications that match your interests and qualifications.

R

Electronic résumés are also organized and designed differently than paper ones. The first lines should include your name, address, and other items with all lines centered. A section of key words should immediately follow the main heading, as shown below.

EXAMPLE

CAROL ANN WALKER
1436 W. Schantz Avenue
Dayton, Ohio 45401
(513) 555-1212
email: caw@hbk.com

Key words: Financial Planner. Research Analyst. Banking Intern. Executive Curriculum. MBA Warton School. University of Pennsylvania. Computer Model Development. Published articles in Finance Journal. Written and oral communication skills. Grant writing for Foundation and Government Funding.

Following the section of key words (which can number as many as fifty), you can include the fairly standard sections shown earlier in this entry, such as "Education," "Employment Experience," and "Special Interests." However, as in the key word section, use nouns such as *designer* and *management* rather than verbs such as *designed* and *managed*.

If you submit a résumé on paper to be scanned, or on a disc, the document needs to be universally readable in any application. Therefore, either format the résumé in a file-transfer format such as ASCII on your disc or use ASCII-compatible features on the paper copy. For example, do not use underlining, **italics,** or bold. Avoid decorative, uncommon, or otherwise fancy typefaces; use simple font styles (sans serif, for example) and sizes between 10 and 14 points. For résumés that are to be scanned, use white space generously; scanners use it to recognize that one topic has ended and another has begun. Don't worry about limiting the document to a single page, but keep the résumé simple, clear, and as concise as possible. Use white or beige paper and do not fold a document to be scanned when you mail it because a folded line may produce a misreading when scanned.

revision

The more natural a work of writing seems to the **reader,** the more effort the writer has probably put into revision. Anyone who has ever had to say of his or her own writing, "Now I wonder what I meant by that," can testify to the importance of revision. The time you invest in review will make the difference between clear writing and unclear writing.

If you have followed the appropriate steps of the writing process, you will have a very rough draft that can hardly be considered finished writing. Revision is the obvious next step.

Allow a day or two, if possible, to go by without looking at your draft before beginning to revise it. Without a cooling period, you are too close to the draft to be objective in evaluating it.

Read and evaluate the draft with deliberation and objectivity, from the **point of view** of a reader. Be anxious to find and correct faults, and be honest. Do not try to do all your revision at once. Read through your rough draft several times, each time searching for and correcting a different set of problems.

Revision Checklist

1. *Completeness.* Your writing should give readers exactly what they need, but no more.
2. *Accuracy.* No matter how careful and painstaking you may have been in conducting your research, compiling your notes, and creating your outline, you could easily have made errors when transferring your thoughts from the outline to the rough draft. Be certain also that contradictory facts have not crept into your draft.
3. *Unity and coherence.* If a paragraph has unity, all its sentences and ideas are closely tied together and contribute directly to the main idea expressed in the topic sentence of the paragraph. Writing that is coherent flows smoothly from one point to another, from one sentence to another, and from one paragraph to another. Where transition is missing, provide it; where transition is weak, strengthen it.
4. *Consistent labeling.* Make sure, for example, that you have not called the same item a "routine" on one page and a "program" on another.
5. *Conciseness.* Tighten your writing so that it says exactly what you mean by pruning unnecessary words, phrases, sentences, and even paragraphs.
6. *Awkwardness.* Look especially for excessive passive voice constructions and excess words.
7. *Word choice.* Delete or replace vague or pretentious words and unnecessary intensifiers. Check for sexist language, especially in pronoun references.
8. *Jargon.* If you have any doubt that all your readers will understand any jargon you may have used, eliminate it.
9. *Clichés.* Replace clichés with fresh figures of speech or direct statements.
10. *Grammar.* Check your draft for possible grammatical errors.
11. *Proofreading.* Finally, check your final draft for typographical errors by proofreading it.

rhetorical questions

A *rhetorical question* is a question to which a specific answer is neither needed nor expected. The question is often intended to make the **reader** think about the subject from a different perspective; the writer then answers the question in the article or essay.

EXAMPLES   Is methanol the answer to the energy shortage?
Does advertising lower consumer prices?

The answer to a rhetorical question may not be a yes or no; in the above examples, it might be a detailed explanation of the pros and cons of the value of methanol as a source of energy or the effect of advertising on consumer prices.

The rhetorical question can be an effective **opening,** and it is often used for a **title.** By its nature, however, it is somewhat informal and should therefore be used judiciously in technical writing. For example, a rhetorical question would not be an appropriate opening for a **report** or **memorandum** addressed to a busy superior in your company. When you do use a rhetorical question, be sure that it is not trivial, obvious, or forced. More than any other writing device, the rhetorical question requires that you know your reader.

run-on sentences

A *run-on sentence,* sometimes called a *fused sentence,* is two or more sentences without **punctuation** to separate them. The term is also sometimes applied to a pair of independent **clauses** separated by only a **comma,** although this variation is usually called a *comma fault* or *comma splice.* Run-on sentences can be corrected by (1) making two sentences, (2) joining the two clauses with a semicolon (if they are closely related), (3) joining the two clauses with a comma and a coordinating **conjunction,** or (4) subordinating one clause to the other.

> CHANGE The new manager instituted several new procedures some were impractical. (run-on sentence)
>
> TO The new manager instituted several new procedures. Some were impractical. (period)
>
> OR The new manager instituted several new procedures; some were impractical. (semicolon)
>
> OR The new manager instituted several new procedures, but some were impractical. (comma plus coordinating conjunction)
>
> OR The new manager instituted several new procedures, some of which were impractical. (one clause subordinated to the other)

(See also **sentence faults.**)

R

S

sales proposals (see **proposals**)

same

When used as a **pronoun,** *same* is awkward and outmoded.

> **CHANGE** Your order has been received, and we will respond to *same* next week.
>
> **TO** Your order has been received, and we will respond to *it* next week.
>
> **OR** We received your order, and we will comply with *it* next week.

schematic diagrams

The *schematic diagram,* which is used primarily in electronics and chemistry and in electrical and mechanical engineering, attempts to show the operation of its subject with lines and **symbols** rather than by a physical likeness to it. The schematic diagram is usually a highly abstract representation of its subject because it emphasizes the relationships among the parts at the expense of precise proportions. Because a schematic diagram is a symbolic representation of the subject, as opposed to a realistic representation of it, the schematic relies heavily on the symbols and **abbreviations** that are common to the subject.

For guidelines on how to use illustrations and incorporate them into your texts, see **illustrations.**

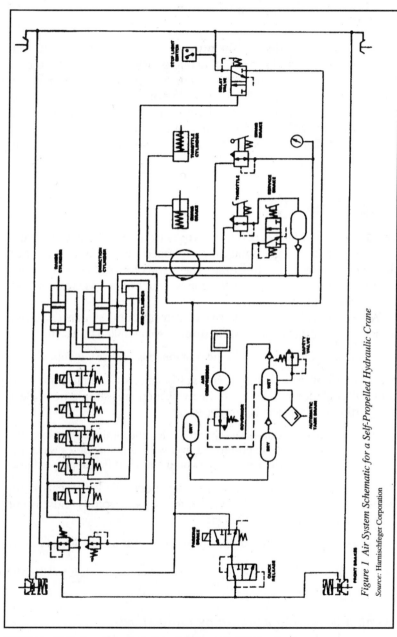

Figure 1 Air System Schematic for a Self-Propelled Hydraulic Crane

Source: Harnischfeger Corporation

S

FIGURE 1   Air System Schematic for a Self-Propelled Hydraulic Crane

scope

If you know your **reader** and the **purpose** of your writing project, you will know the type and amount of detail to include in your document. This is *scope,* which may be defined as the depth and breadth to which you need to cover your subject. If you do not determine your scope of coverage in the planning stage of your writing project, you will not know how much or what kind of information to include. Your scope should be designed to satisfy the needs of your objective and your reader.

semicolons

The semicolon links independent **clauses** or other **sentence** elements of equal weight and grammatical rank, especially **phrases** in a series that have **commas** within them. The semicolon indicates a greater pause between clauses than a comma would, but not as great a pause as a **period** would.

When the independent clauses of a compound sentence are not joined by a comma and a **conjunction,** they are linked by a semicolon.

> **EXAMPLE**   No one applied for the position; the job was too difficult.

Make sure, however, that such **clauses** balance or contrast with each other. The relationship between the two statements should be so clear that further explanation is not necessary.

> **EXAMPLE**   It is a curious fact that there is little similarity between the chemical composition of river water and that of sea water; the various elements are present in entirely different proportions.

Do not use a semicolon between a dependent clause and its main clause. Remember that elements joined by semicolons must be of equal grammatical rank or weight.

> **CHANGE**   No one applied for the position; even though it was heavily advertised.
> **TO**   No one applied for the position, even though it was heavily advertised.

Using a Semicolon with Strong Connectives

In complicated sentences, a semicolon may be used before transitional words or phrases *(that is, for example, namely)* that introduce examples or further explanation.

EXAMPLE   The study group was aware of his position on the issue; that is, federal funds should not be used for the housing project.

A semicolon should also be used before conjunctive **adverbs** (such as *therefore, moreover, consequently, furthermore, indeed, in fact, however*) that connect independent clauses.

EXAMPLE   I won't finish today; *moreover,* I doubt that I will finish this week.

The semicolon in this example shows that *moreover* belongs to the second clause.

Using a Semicolon for Clarity in Long and Complicated Sentences

Use a semicolon between two independent clauses connected by a coordinating conjunction *(and, but, for, or, nor, yet)* if the clauses are long and contain other **punctuation.**

EXAMPLE   In most cases these individuals are corporate executives, bankers, Wall Street lawyers; but they do not, as the economic determinists seem to believe, simply push the button of their economic power to affect fields remote from economics.

A semicolon may also be used if items in a series contain commas within them.

EXAMPLE   Among those present were John Howard, president of the Omega Paper Company; Carol Martin, president of Alpha Corporation; and Larry Stanley, president of Stanley Papers.

Do not, however, use semicolons to enclose a parenthetical element that contains commas. Use **parentheses** or **dashes** for this purpose.

CHANGE   All affected job classifications; typists, secretaries, clerk-stenographers, and word processors; will be upgraded this month.
TO   All affected job classifications (typists, secretaries, clerk-stenographers, and word processors) will be upgraded this month.
OR   All affected job classifications—typists, secretaries, clerk-stenographers, and word processors—will be upgraded this month.

Do not use a semicolon as a mark of anticipation or enumeration. Use a **colon** for this purpose.

CHANGE   Three decontamination methods are under consideration; a zeolite-resin system, an evaporation and resin system, and a filtration and storage system.
TO   Three decontamination methods are under consideration: a zeolite-resin system, an evaporation and resin system, and a filtration and storage system.

The semicolon always appears outside closing **quotation marks.**

> EXAMPLE   The attorney said, "You must be accurate"; the client said, "I will."

sentence construction

Word order can make a great difference in the meaning of a sentence.

> EXAMPLES   He was *only* the engineer.
> He was the *only* engineer.

Consider the many different ways the same idea can be expressed to achieve different **emphasis.**

> EXAMPLES   Every person in the department, from manager to typists, must work overtime this week to ensure that we meet the deadline.
>
> To ensure that we meet the deadline, every person in the department, from manager to typists, must work overtime this week.
>
> From manager to typists, every person in the department must work overtime this week to ensure that we meet the deadline.
>
> To ensure that we meet the deadline, every person in the department must work overtime this week, from manager to typists.
>
> This week, every person in the department, from manager to typists, must work overtime to ensure that we meet the deadline.

At the most basic, a sentence must have a *subject* and *predicate.*

Subject

Not only every sentence, but every **clause** in a sentence, must have a subject and a predicate.

The subject of a sentence is a word or group of words about which the sentence or clause makes a statement. It indicates the topic of the sentence, telling what the predicate is about. It may appear anywhere in a sentence but most often appears at the beginning.

> EXAMPLES   *To increase sales* is our goal.
> *The wiring* is defective.
> *We* often work late.
> *That he will be fired* is now doubtful.

The simple subject is a **substantive;** the complete subject is the simple subject and its **modifiers.** In the following sentence, the simple subject is *procedures;* the complete subject is *the procedures that you instituted.*

> EXAMPLE   The procedures that you instituted have increased efficiency.

Grammatically, a subject must agree with its **verb** in **number.**

> EXAMPLES   The *departments have* much in common.
> The *department has* several advantages.

The subject is the actor in active-**voice** sentences.

> EXAMPLE   The *aerosol bomb* propels the liquid as a mist.

A compound subject has two or more elements as the subject of one verb.

> EXAMPLE   *The president* and *the treasurer* agreed to withhold the information.

Alternative subjects are joined by *or* and *nor.*

> EXAMPLE   Either *cash* or a *check* is acceptable.

Be careful not to shift subjects in a sentence, as doing so may confuse your reader.

> CHANGE   *Radio amateurs* stay on duty during emergencies, and *sending messages* to and from disaster areas is their particular job.
> TO   *Radio amateurs,* who stay on duty during emergencies, are responsible for sending messages to and from disaster areas.
> OR   *Radio amateurs* stay on duty during emergencies, sending messages to and from disaster areas.

Predicate

The *predicate* is that part of a sentence that contains the main verb and any other words used to complete the thought of the sentence (the verb's modifiers and **complements**). The principal part of the predicate is the verb, just as a **noun** (or noun substitute) is the principal part of the subject.

> EXAMPLE   Bill *piloted the airplane.*

The simple predicate is the verb (or verb phrase) alone; the complete predicate is the verb and its modifiers and complements.

A compound predicate consists of two or more verbs with the same subject. It is an important device for economical writing.

> EXAMPLE   The company *tried* but *did not succeed* in that field.

A predicate nominative is a noun construction that follows a linking verb and renames the subject.

> **EXAMPLES**   He is my *lawyer.* (noun)
> His excuse was *that he had been sick.* (noun clause)

The predicate nominative is one kind of subjective complement; the other is the predicate adjective.

Sentence Types

Sentences may be classified according to *structure* (simple, compound, complex, compound-complex); *intention* (declarative, interrogative, imperative, exclamatory); and *stylistic use* (loose, periodic, minor).

By Structure. A *simple sentence* has one clause. At its most basic, the simple sentence contains only a subject and a predicate.

> **EXAMPLES**   Profits (subject) rose (predicate).
> The strike (subject) finally ended (predicate).

Both the subject and the predicate may be compounded without changing the basic structure of the simple sentence.

> **EXAMPLES**   *Bulldozers and road graders* have blades. (compound subject)
> Bulldozers *strip, ditch, and backfill.* (compound predicate)

Although modifiers may lengthen a simple sentence, they do not alter its basic structure.

> **EXAMPLE**   The *recently introduced* procedure works *very well.*

Inverting the subject and predicate does not alter the basic structure of the simple sentence.

> **EXAMPLE**   A better job I never had.

A simple sentence may contain **noun** phrases and prepositional **phrases** in either the subject or the predicate.

> **EXAMPLES**   The man *in the blue suit* is my boss. (prepositional phrase in the subject)
> I wrote the report *in a day.* (prepositional phrase in the predicate)

A simple sentence may include modifying phrases in addition to the independent clause. This fact causes most of the confusion about simple sentences. The following sentence, for example, is a simple sentence because the introductory phrase is an adjective phrase and not a dependent clause:

> **EXAMPLE**   *Hard at work in my office,* I did not realize how late it was.

A *complex sentence* provides a means of subordinating one thought to another (or, put another way, of emphasizing one thought over another) because it contains one independent clause and at least one dependent clause that expresses a subordinate idea.

> **EXAMPLE** The generator will shut off automatically (independent clause) if the temperature rises above a specified point (dependent clause).

The independent clause carries a main point, and the dependent clause carries a related subordinate point.

Complex sentences offer more variety than simple sentences. And frequently the meaning of a compound sentence can be made more precise by subordinating one of the two independent clauses to the other to create a complex sentence (thereby establishing the relationship of the two parts more clearly). (See also **subordination.**)

> **EXAMPLES** We moved and we lost some of our efficiency. (compound sentence with coordinating conjunction)
>
> When we moved, we lost some of our efficiency. (complex sentence with subordinating conjunction)

Learn to handle the complex sentence well; it can be a useful tool with which to express your thoughts clearly and exactly.

A *compound sentence* combines two or more related independent clauses that are of equal importance.

> **EXAMPLE** Drilling a well is the only sure way to determine the presence of oil, *but* it is a very costly endeavor.

The independent clauses of a compound sentence may be joined by a comma and a coordinating conjunction, by a semicolon, or by a conjunctive adverb preceded by a semicolon and followed by a comma.

> **EXAMPLES** People deplore violence, *but* they have an insatiable appetite for it on television and in films. (coordinating conjunction)
>
> There is little similarity between the chemical composition of sea water and that of river water; the various elements are present in entirely different proportions. (semicolon)
>
> It was 500 miles to the site; *therefore,* we made arrangements to fly. (semicolon and conjunctive adverb)

A *compound-complex sentence* consists of two or more independent clauses and at least one dependent clause.

> **EXAMPLE** At the same time *that it cools and lubricates the bit and brings the cuttings to the surface,* the "drilling mud" deposits a sheath of mud cake on the wall of the hole to prevent cave-ins; it reduces the friction *which is created by the drill string's rubbing against the*

S

> *wall of the hole; and since the weight of the mud column bears against the wall of the hole,* it helps to contain formation pressures and prevent a blowout.

Here, the three parallel independent clauses are joined in a series linked by semicolons and the coordinating **conjunction** *and;* each independent clause contains a dependent clause (in **italics**). The first two dependent clauses are adjective clauses modifying *time* and *friction,* and the third is an adverb clause that modifies its independent clause as a whole.

The compound-complex sentence offers a vehicle for a more elaborate grouping of ideas than other sentence types permit. It is not really recommended for inexperienced writers because it is difficult to handle skillfully and often leads them into sentences that become so complicated that they lose all logic and coherence. The result can be a **garbled sentence.** The more technically complex the subject, the more difficult the compound-complex sentence is to control effectively.

By Intention.   By intention, a sentence may be *declarative, interrogative, imperative,* or *exclamatory.*

A declarative sentence conveys information or makes a factual statement.

>   **EXAMPLE**   This motor powers the conveyor belt.

An interrogative sentence asks a direct question.

>   **EXAMPLE**   Does the conveyor belt run constantly?

An imperative sentence issues a command.

>   **EXAMPLE**   Start the generator.

An exclamatory sentence is an emphatic expression of feeling, fact, or opinion. It is a declarative sentence that is stated with great feeling.

>   **EXAMPLE**   The heater exploded!

By Stylistic Use.   A *loose* sentence makes its major point at the beginning and then adds subordinate phrases and clauses that develop the major point. It is the pattern in which we express ourselves most naturally and easily. A loose sentence could be ended at one or more points before it actually does.

>   **EXAMPLE**   It went up (.), a great ball of fire about a mile in diameter (.), changing colors as it kept shooting upward (.), an elemental force freed from its bonds (.) after being chained for billions of years.

A *periodic* sentence delays its main idea until the end by presenting subordinate ideas or modifiers first, thus holding the reader's interest until the end.

> EXAMPLE   During the last decade or so, the attitude of the American citizen toward automation has undergone a profound change.

A *minor* sentence is an incomplete sentence. It makes sense in its context because the missing element is clearly implied by the preceding sentence.

> EXAMPLE   In view of these facts, is automation really useful? *Or economical?* There is no question that it has made our country one of the most wealthy and technologically advanced nations the world has ever known.

Minor sentences are common in advertising copy and fictional dialogue; they are not normally appropriate to technical writing.

Constructing Sentences

Most sentences follow the subject-verb-object pattern. In "The company dismissed Joe," we know the subject and the object by their positions relative to the verb. The knowledge that the usual sentence order is subject-verb-object helps readers interpret what they read. The fact that readers tend to expect this order explains why the writer's departures from it can be effective if used sparingly for emphasis and variety, but annoying if overdone.

An inverted sentence places the elements in other than normal order.

> EXAMPLES   A better job I never had. (direct object-subject-verb)
>
> More optimistic I have never been. (subjective complement-subject-linking verb)

Inverted sentence order may be used in questions and exclamations and also to achieve emphasis.

> EXAMPLES   Have you a pencil? (verb-subject-complement)
> How heavy your book feels! (complement-subject-verb)
> A sorry sight we presented! (complement-subject-verb)

In sentences introduced by **expletives** (there, it), the subject generally follows the verb because the expletive precedes the verb.

> EXAMPLES   *There* (expletive) *are* (verb) certain *principles* (subject) of drafting that must not be ignored.
>
> *It* (expletive) *is* (verb) *difficult* (complement) *to work* (subject) in a noisy office.

S

Beginning sentences with expletives is often wordy.

> **CHANGE**   *There are* certain principles of drafting that must be followed.
> **TO**   Certain principles of drafting must be followed.

Use uncomplicated sentences to state complex ideas. If readers must cope with a complicated sentence in addition to a complex idea, they are likely to become confused.

> **CHANGE**   When you are purchasing parts, remember that although an increase in the cost of aluminum forces all the vendors to increase their prices, some vendors will have a supply of aluminum purchased at the old price, and they may be willing to sell parts to you at the old price in order to get your business.
> **TO**   Although an increase in the cost of aluminum forces all vendors to increase their prices, some vendors will have a supply of aluminum purchased at the old price. When you are purchasing aluminum parts, remember that these vendors may be willing to sell you the parts at the old price in order to get your business.

Just as simpler sentences make complex ideas more digestible, a complex sentence construction makes a series of simple ideas more palatable.

> **CHANGE**   The computer is a calculating device. It was once known as a mechanical brain. It has revolutionized industry.
> **TO**   The computer, a calculating device once known as a mechanical brain, has revolutionized industry.

Do not string together in a series a number of thoughts that should be written as separate sentences or some of which should be subordinated to others. Sentences carelessly tacked together this way are monotonous and hard to read because all ideas seem to be of equal importance.

> **CHANGE**   We started the program three years ago, there were only three members on the staff, and each member was responsible for a separate state, but it was not an efficient operation.
> **TO**   When we started the program three years ago, there were only three members on the staff, each having responsibility for a separate state; however, that arrangement was not efficient.

Sentences can often be improved by eliminating trailing constructions and ineffective **repetition.**

> **CHANGE**   We conducted a new experiment last month and learned much from it.
> **TO**   We learned much from a new experiment last month.

Express coordinate ideas in similar form. The very construction of the sentence helps the reader grasp the similarity of its components. (See also **parallel structure.**)

EXAMPLE Similarly, atoms come and go in a molecule, but the molecule remains; molecules come and go in a cell, but the cell remains; cells come and go in a body, but the body remains; persons come and go in an organization, but the organization remains.

Subordinate your minor ideas to emphasize your more important ideas. (See also **subordination.**)

CHANGE We all had arrived, and we began the meeting early.
TO Since we all had arrived, we began the meeting early.

The most emphatic positions within a sentence are at the beginning and the end. Do not waste them by tacking on phrases and clauses almost as an afterthought or by burying the main point in the middle of a sentence between less important points. For example, consider the following original and revised versions of a statement written for a company's annual report to its stockholders:

CHANGE Sales declined by 3 percent in 19—, but nevertheless the Company had the most profitable year in its history, thanks to cost savings that resulted from design improvements in several of our major products; and we expect 19— to be even better, since further design improvements are being made.
TO Cost savings from design improvements in several major products not only offset a 3-percent sales decline but made 19— the most profitable year in the Company's history. Further design improvements now in progress promise to make 19— even more profitable.

Reversing the normal word order is also used to achieve emphasis.

EXAMPLES I will never agree to that (normal word order).
That I will never agree to.
Never will I agree to that.

There is no rule against beginning a sentence with a coordinating conjunction; in fact, coordinating conjunctions can be strong transitional words and at times provide emphasis.

EXAMPLE I realize the project was more difficult than expected and that you have also encountered personnel problems. *But* we must meet our deadline.

sentence faults 🔄

A number of problems can create sentence faults.

Faulty Subordination

Faulty **subordination,** one of the most common sentence faults, occurs (1) when a grammatically subordinate element, such as a dependent **clause,** actually contains the main idea of the sentence or (2) when a subordinate element is so long or detailed that it dominates or obscures the main idea. Avoiding the first problem (main idea expressed in a subordinate element) depends on the writer's knowing which idea is *meant* to be the main idea. Note that both of the following sentences appear logical; which one really is logical depends on which of two ideas the writer means to emphasize.

> **EXAMPLES** Although the new filing system saves money, many of the staff are unhappy with it.
>
> The new filing system saves money, although many of the staff are unhappy with it.

In this example, if the writer's main point is that *the new filing system saves money,* the second sentence is correct. If the main point is that *many of the staff are unhappy,* then the first sentence is correct.

The other major problem with subordination is the loading of so much detail into a subordinate element that it "crushes" the main point by its sheer size and weight.

> **CHANGE** Because the noise level on a typical street in New York City on a weekday is as loud as an alarm clock ringing three feet away, New Yorkers often have hearing problems.
>
> **TO** Because the noise level in New York City is so high, New Yorkers often have hearing problems.

Clauses with No Subjects

Writers sometimes inappropriately assume a subject that is not stated in a clause.

> **CHANGE** Your application program can request to end the session after the next command. (Request *who* or *what* to end the session? The reader doesn't know.)
>
> **TO** Your application program can request *the host program* to end the session after the next command.

S

CHANGE   This command enables sending the entire message again if an incomplete message transfer occurs. (The sentence contains no subject for the verb *sending*. The reader doesn't know *who* or *what* is doing the *sending*.)

TO   This command enables *you to send* the entire message again if an incomplete message occurs.

(See also **telegraphic style.**)

Rambling Sentences

Sentences that contain more information than the reader can comfortably absorb in one reading are known as *rambling sentences*. The obvious remedy for a rambling sentence is to divide it into two or more sentences. In doing so, however, put the main message of the rambling sentence into the first of the revised sentences.

CHANGE   The payment to which a subcontractor is entitled should be made promptly in order that in the event of a subsequent contractual dispute we, as general contractors, may not be held in default of our contract by virtue of nonpayment.

TO   Pay subcontractors promptly. Then if a contractual dispute should occur, we cannot be held in default of our contract because of nonpayment.

Miscellaneous Sentence Faults

The assertion made by a sentence's predicate about its subject must be logical. "Mr. Wilson's *job* is a salesman" is not logical, but "*Mr. Wilson* is a salesman" is. "Jim's *height* is six feet tall" is not logical, but "*Jim* is six feet tall" is.

Do not omit a required **verb.**

CHANGE   The floor is swept and the lights out.
TO   The floor is swept and the lights *are* out.

CHANGE   I never have and probably never will write the annual report.
TO   I never have *written* and probably never will *write* the annual report.

Do not omit a subject.

CHANGE   He regarded price fixing as wrong, but until abolished by law, he engaged in it, as did everyone else.
TO   He regarded price fixing as wrong, but until *it was* abolished by law, he engaged in it, as did everyone else.

Avoid compound sentences containing clauses that have little or no logical relationship to one another.

CHANGE   The reactor oxidizes the harmful exhaust ingredients into harmless water vapor and carbon dioxide, and it is housed in a double-walled metal shell.

TO   The reactor oxidizes the harmful exhaust ingredients into harmless water vapor and carbon dioxide. It is housed in a double-walled metal shell.

(See also **run-on sentences** and **sentence fragments.**)

sentence fragments

A sentence fragment is an incomplete grammatical unit that is punctuated as a sentence.

EXAMPLES   He quit his job. (sentence)
And quit his job. (fragment)

Sentence fragments are often introduced by relative **pronouns** *(who, which, that)* or subordinating **conjunctions** (such as *although, because, if, when,* and *while*). This knowledge can tip you off that what follows is a dependent clause and must be combined with a main clause.

CHANGE   The new manager instituted several new procedures. *Many of which are impractical.* (The last is an adjective clause modifying *procedures* and linked to it by the relative pronoun *which*.)

TO   The new manager instituted several new procedures, many of which are impractical.

A sentence must contain a finite verb; **verbals** will not do the job. The following examples are sentence fragments because their verbals *(providing, to work, waiting)* cannot perform the function of a finite verb.

EXAMPLES   *Providing* all employees with hospitalization insurance.
To work a forty-hour week.
The customer *waiting* to see you.

Fragments usually reflect incomplete and sometimes confused thinking. The most common type of fragment occurs because of the careless addition of an afterthought. Such fragments should either be a part of the preceding sentence or converted into their own sentences.

CHANGE   These are my coworkers. *A fine group of people.*
TO   These are my coworkers, a fine group of people.

CHANGE   The field tests showed the prototype to be extremely rugged. *The most durable we've tested this year.* (appositive phrase)
TO   The field tests showed the prototype to be extremely rugged, the most durable we've tested this year.

CHANGE   We have one major goal this month. *To increase the strength of the alloy without reducing its flexibility.* (infinitive phrase)

TO   We have one major goal this month: to increase the strength of the alloy without reducing its flexibility.

Explanatory **phrases** beginning with *such as, for example,* and similar terms often lead writers to create sentence fragments.

CHANGE   The staff wants additional benefits. For example, the use of company automobiles. (fragment)

TO   The staff wants additional benefits, such as the use of company automobiles. (one sentence)

OR   The staff wants additional benefits. For example, one possible benefit is the use of company automobiles. (two sentences)

A hopelessly snarled fragment simply has to be rewritten. This rewriting is best done by pulling the main points out of the fragment, listing them in the proper sequence, and then rewriting the sentence. (See also **garbled sentences.**)

CHANGE   Removing the protection cap and the piston secured in the housing by means of the spring placed between the piston and the housing.

MAIN POINTS   1. Remove the protection cap.
2. The piston is held in the housing by a spring.
3. The spring is connected to the piston at one end and the housing at the other.
4. To remove the piston, disconnect the spring.

TO   Removing the protection cap enables you to remove the piston by disconnecting a spring that connects to the piston at one end and the housing at the other.

(See also **sentence construction, sentence faults,** and **run-on sentences.**)

sentence types (see **sentence construction**)

sentence variety

Sentences may be long or short; loose or periodic; simple, compound, complex, or compound-complex; declarative, interrogative, exclamatory, or imperative—even elliptical. There is never a legitimate excuse for letting your sentences become tiresomely alike.

Sentence Length

Varying sentence length makes writing more interesting to the **reader** because a long series of sentences of the same length is monotonous. For example, avoid stringing together a number of short independent clauses. Either connect them with subordinating connectives, thereby making some dependent, or make some clauses into separate sentences. (See also **emphasis** and **subordination.**)

> CHANGE The river is 60 miles long, and it averages 50 yards in width, *and* its depth averages 8 feet.
>
> TO This river, *which* is 60 miles long and averages 50 yards in width, has an average depth of 8 feet.
>
> OR This river is 60 miles long. It averages 50 yards in width and 8 feet in depth.

Short sentences can often be effectively combined by converting **verbs** into **adjectives.**

> CHANGE The steeplejack *fainted.* He collapsed on the scaffolding.
>
> TO The *fainting* steeplejack collapsed on the scaffolding.

Although too many short sentences make your writing sound choppy and immature, a short sentence can be effective at the end of a passage of long ones.

> EXAMPLE During the past two decades, many changes have occurred in American life, the extent, durability, and significance of which no one has yet measured. *No one can.*

In general terms, short sentences are good for emphatic, memorable statements. Long sentences are good for detailed explanations and support. There is nothing inherently wrong with a long sentence, or even with a complicated one, as long as its meaning is clear and direct. Sentence length becomes an element of **style** when varied for **emphasis** or contrast; a conspicuously short or long sentence can be used to good effect.

Word Order

When a series of sentences all begin in exactly the same way, the result is likely to be monotonous. You can make your sentences more interesting by occasionally starting with a modifying word, phrase, or clause.

> EXAMPLES *Exhausted,* the project director slumped into a chair. (**adjective**)
>
> *Reading the report,* she found several errors. (participial phrase)
>
> *Because we now know the result of the survey,* we may proceed with certainty. (**adverb** clause)

S

Inverted sentence order can be an effective way to achieve variety.

EXAMPLES   Then occurred the event that gained us the contract.
Never have sales been so good.

Too many inverted sentences, however, can be distracting.

Be careful in your sentences to avoid the unnecessary separation of subject and verb, **preposition** and **object,** and parts of a verb phrase.

CHANGE   Electrical equipment can, if not carefully handled, cause serious accidents. (parts of verb phrase separated)

TO   Electrical equipment can cause serious accidents if not carefully handled.

This advice, however, does not mean that a subject and verb should never be separated by a modifying phrase or clause.

EXAMPLE   John Stoddard, who founded the firm in 1943, is still an active partner.

Loose/Periodic/Insertion Sentences

A loose sentence makes its major point at the beginning and then adds subordinate phrases and clauses that develop or modify the point. A periodic sentence delays its main idea until the end by presenting modifiers or subordinate ideas first, thus holding the reader's interest until the end.

EXAMPLES   The attitude of the American citizen toward automation has undergone a profound change during the last decade or so. (loose)

During the last decade or so, the attitude of the American citizen toward automation has undergone a profound change. (periodic)

Avoid the singsong monotony of a long series of loose sentences, particularly a series containing coordinate clauses joined by conjunctions. Subordinating some thoughts to others makes your sentences more interesting.

S

CHANGE   The auditorium was filled to capacity, *and* the chairman of the board came onto the stage. The meeting started at eight o'clock, *and* the president made his report of the company's operations during the past year. The audience of stockholders was obviously unhappy, *but* the members of the board of directors all were reelected.

TO   By eight o'clock, *when* the chairman of the board came onto the stage and the meeting began, the auditorium was filled to capacity. *Although* the audience of stockholders was obviously unhappy with the president's report of the company's operations during the past year, the members of the board of directors all were reelected.

For variety, you may also alter the normal sentence order with an inserted phrase or clause.

EXAMPLE   Titanium fills the gap, both in weight and strength, between aluminum and steel.

The technique of inserting such a phrase or clause is good for emphasis, for providing detail, for breaking monotony, and for regulating **pace.** (See also **sentence construction,** and **sentence faults.**)

sequential method of development

The sequential, or step-by-step, method of development is especially effective for explaining a process or describing a mechanism in operation. It would also be the logical method for writing **instructions** (Figure 1).

The main advantage of the sequential method of development is that it is easy to follow because the steps correspond to the elements of the process or operation being described.

PROCESSING FILM

Developing. In total darkness, load the film on the spindle and enclose it in the developing tank. Be careful not to allow the film to touch the tank walls or other film. Add the developing solution, turn the lights on, and set the timer for seven minutes. Agitate for five seconds initially and then every half minute.

Stopping. When the timer sounds, drain the developing solution from the tank and add the stop bath. Agitate continually for 30 seconds.

Fixing. Drain the stop bath and add the fixing solution. Allow the film to remain in the fixing solution for two to four minutes. Agitate for five seconds initially and then every half minute.

Washing. Remove the tank top and wash the film for at least 30 minutes under running water.

Drying. Suspend the film from a hanger to dry. It is generally advisable to place a drip pan below the rack. Sponge the film gently to remove excess water. Allow the film to dry completely.

FIGURE 1 Sequential Method of Development

The disadvantages of the sequential method are that it can become monotonous and does not lend itself very well to achieving **emphasis.**

Practically all **methods of development** have elements of sequence within them. The **chronological method of development,** for example, is also sequential: to describe a trip chronologically, from beginning to end, is also to describe it sequentially.

service

When used as a **verb,** *service* means "keep up or maintain" as well as "repair."

> EXAMPLE   Our company will *service* your equipment.

To mean providing a more general benefit, use *serves.*

> CHANGE   Our company services the northwest area of the state.
> TO   Our company *serves* the northwest area of the state.

set/sit

Sit is an intransitive **verb;** it does not, therefore, require an **object.** Its past **tense** is *sat.*

> EXAMPLES   I *sit* by a window in the office.
>   We *sat* around the conference table.

Set is usually a transitive verb, meaning "put or place," "establish," or "harden." Its past tense is *set.*

> EXAMPLES   Please *set* the trophy on the shelf.
>   The jeweler *set* the stone beautifully.
>   Can we *set* a date for the tests?
>   The high temperature *sets* the epoxy quickly.

Set is occasionally intransitive.

> EXAMPLES   The sun *sets* a little earlier each day.
>   The glue *sets* in 45 minutes.

S

sexism in language

See the following entries:

- **agreement of pronouns and antecedents**
- **chair/chairperson**

- **correspondence**
- **everybody/everyone**
- **female**
- **gender**
- **he/she**
- **male**
- **Ms./Miss/ Mrs.**

shall/will

It was at one time fairly common to use *shall* for first-person constructions and *will* for second- and third-person constructions.

> EXAMPLES   I *shall* go.
> You *will* go.
> He *will* go.

This distinction is unnecessary, however, because no one could be confused by *I will go* to express an action in the near future. *Shall* is sometimes used in all **persons,** nonetheless, to emphasize determination.

> EXAMPLE   Employees *shall* submit a written reason for a leave of absence.

sic

Latin for "thus," *sic* is used in **quotations** to indicate that the writer has quoted the material exactly as it appeared in the original source. It is most often used when the original material contains an obvious error or might in some other way be questioned. *Sic* is placed within **brackets.**

> EXAMPLE   In *Basic Astronomy,* Professor Jones noted that the "earth does not revolve around the son *[sic]* at a constant rate."

S

simile (see **figures of speech**)

simple sentences (see **sentence construction**)

-size/-sized

As **modifiers,** the **suffixes** *-size* and *-sized* are more common to advertising copy than to general writing.

EXAMPLES   a *king-sized* bed, an *economy-sized* carton

However, they are usually redundant unless they are part of the name of a product and should generally not be used.

CHANGE   It was a *small-size* desk.
TO   It was a *small* desk.

slashes

Although not always considered a mark of **punctuation,** the slash performs punctuating duties by separating and showing omission. The slash is called a variety of names, including slant line, virgule, bar, solidus, shilling sign.

The slash is often used to separate parts of addresses in continuous writing.

EXAMPLE   The return address on the envelope was Ms. Rose Howard/62 W. Pacific Court/Dalton/Ontario/Canada.

The slash can indicate alternative items.

EXAMPLE   David's telephone number is 549-2278/2335.

The slash often indicates omitted words and letters.

EXAMPLES   miles/hour for "miles per hour"
c/o for "in care of"
w/o for "without"

In fractions the slash separates the numerator from the denominator.

EXAMPLES   2/3 (2 of 3 parts), 3/4 (3 of 4 parts), 27/32 (27 of 32 parts)

In informal writing, the slash is also used to separate day from month and month from year in **dates.**

EXAMPLE   12/29/87

S

so/so that/such

So is often vague and should be avoided if another word would be more precise.

CHANGE   She writes faster, *so* she finished before I did.
TO   *Because* she writes faster, she finished before I did.

Another problem occurs with the **phrase** *so that,* which should never be replaced with *so* or *such that.*

CHANGE	The report should be written *such that* it can be copied.
TO	The report should be written *so that* it can be copied.

Such, an **adjective** meaning "of this or that kind," should never be used as a **pronoun.**

CHANGE	Our company does not need computers, and we do not anticipate using *such.*
TO	Our company does not need computers, and we do not anticipate using *any.*

some

When *some* functions as an indefinite **pronoun** for plural count **nouns,** or as an indefinite **adjective** modifying plural count nouns, use a plural **verb.**

EXAMPLES	*Some* people *are* kinder than others.
	Some of us *are* prepared to leave.

Some is singular, however, when used with mass nouns.

EXAMPLES	*Some* sand *has* trickled through the crack.
	Some oil *was* spilled on the highway.
	Some stationery *is* missing from the supply cabinet.
	Most of the water evaporated, but *some remains.*

some/somewhat

Some, an **adjective** or **pronoun** meaning "an undetermined quantity" or "certain unspecified persons," should not replace the **adverb** *somewhat,* which means "to some extent."

CHANGE	His writing has improved *some.*
TO	His writing has improved *somewhat.*
OR	His writing is *somewhat* improved.

some time/sometime/sometimes

Some time refers to a duration of time.

EXAMPLE	We waited for *some time* before calling the customer.

Sometime refers to an unknown or unspecified time.

EXAMPLE	We will visit with you *sometime.*

Sometimes refers to occasional occurrences at unspecified times.

EXAMPLE   He *sometimes* visits the branch offices.

spatial method of development

In a spatial sequence, you describe an object or a process according to the physical arrangement of its features. Depending on the subject,

---

Conducting a Methodical Room Search

First, look around the room to decide how the room should be divided for the search and to what height the first searching sweep should extend. The first sweep should include all items resting on the floor up to the selected height.

As nearly as possible, divide the room into two equal parts. Base the division on the number and type of objects in the room, not on the size of the room. Divide the room with an imaginary line that extends from one object to another—for example, the window on the north wall to the floor lamp next to the south wall.

Next, select a search height for the first sweep. Base this height on the average height of the majority of objects resting on the floor. In a typical room, this height is established by such objects as table and desk tops, chair backs, and so on. As a rule, the first sweep will be made hip high and below.

After dividing the room and establishing the first search height, go to one end of the agreed upon room division. This point will be the starting point for the first and all subsequent search sweeps. Beginning back to back, work your way around the room along the walls toward the other team member at the other end of the imaginary dividing line. Check all items resting on the floor adjacent to the walls; be sure to check the floor under the rugs. When you meet your partner at the other end of the room, return to the starting point and search all items in the middle of the room up to the first search height. Include all items mounted in or on the walls in the first sweep, such as air conditioning ducts, baseboards, heaters, built-in storage units, and so on. During this and all subsequent searches, use an electronic or a medical stethoscope.

Then determine the search height for the second search sweep. This height is usually set at the chin or at the top of the head of the searchers. Return to the starting point and repeat the searching technique up to the second search height. This sweep typically covers objects hanging on walls, built-in storage units, tall standing items on the floor, and the like.

Next, establish the third search height. This area usually includes everything in the room from the top of the searcher's head to the ceiling. In this sweep, features like hanging light fixtures and mounted air conditioning units are examined.

Finally, if the room has a false or suspended ceiling, perform a fourth sweep. Check flush or ceiling light fixtures, air conditioning or ventilation ducts, speaker systems, structural frames, and so forth.

S

FIGURE 1   Spatial Sequencing Method of Development

you may describe the features from top to bottom, side to side, east to west (or west to east), inside to outside, and so on. Descriptions of this kind rely mainly on dimension (height, width, length), direction (up, down, north, south), shape (rectangular, square, semicircular), and proportion (one-half, two-thirds). Features are described in relation to one another.

> **EXAMPLE**   One end is raised six to eight inches higher than the other end to permit the rain to run off.

Features are also described in relation to their surroundings.

> **EXAMPLE**   The lot is located on the east bank of the Kingman River.

The spatial **method of development** is commonly used in descriptions of laboratory equipment, **proposals** for landscape work, construction-site **progress and activity reports,** and, in combination with a step-by-step sequence, in many types of **instructions.**

The preceding **instructions** (Figure 1 on p. 559), which explain how a two-person security team should conduct a methodical room search, use spatial sequencing.

specie/species

Specie means "coined money" or "in coin."

> **EXAMPLE**   Paper currency was virtually worthless, and creditors began to demand payment in *specie.*

Species means a category of animals, plants, or things having some of the same characteristics or qualities. *Species* is the correct spelling for both singular and plural.

> **EXAMPLES**   The wolf is a member of the canine *species.*
> Many animal *species* are represented in the Arctic.

specific-to-general method of development

The *specific-to-general method of development* begins with a specific statement and builds to a general **conclusion.** It is somewhat like the **increasing-order-of-importance method of development** in that it carefully builds its case, often with examples and analogies in addition to facts or statistics, and does not actually make its point until the end. For example, if your subject were highway safety, you might begin with a specific highway accident and then go on to generalize about how

---

The Facts about Seat Belts

Statistic
Recently the Highway Safety Foundation studied the use of seat belts in 4,500 accidents involving nearly 13,000 people. Nearly all these accidents occurred on routes which had a speed limit of at least 40 mph. Only 20 percent of all the vehicle occupants were wearing any kind of seat belt.

Statistic
The shoulder-type belts were even more unpopular than the lap belts, and only 4 percent of the occupants who had shoulder belts were wearing them.

Statistic
In this study, as in other studies, it was found that vehicle passengers not using seat belts were more than four times as likely to be killed as those using them. The driver in some cases escaped a more serious injury by being thrown against the steering wheel.

General conclusion
A conservative estimate is that 40 percent of the front-seat passenger car deaths could be prevented if everyone used the seat belts, which the law requires the manufacturer to put into each automobile. If you are in an accident, your chances of survival are far greater if you are using your seat belt.

—*The Safe Driving Handbook* (New York: Grosset, 1970) 84–85.

FIGURE 1   Specific-to-General Method of Development

details of the accident were common enough to many similar accidents that recommendations could be made to reduce the probability of such accidents. Figure 1 is an example of the specific-to-general method of development.

specifications

By definition, a *specification* is "a detailed and exact statement of particulars; especially a statement prescribing materials, dimensions, and workmanship for something to be built, installed, or manufactured." The most significant words in this definition are *detailed* and *exact*.

Although there are two broad categories of specifications—government specifications and industrial specifications—both require the writer of the specification to be both accurate and exact. A specification must be written so clearly that no one could misinterpret any statement contained in it; therefore, do not imply or suggest—state explicitly what is needed. **Ambiguity** in a specification not only can waste money but can result in a lawsuit. Because of the stringent requirements for completeness and exactness of detail in writing specifications, careful **research** and **preparation** are especially important before you begin to write, as is careful **revision** after you have completed the draft of your specification.

Government Specifications

Government agencies are required by law to contract for equipment strictly according to definitions provided in formal specifications. Government specifications are contractual documents that protect both the procuring government agency and the contractor.

A government specification is a precise definition of exactly what the contractor is to provide for the money he or she is paid. In addition to a technical description of the device to be purchased, the specification normally includes an estimated cost, an estimated delivery date, and the standards for the design, manufacture, workmanship, testing, training of government personnel, governing codes, inspection, and delivery of the item to be purchased. Government specifications are often used to prescribe the content and deadline for a government **proposal** submitted by a vendor or a company that wishes to bid on a project.

Government specifications contain details on (1) the **scope** of the project, (2) the documents that the contractor is required to furnish with the purchased device, (3) the required product characteristics and functional performance of the purchased device, (4) the required tests, test equipment, and test procedures, (5) the required preparations for delivery, (6) notes, and (7) an **appendix.**

The "characteristics and performance" section of the specification (item 3 in the previous paragraph) must precisely define every product characteristic not covered by the engineering drawings, and it must precisely define every functional performance requirement the device must meet when it is operational. The "test" section of the specification (item 4 in the previous paragraph) must specify how the device is to be tested to verify that it meets all requirements. The test section includes the precise tests that are to be performed, the procedure to be used in conducting the tests, and the test equipment to be used.

Industrial Specifications

The industrial specification is used in areas like computer software in which there are no engineering drawings or other means of documentation. It is a permanent document, whose purpose is twofold: (1) to document the item being implemented so that it can be maintained if the person who designed it is promoted, transferred, or leaves the company and (2) to provide detailed technical information on the item being implemented to all those in the company who need it (this includes other engineers and technicians, technical writers, technical instructors, and possibly salespeople and purchasing agents).

The industrial specification describes (1) a planned project, (2) a newly completed project, or (3) an old project. The specification for a planned project describes how it is going to be implemented; the specification for a newly completed project describes how the newly completed project was implemented; and the specification for an old project describes the project as it finally exists after it has been operational long enough for all the problems to have been discovered and corrected. All three types of industrial specifications contain a detailed technical description of all aspects of the item being described, including what was done, how it was done, what is required to use the item, how it is used, what its function is, who would use it, and so on. In addition, the specification for a planned project often estimates the time to complete the project (in man-hours or man-months in addition to calendar months) and estimated costs.

The industrial specification differs from the **technical manual** that is given to the customer for the same project in that the specification contains detailed information that the customer does not need.

spelling

The use of a computer spell checker helps enormously with spelling problems; however, it does not solve all spelling problems. It cannot detect a spelling error if the error spells a valid word; for example, if you mean *to* but inadvertently write *too*, the spell checker cannot detect the error. You still must carefully proofread your document.

Whether or not you have access to a personal computer, the following system will help you learn to spell correctly.

1. Keep your **dictionary** handy, and use it regularly. If you are unsure about the spelling of a word, don't rely on memory or guesswork—consult the dictionary. When you look up a word, focus on both its spelling and its meaning.
2. After you have looked in the dictionary for the spelling of the word, write the word from memory several times. Then check the accuracy of your spelling. If you have misspelled the word, repeat this step. If you do not follow through by writing the word from memory, you lose the chance of retaining it for future use. Practice is essential.
3. Keep a list of the words you commonly misspell, and work regularly at whittling it down. Do not load the list with exotic words; many of

us would stumble over *asphyxiation* or *pterodactyl.* Concentrate instead on words like *calendar, maintenance,* and *unnecessary.* These and other frequently used words should remain on your list until you have learned to spell them.

4. Check all your writing for misspellings by **proofreading.**

spin-off

In technical usage, *spin-off* usually refers to benefits that come about in one area (for example, housing materials) as the result of achievements in another area (for example, space technology **research**). Because it is **jargon,** do not use the term unless you are certain that all your **readers** understand it.

>   EXAMPLE   The Teflon coating on cookware is a *spin-off* from the space
>   program.

strata

Strata is the plural form of *stratum,* meaning a "layer of material."

>   EXAMPLES   The land's *strata are* exposed by erosion.
>   Each *stratum is* clearly visible in the cliff.

style

A **dictionary** definition of *style* is "the way in which something is said or done, as distinguished from its substance." Writers' styles are determined by the way they think and transfer their thoughts to paper—the way they use words, sentences, images, **figures of speech,** and so on.

A writer's style is the way his or her language functions in particular situations. For example, a letter to a friend would be relaxed, even chatty in **tone,** whereas a job **application letter** would be more restrained and deliberate. Obviously, the style appropriate to the one letter would not be appropriate to the other. In both letters, the **reader** and situation determine the manner or style the writer adopts.

Standard English can be divided into two broad categories of style—formal and informal—according to how it functions in certain situations. Understanding the distinction between formal and informal writing styles helps writers use the appropriate style in the appropriate place. We must recognize, however, that no clear-cut line divides the two categories, and that some writing may call for a combination of the two.

A formal writing style can perhaps best be defined by pointing to certain material that is clearly formal, such as scholarly and scientific articles in professional journals, lectures read at meetings of professional societies, and legal documents.

Material written in a formal style is usually the work of a specialist writing to other specialists, or writing that embodies laws or regulations. As a result, the vocabulary is specialized and precise. The writer's tone is impersonal and objective because the subject matter looms larger in the writing than does the author's personality. (See **point of view.**) Unlike an informal writing style, a formal writing style does not use **contractions,** slang, or dialect. (See **English, varieties of.**) **Sentences** may be elaborate because complex ideas are generally being examined.

Formal writing need not be dull and lifeless, however. By using such techniques as the active **voice** whenever possible, **sentence variety,** and **subordination,** a writer can make formal writing lively and even interesting, especially if the subject matter is inherently interesting to the reader.

EXAMPLE   Although a knowledge of the morphological chemical constitution of cells is necessary to the proper understanding of living things, in the final analysis it is the activities of their cells that distinguish organisms from all other objects in the world. Many of these activities differ greatly among the various types of living things, but some of the basic sorts are shared by all, at least in their essentials. It is thse fundamental actions with which we are concerned here. They fall into two major groups—those that are characteristic of the cell in the *steady state,* that is, in the normally functioning cell not engaged in reproducing itself, and those that occur during the process of *cellular reproduction.*

—Lawrence S. Dillon, *The Principles of Life Sciences*

S

Whether you should use a formal style in a particular instance depends on your reader and **purpose.** When writers attempt to force a formal style when it should not be used, they are likely to fall into **affectation, awkwardness,** and **gobbledygook.**

An informal writing style is a relaxed and colloquial way of writing standard English. It is the style found in most private letters and in some **correspondence, memorandums,** nonfiction books of general interest, and mass-circulation magazines. There is less distance between the writer and reader because the tone is more personal than in a formal writing style. Contractions and elliptical constructions are commonplace. Consider the following passage, written in an informal style, from a nonfiction book:

EXAMPLE   All you need is a talent for spotting the idiocies now built into the system. But you'll have to give up being an administrator who loves to run others and become a manager who carries water for his people so they can get on with the job. And you'll have to keep a suspicious eye on the phonies who cater to your uncertainties or feed your trembling ego on press releases, office perquisites, and optimistic financial reports. You'll have to give substance to such tired rituals as the office party. And you'll certainly have to recognize, once you get a hunk of your company's stock, that you aren't the last man who might enjoy the benefits of shareholding. These elegant simplicities require a sense of justice that won't be easy to hang on to.

—Robert Townsend, *Up the Organization*

As this example illustrates, the vocabulary of an informal writing style is made up of generally familiar rather than unfamiliar words and expressions, although slang and dialect are usually avoided. An informal style approximates the cadence and structure of spoken English, while conforming to the grammatical conventions of written English.

When we consciously attempt to create a "style," we usually defeat our purpose. One writer may attempt to impress the reader with a flashy writing style, which can lead to affectation. Another may attempt to impress the reader with scientific objectivity and produce a style that is dull and lifeless. Technical writing need be neither affected nor dull. It can and should be simple, clear, direct, and even interesting—the key is to master the basic writing skills and always to keep your reader in mind. What will be both informative and interesting to your reader? When this question is uppermost in your mind as you apply the steps of the writing process, you will achieve an interesting and informative writing style. (See the Checklist of the Writing Process, at the beginning of this book.)

The following guidelines will help you produce a brisk, interesting style. Concentrate on them as you revise your rough draft.

1. Use the active **voice**—not exclusively but as much as possible without becoming awkward or illogical.
2. Use **parallel structure** whenever a sentence presents two or more thoughts that are of equal importance.
3. Avoid the monotony of a singsong style by using a variety of sentence structures. (See **sentence variety.**)

4. Avoid stating positive thoughts in negative terms (write *40 percent responded* instead of *60 percent failed to respond*). (See also **positive writing.**)
5. Concentrate on achieving the proper balance between **emphasis** and subordination.

Beyond an individual's personal style, there are various kinds of writing that have distinct stylistic traits, such as **technical writing style.**

subjective complements (see complements)

subjects of sentences (see sentence construction)

subordinating conjunctions (see conjunctions)

subordination

Subordination is a technique used by writers to show, in the structure of a sentence, the appropriate relationship between ideas of unequal importance by subordinating the less important ideas to the more important ideas. The skillful use of subordination is a mark of mature writing.

CHANGE	Beta Corp. now employs 500 people. It was founded just three years ago.
TO	Beta Corp., *which now employs 500 people,* was founded just three years ago.
OR	Beta Corp., *which was founded just three years ago,* now employs 500 people.

Effective subordination can be used to achieve **sentence variety, conciseness,** and **emphasis.** For example, consider the following sentence. "The city manager's report was carefully illustrated, and it covered five typed pages." See how it might be rewritten, using subordination, in any of the following ways:

EXAMPLES	The city manager's report, *which covered five typed pages,* was carefully illustrated. (dependent clause)
	The city manager's report, *covering five typed pages,* was carefully illustrated. (phrase)

S

>The city manager's report, *five typed pages,* was carefully illus-
>trated. (phrase)
>
>*The carefully illustrated report* of the city manager covered five
>typed pages. (phrase)
>
>The city manager's *five-page* report was carefully illustrated.
>(single modifier)

We sometimes use a coordinating **conjunction** to concede that an
opposite or balancing fact is true; however, a subordinating connective
can often make the point more smoothly.

>CHANGE   Their bank has a lower interest rate on loans, *but* ours provides
>a fuller range of essential services.
>
>TO   *Although* their bank has a lower interest rate on loans, ours pro-
>vides a fuller range of essential services.

The relationship between a conditional statement and a statement of
consequences will be clearer if the condition is expressed as a subordinate clause.

>CHANGE   The bill was incorrect, *and* the customer was angry.
>
>TO   The customer was angry *because* the bill was incorrect.

Subordinating connectives (such as *because, if, while, when, though*)
and relative **pronouns** *(who, whom, which, that)* achieve subordination effectively when the main **clause** states a major point and the
dependent clause establishes a relationship of time, place, or **logic**
with the main clause.

>EXAMPLE   A buildup of deposits is impossible *because* the apex seals are
>constantly sweeping the inside chrome surface of the rotor
>housing.

Relative pronouns *(who, whom, which, that)* can be used effectively to
combine related ideas that would be stated less smoothly as independent clauses or sentences.

>CHANGE   The generator is the most common source of electric current. It
>uses mechanical energy to produce electricity.
>
>TO   The generator, *which* is the most common source of electric cur-
>rent, uses mechanical energy to produce electricity.

Avoid overlapping subordinate constructions, with each depending
on the last. Often the relationship between a relative pronoun and its
antecedent will not be clear in such a construction.

>CHANGE   Shock, *which* often accompanies severe injuries, severe infec-
>tions, hemorrhages, burns, heat exhaustion, heart attacks, food

or chemical poisoning, and some strokes, is a failure of the circulation, *which* is marked by a fall in blood pressure *that* initially affects the skin (*which* explains pallor) and later the vital organs of kidneys and brain; there is a marked fall in blood pressure.

TO Shock often accompanies severe injuries, severe infections, hemorrhages, burns, heat exhaustion, heart attacks, food or chemical poisoning, and some strokes. It is a failure of the circulation, initially to the skin (this explains pallor) and later to the vital organs of kidneys and brain; there is a marked fall in blood pressure.

substantives

A substantive is a word, or a group of words, that functions in its **sentence** as a subject. It may be a **noun,** a **pronoun,** or a **verbal** (gerund or infinitive), or it may be a **phrase** or even a **clause** that is used as a noun.

EXAMPLES The *report* is due today. (noun)
We must finish the project on schedule. (pronoun)
Several college graduates applied for the job. (noun phrase)
Drilling is expensive. (gerund)
To succeed will require hard work. (infinitive)
What I think is unimportant. (noun clause)

suffixes

A *suffix* is a letter or letters added to the end of a word to change its meaning in some way. Suffixes can change the **part of speech** of a word.

EXAMPLES The *wire* is on the truck. (noun)
The *wirelike* tubing is on the truck. (adjective)
What is the *length* of the unit? (noun)
Move the unit *lengthwise* through the conveyor. (adverb)

S

sweeping generalizations (see **logic**)

symbols

From highway signs to **mathematical equations,** people communicate in written symbols. When a symbol seems appropriate in your writing, either be certain that your **reader** understands its meaning or place an explanation in **parentheses** following the symbol the first time it appears. However, never use a symbol when your reader would more readily understand the full term.

synonyms

A *synonym* is a word that means nearly the same thing as another word does.

> **EXAMPLES**   purchase, acquire, buy;
> seller, vendor, supplier

The **dictionary** definitions of synonyms are very similar. The **connotations** of such words, however, may differ. (A *seller* may be the same thing as a *supplier*, but the term *supplier* does not suggest a commercial transaction as strongly as *seller* does.)

Do not try to impress your **reader** by finding fancy or obscure synonyms in a **thesaurus;** the result is likely to be **affectation.** (See also **connotation/denotation** and **antonyms.**)

syntax

Syntax refers to the way that words, **phrases,** and **clauses** are combined to form sentences. In English, the most common structure is the subject-**verb-object** pattern. (For more information about the word order of sentences, see **sentence construction, sentence faults, sentence fragments,** and **sentence variety.**)

S

T

table of contents

A *table of contents* is a list of **headings** in a **report** or chapters in a book. It lists them in their order of appearance and cites their page numbers. Appearing at the front of a work, a table of contents permits **readers** to preview what is in the work and assess the work's value to them. It also aids readers who may want to read only certain sections.

The length of your report should determine whether it needs a table of contents. If it does, use the major heads and subheads of your **outline** to create your table of contents, as shown in Figure 1 (p. 572).

To punctuate a table of contents, see "leaders" in **periods,** and for guidance on the placement of the table of contents in a report, see **formal reports.**

tables

A table is useful for showing large numbers of specific, related facts or statistics in a small space. A table can present **data** more concisely than text can, and it is more accurate than graphic presentations are because it provides numerous facts that a **graph** cannot convey. A table facilitates **comparisons** among data because of the arrangement of the data into rows and columns. Overall trends about the information, however, are more easily seen in charts and graphs. But do not rely on a table (or any **illustration**) as the only method of conveying significant information.

Guidelines for Creating Tables

Tables typically include the following elements, as shown in Figure 1 (p. 573).

TABLE OF CONTENTS

ABSTRACT	ii
LIST OF FIGURES	iv
LIST OF TABLES	v
PREFACE	vii
1 SUMMARY	1
2 INTRODUCTION	6
3 METHODOLOGY	10
3.1 Selection of Study Areas	10
3.2 Descriptions of Study Sites	11
3.2.1 Pilgrim 1	12
3.2.2 Millstone 1	12
3.2.3 Oyster Creek	13
3.2.4 R. E. Ginna	14
3.3 Time Series Analysis	14
3.4 Cross-Section Analysis	17
3.4.1 Real Selling Price	17
3.4.2 House Characteristics	18
3.4.3 Lot Characteristics	19
3.4.4 Accessibility Characteristics	19
3.4.5 Locational Characteristics	20
3.5 Collection of Data	22
3.5.1 Time Series Data	23
3.5.2 Cross-Section Data	24
4 DATA ANALYSIS AND INTERPRETATION	27
4.1 Time Series Studies	27
4.1.1 Data Analysis	28
4.1.2 Interpretation of Results	28
4.2 Cross-Section Studies	30
4.2.1 Functional Forms of the Model	31
4.2.2 Statistical Package	33
4.2.3 Regression Results: Individual Study Areas	35
4.2.4 Regression Results: Pooled Data	37
4.2.4.1 Linear Model	37
4.2.4.2 Log-Log Model	39
4.3 References	41
5 CONCLUSIONS	43
APPENDIX	A-1

iii

FIGURE 1 Sample Table of Contents

Table Number. If you are using several tables, assign each a number, and then center the number and title above the table. The numbers are usually Arabic, and they should be assigned sequentially to the tables throughout the text. Tables should be referred to in the text by table number rather than by direction ("Table 4" rather than "the above

Table number ——→ Table 1 Recreational fresh-water angling by water-body type and geographical region* ←—— Caption

Column captions ↓   Body

Geographical Regions	Reservoirs	Man-Made Ponds	Natural Lakes & Ponds	Rivers & Streams	Farms Ponds
New England	130	40	570	410	410
Middle Atlantic	710	290	780	1200	630
East North Central	1200	760	3100	1600	1300
West North Central	810	550	1200	970	980
South Atlantic	1100	760	640	1500	1600
East South Central	890	630	190	670	1200
West South Central	1700	610	430	880	1300
Mountain	820	50	280	600	230
Pacific	950	200	820	1400	470
Totals	8300	39000	8000	9200	7800

Boxhead ↗↖   Stub ↗↖   Rule ↑

Footnote ——→ *In thousands of anglers. Anglers who fished in more than one water body or region are represented in more than one category.

Source line ——→ Source: U.S. Department of the Interior

FIGURE 1 Table of Data

table"). If there are more than five tables in your report or paper, give them a separate heading ("List of Tables"), and list them by title and page number, together with any figure numbers, on a separate page immediately after the **table of contents.** The first text reference to the table should precede the table. If a document contains several chapters or sections, tables can be numbered by chapter or section (Table 1-1, 1-2, . . . 3-1, 3-2).

Title. The **title,** which is placed just above the table, should describe concisely what the table represents.

Boxhead. The boxhead carries the column headings. These should be kept concise but descriptive. Units of measurement, where necessary, should be specified either as part of the heading or enclosed in **parentheses** beneath the heading. Avoid vertical lettering if possible.

Stub. The left-hand vertical column of a table is the stub. It lists the items about which information is given in the body of the table.

Body. The body comprises the data below the boxhead and to the right of the stub. Within the body, columns should be arranged so that the terms to be compared appear in adjacent rows and columns. Leaders are sometimes used between figures to aid the eye in following data from column to column. Where no information exists for a specific item, leave an empty space to acknowledge the gap.

Rules. These are the lines that separate the table into its various parts. Horizontal rules are placed below the title, below the body of the table, and between the column headings and the body of the table. They should not be closed at the sides. The columns within the table may be separated by vertical rules if they aid clarity.

Footnotes. Footnotes are used for explanations of individual items in the table. Symbols (*, #) or lowercase letters, rather than numbers, are ordinarily used to key table footnotes because numbers might be mistaken for the data in a numerical table.

Source Line. The source line, which identifies where the data were obtained, appears below any footnotes (when a source line is appropriate).

Continued Lines. When a table must be divided so that it can be continued on another page, repeat the boxhead and give the table number at the head of each new page with a "continued" label ("Table 3, continued").

Informal Tables

To list relatively few items that would be easier for the reader to grasp in tabular form, use an informal table; be sure you introduce it properly.

> The sound intensity levels (decibels) for the three frequency bands (in Hertz) were determined to be:

Frequency Band (Hz)	Decibels
600–1200	68
1200–2400	62
2400–4800	53

Although informal tables do not need titles or table numbers to identify them, they do require column headings that accurately describe the information listed.

technical information letters (see **correspondence**)

technical manuals

Technical manuals help technical specialists and customers use and maintain products, from word processors purchased by general consumers to multimillion-dollar aircraft or sophisticated mainframe computer systems purchased by manufacturers, businesses, and governments. Technical manuals are normally written by professional technical writers, although in smaller companies, engineers and technicians may write them. They are sometimes delivered on CD-ROM rather than as bound manuals.

Today, many manufacturers consider good technical manuals to be an important marketing tool; if a manual makes equipment easy to operate or repair, consumers are more likely to buy it. The following **list** describes typical technical manuals and the **readers** and **purposes** they serve.

User Manuals

User manuals are aimed at skilled or unskilled consumers of such equipment as word processors. For example, in the case of word processors, these manuals enable users to set up, operate, and care for both the hardware and software of their systems.

Tutorials

Tutorials are self-study, workbook-like teaching guides for users of a product or system. Either packaged with user manuals or published separately, tutorials walk the novice user through the operation of the product or system. Both user manuals and tutorials are written for home and office applications.

Training Manuals
Training manuals are textbooks used in the classroom training of individuals in some procedure or skill, such as flying an airplane or processing an insurance claim. In many technical and vocational fields, they are the primary teaching device. Training manuals are often accompanied by slides, tapes, or other audiovisual material.

Operators' Manuals
Written for skilled operators of construction, manufacturing, computer, or military equipment, operators' manuals contain essential instructions and safety warnings.

Service Manuals
Service manuals help trained technicians repair equipment or systems, usually at the customer's location. Such manuals often contain elaborate troubleshooting guides for locating technical problems.

Maintenance Manuals
Maintenance manuals are written for semiskilled operators or technicians who must keep such equipment as aircraft and manufacturing machinery operating efficiently.

Repair Manuals
Repair manuals guide skilled technicians as they perform extensive repairs or rebuild equipment, usually at the manufacturing site. Such manuals contain detailed illustrations and sophisticated technical information, sometimes including the theory or operation for highly technical electronic and computer equipment.

Special-Purpose Manuals
Programmer reference manuals provide computer programmers with definitions, syntax rules, and other technical information about specific programming languages, such as FORTRAN and COBOL.

Overhaul manuals are guides for rebuilding items at the factory. Overhaul manuals are often used in the heavy-equipment industry, in which rebuilding a huge crane, for example, is much less costly than manufacturing a new one.

Handling and setup manuals are guides for skilled and unskilled employees in the safe handling, installation, or setup of various kinds of equipment.

Safety manuals are guides for operators of potentially dangerous equipment, illustrating safe operating practices.

Some manuals, of course, are combinations or variations of those described.

Technical manuals often use extensive **illustrations,** such as exploded-view **drawings, schematic diagrams, flowcharts, photographs,** and **tables.** They include such sections as parts lists, troubleshooting guides, warning statements, the standard symbols for potential dangers, and **indexes.**

The packaging of manuals is important. Although they may be either bound or looseleaf, manuals are more often looseleaf to enable easy revision and updating. They are sized and packaged for easy use. A word-processing manual, for example, may use a small ($7'' \times 9''$) three-ring binder that can be easily held or that lays flat open on a desk while the user views the screen. Or an aircraft training manual may be spiral bound to allow it to fit neatly in the cockpit of the aircraft.

Writing a technical manual requires careful adherence to the steps outlined in the Checklist of the Writing Process. **Preparation** and **organization** are especially important because of the complexity of the products and systems for which the manuals are written. Before beginning to write a technical manual, give particular care to **outlining,** and review the entries on **format, instructions, process explanation,** and **technical writing style.** These elements are central to writing a manual, regardless of its purpose and **scope.** Above all, pay particular attention to the readers' needs and their level of technical knowledge of the product or system you are documenting.

To ensure that manuals are helpful and accurate, you should submit them to technical, legal, and peer reviews. Some companies test the manuals by using prototypes with typical manual readers. (See also **proofreading.**)

technical writing style

Technical writing style is standard **exposition** in which the **tone** is objective, with the author's voice taking a back seat to the subject matter. Because the focus is on helping the **reader** use a device or understand a process or concept, the language is utilitarian—emphasizing exactness rather than elegance for its own sake. Thus the writing is usually not adorned with figurative language, except when a **figure of speech** would facilitate understanding. Technical writing requires an effective **introduction** or **opening,** good **organization,** and **sentence variety,** all of which benefit greatly from **headings** and **illustrations.**

Good technical writing avoids overusing the passive **voice.** Its vocabulary is appropriately technical, although the general word is preferable to the technical word for most readers. **Gobbledygook** and **jargon** are poor substitutes for clear and direct writing. Do not use a big or technical term merely because you know it—make sure that your readers also understand it. (See also **affectation.**)

Technical writing is direct and often is aimed at readers who are not experts in the subject—such as consumers needing to learn how to operate unfamiliar equipment or managers trained in business rather than technical methods. Figure 1 is an excerpt from a manual that teaches customers how to operate eye test equipment. As this opening for a **technical manual** illustrates, its **format** (or the page design) is important because it keeps the reader oriented and helps make the writing readable. (See also **style.**)

MONOCULAR RECORDING

A monocular electrode configuration might be useful to measure the horizontal eye movement of each eye separately. This configuration is commonly used during the voluntary saccade tests.

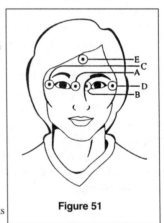

Figure 51

1. Place electrodes at the inner canthi of both eyes (**A and B,** Figure 51).
2. Place electrodes at the outer canthi of both eyes (**C and D**).
3. Place the last electrode anywhere on the forehead (**E**).
4. Plug the electrodes into the jacks described in Figure 52 on the following page.

With this configuration, the Channel A display moves upward when the patient's left eye moves nasally and moves downward when the patient's left eye moves temporally. The Channel B display moves upward when the patient's right eye moves temporally and moves downward when the patient's right eye moves nasally.

NOTE: *During all testing, the Nystar Plus identifies and eliminates eye blinks without vertical electrode recordings.*

FIGURE 1   Example of Technical Writing (Reprinted with permission from Nicolet Biomedical)

telegraphic style

Telegraphic style condenses writing by omitting articles, **pronouns, conjunctions,** and **transitional** expressions. Although **conciseness** is important in writing, writers sometimes make their **sentences** too brief by omitting these words. Telegraphic style forces the **reader** to mentally supply the missing words. Compare the following two passages, and notice how much easier the revised version reads (the added words are italicized).

> CHANGE   Take following action when treating a serious burn. Remove loose clothing on or near burn. Cover injury with clean dressing and wash area around burn. Secure dressing with tape. Separate fingers/toes with gauze/cloth to prevent sticking. Do not apply medication unless doctor prescribes.
>
> TO   Take *the* following action when treating *a* serious burn. Remove *any* loose clothing on or near *the* burn. Cover *the* injury with *a* clean dressing, and wash *the* area around *the* burn. *Then* secure *the* dressing with tape. Separate fingers *or* toes with gauze *or* cloth to prevent *them from* sticking *together*. Do not apply medication unless *a* doctor prescribes *it*.

Telegraphic style can also produce **ambiguity,** as the following example demonstrates:

> CHANGE   Grasp knob and adjust lever before raising boom.

Does this sentence mean that the reader should *adjust the lever* or *grasp an adjust lever?*

> TO   Grasp the knob and the adjust lever before raising the boom.
> OR   Grasp the knob and adjust the lever before raising the boom.

Some writers excuse telegraphic style with the claim that their readers will "understand." Maybe so, but they will probably have to work hard to do so—and may therefore even misunderstand. Though telegraphic writing can save space, it never really saves time. As a writer, remember that although you may save yourself work by writing telegraphically, your readers will have to work that much harder to read—or decipher— your writing. (See also **transitions.**)

tenant/tenet

A *tenant* is one who holds or temporarily occupies a property owned by another person.

> EXAMPLE   The *tenants* were upset by the increased rent.

A *tenet* is an opinion or belief held by a person or an organization.

> **EXAMPLE**   The idea that competition will produce adequate goods and services for a society is a central *tenet* of capitalism.

tense

Tense is the grammatical term for **verb** forms that indicate time distinctions. There are six tenses in English: past, past perfect, present, present perfect, future, and future perfect. Each of these has a corresponding progressive form.

Tense	Basic	Progressive
Past	I began	I was beginning
Past Perfect	I had begun	I had been beginning
Present	I begin	I am beginning
Present Perfect	I have begun	I have been beginning
Future	I will begin	I will be beginning
Future Perfect	I will have begun	I will have been beginning

Perfect tenses allow you to express a prior action or condition that continues in a present, past, or future time.

> **EXAMPLES**   I *have begun* to write the annual report, and I will work on it for the rest of the month. (present perfect)
>
> I *had begun* to read the manual when the lights went out. (past perfect)
>
> I *will have begun* this project by the time funds are allocated. (future perfect)

Note from the **table** that the progressive forms are created by combining the helping **verb** *be,* in the appropriate tenses, with the present participle *(-ing)* form of the main verb.

Past Tense

The simple past tense indicates that an action took place in its entirety in the past. The past tense is usually formed by adding *-d* or *-ed* to the root form of the verb.

> **EXAMPLE**   We *closed* the office early yesterday.

Past Perfect Tense

The past perfect tense indicates that one past event preceded another. It is formed by combining the helping verb *had* with the past participle form of the main verb.

> EXAMPLE   He *had finished* by the time I arrived.

Present Tense

The simple present tense represents action occurring in the present, without any indication of time duration.

> EXAMPLE   I *use* the beaker.

A general truth is always expressed in the present tense.

> EXAMPLE   He learned that the saying "time *heals* all wounds" is true.

The present tense can be used to present actions or conditions that have no time restrictions.

> EXAMPLE   Water *boils* at 212°F.

The present tense can be used to indicate habitual action.

> EXAMPLE   I *pass* the paint shop on the way to my department every day.

The present tense can be used as the "historical present" to make things that occurred in the past more vivid.

> EXAMPLE   He *asks* for more information on production statistics and *receives* a detailed report on every product manufactured by the company. Then he *asks,* "Is each department manned at full strength?" In his office, surrounded by his staff, he goes over the figures and *plans* for the coming year.

Present Perfect Tense

The present perfect tense describes something from the recent past that has a bearing on the present—a period of time before the present but after the simple past. The present perfect tense is formed by combining a form of the helping verb *have* with the past principle form of the main verb.

> EXAMPLES   He *has retired,* but he visits the office frequently.
> We *have finished* the draft and are ready to begin revising it.

T

Future Tense

The simple future tense indicates a time that will occur after the present. It uses the helping verb *will* (or *shall*) plus the main verb.

> **EXAMPLE**   I *will finish* the job tomorrow.

Do not use the future tense needlessly.

> **CHANGE**   This system *will be* explained on page 3.
> **TO**   This system *is* explained on page 3.

Future Perfect Tense

The future perfect tense indicates action that will have been completed at a future time. It is formed by linking the helping verbs *will have* to the past participle form of the main verb.

> **EXAMPLE**   He *will have driven* the test car 40 miles by the time he returns.

Tense Agreement of Verbs

The verb of a subordinate **clause** should usually agree in tense with the verb of the main clause. (See also **agreement.**)

> **EXAMPLES**   When the supervisor *presses* the starter button, the assembly line *begins* to move.
>
> When the supervisor *pressed* the starter button, the assembly line *began* to move.

Shift in Tense.   Be consistent. The only legitimate shift in tense records a real change in time. When you choose a tense in telling a story or discussing an idea, stay with that tense. Illogical shifts in tense will only confuse your **reader.**

> **CHANGE**   Before he *installed* the printed circuit, the technician *cleans* the contacts.
> **TO**   Before he *installed* the printed circuit, the technician *cleaned* the contacts.

T

test reports

The *test report* differs from the more formal **laboratory report** in both size and **scope.** Considerably smaller and less formal than the laboratory report, the test report can be a **memorandum** or a formal business letter, depending on its recipient. Either way, the report should have a subject line at the beginning to identify the test being discussed.

Biospherics, Inc.
4928 Wyaconda Road
Rockville, MD 20852
(301) 962-1332
(301) 962-4219 (fax)

March 14, 19—

Mr. John Sebastiani, General Manager
Midtown Development Corporation
114 West Jefferson Street
Milwaukee, WI 53201

Subject: Results of Analysis of Soil Samples for Arsenic

Dear Mr. Sebastiani:

Following are the results of the analysis of 22 soil samples for arsenic. The arsenic values listed are based on a wet-weight determination. The moisture content of the soil is also given to allow conversion of the results to a dry-weight basis if needed.

Hole Number	Depth	Percent of Moisture	Arsenic Total As ppm
1	12"	19.0	312.0
2	Surface	11.2	737.0
3	12"	12.7	9.5
4	12"	10.8	865.0
5	12"	17.1	4.1
6	12"	14.2	6.1
7	12"	24.2	2540.0
8	Surface	13.6	460.0

I noticed that some of the samples contained large amounts of metallic iron coated with rust. Arsenic tends to be absorbed into soils high in iron, aluminum, and calcium oxides. The large amount of iron present in some of these soil samples is probably responsible for retaining high levels of arsenic. The soils highest in iron, aluminum, and calcium oxides should also show the highest levels of arsenic, provided the soils have had approximately equal levels of arsenic exposure.

If I can be of further assistance, please do not hesitate to contact me.

Yours truly,

Gunther Gottfried
Chemist

GG/jrm

FIGURE 1 Test Report

The **opening** of a test report should state the test's purpose, unless it is obvious. The body of the report presents the data. A report on the tensile strength of metal, for example, includes the readings from the test equipment. If the procedure used to conduct the test would interest the **reader,** it should be described. The results of the test should be

---

Biospherics, Inc.
4928 Wyaconda Road
Rockville, MD 20852
(301) 962-1332
(301) 962-4219 (fax)

September 9, 19—

Mr. Leon Hite, Administrator
The Angle Company, Inc.
1869 Slauson Boulevard
Waynesville, VA 23927

Dear Mr. Hite:

On Tuesday, 30 August, Biospherics, Inc., performed asbestos-in-air monitoring at your Route 66 construction site, near Front Royal, Virginia. Six persons and three construction areas were monitored.

All monitoring and analyses were performed in accordance with "Occupational Exposure to Asbestos," U.S. Department of Health, Education and Welfare, Public Health Service, National Institute for Occupational Safety and Health, 1972. Each worker or area was fitted with a battery-powered personal sampler pump operating at a flow rate of approximately one liter per minute. We collected the airborne asbestos on a 37-mm Millipore type AA filter mounted in an open-face filter holder over an 8-hour period.

We mounted a wedge-shaped piece of each filter on a microscope slide with a drop of 1:1 solution of dimethyle phthalate and diethyl oxalate and then covered it with a cover clip. We counted samples within 24 hours after mounting, using a microscope with phase contrast option.

In all cases, the workers and areas monitored were exposed to levels of asbestos fibers well below the NIOSH standard. The highest exposure we found was that of a driller who was exposed to 0.21 fibers per cubic centimeter. We analyzed the driller's sample by scanning electron microscopy followed by energy dispersive X-ray techniques which identify the chemical nature of each fiber, thereby verifying the fibers as asbestos or identifying them as other fiber types. Results from these analyses show that the fibers present are tremolite asbestos. We found no nonasbestos fibers.

If you have any questions about the tests, please call or write me.

Yours truly,

Gary Willis

Gary Willis
Chemist

GW/jrm

FIGURE 2   Test Report

---

stated and, if necessary, interpreted. Often there is reason to discuss their significance. The report should conclude with any recommendations made as a result of the test.

The test report in Figure 1 (page 583) does not explain how the tests were performed because such an explanation is unnecessary. The example in Figure 2, however, does explain this.

that

Avoid the unnecessary repetition of *that*.

CHANGE   You will note *that* as you assume greater responsibility and as your years of service with the company increase, *that* your benefits will increase accordingly.

TO   You will note *that* as you assume greater responsibility and as your years of service with the company increase, your benefits will increase accordingly.

OR   You will note *that* your benefits will increase as you assume greater responsibility and as your years of service with the company increase.

However, do not delete *that* from a sentence in which it is necessary for the **reader's** understanding.

CHANGE   Some engineers fail to recognize sufficiently the human beings who operate the equipment constitute an important safety system.

TO   Some engineers fail to recognize sufficiently *that* the human beings who operate the equipment constitute an important safety system.

that/which/who

Who refers to persons, whereas *that* and *which* refer to animals and things.

EXAMPLES   John Brown, *who* is retiring tomorrow, has worked for the company for twenty years.

Companies *that* fund basic research must not expect immediate results.

The jet stream, *which* is approximately 8 miles above the earth, blows at an average of 64 miles per hour from the west.

That is often overused. (See **that.**) However, do not eliminate it if to do so would cause **ambiguity** or problems with **pace.**

CHANGE   On the file specifications input to the compiler for any chained file, the user must ensure the number of sectors per main file section is a multiple of the number of sectors per bucket.

TO   On the file specifications input to the compiler for any chained file, the user must ensure *that* the number of sectors per main file section is a multiple of the number of sectors per bucket.

Which, rather than *that,* should be used with nonrestrictive **clauses** (clauses that do not change the meaning of the basic **sentence**).

> EXAMPLES   After John left the restaurant, *which* is one of the best in New York, he came directly to my office. (nonrestrictive)
>
> A company *that* diversifies often succeeds. (restrictive)

(See also **who/whom,** relative **pronouns,** and **restrictive and non-restrictive elements.**)

there/their/they're

There, their, and *they're* are often confused because they sound alike. *There* is an **expletive** or an **adverb.**

> EXAMPLES   *There* were more than 1,500 people at the conference. (expletive)
> More than 1,500 people were there. (adverb)

Their is the possessive form of *they.*

> EXAMPLE   Our employees are expected to keep *their* desks neat.

They're is a contraction of *they are.*

> EXAMPLE   If *they're* right, we should change the design.

thesaurus

A *thesaurus* is a book of words with their **synonyms** and **antonyms,** arranged by categories. Thoughtfully used, it can help you with **word choice** during the **revision** phase of the writing process. However, this variety of words may tempt you to choose inappropriate or obscure synonyms just because they are available. Use a thesaurus only to clarify your meaning, not to impress your **reader.** Never use a word unless you are sure of its meanings; **connotations** of the word that might be unknown to you could mislead your reader.

thus/thusly

Thus is an **adverb** meaning "in this manner" or "therefore." The *-ly* in *thusly* is superfluous and should be omitted.

> CHANGE   The committee's work is done. *Thusly* we should have the report by the end of the week.
>
> TO   The committee's work is done. *Thus* we should have the report by the end of the week.

'til/until

'Til is a nonstandard **spelling** of *until,* meaning "up to the time of."
Use *until.*

> **CHANGE**   We worked *'til* eight o'clock.
> **TO**   We worked *until* eight o'clock.

titles

The *title* of a document should indicate its **topic** and announce its
scope and **purpose.** Often the title is the only basis on which readers
can decide whether they should read something. Titles are also
increasingly used by information specialists as a source of key terms
by which documents can be indexed for automated information storage and retrieval systems. Imprecise titles thus defeat human and automated information retrieval.

   To create accurate titles, keep the following guidelines in mind:

1. Be specific. Do not use general or vague terms when specific ones will
   pinpoint the topic and its scope. Although the title "Electric Fields
   and Living Organisms" announces the topic of a report, it leaves the
   reader with important unanswered questions: What is the relationship between electric fields and living organisms? What kind of
   organisms does it refer to? What is the intensity of the electric field?
   What stage in the organism's life cycle does it discuss? The addition
   of details identifies not only the topic but its scope and objective as
   well: *Effects of 60-Hz Electric Fields on Embryo and Chick Development,
   Growth, and Behavior.*
2. Be concise, but don't make your title short just for the sake of being
   short. As the example above indicates, making a title accurate may
   mean adding a few terms. Conciseness, instead, means eliminating
   words that do not contribute to accuracy. Avoid titles that begin
   "Notes on . . .," "Studies on . . .," "A Report on . . .," or "Observations
   on. . . ." These notions are self-evident to the reader. On the other
   hand, certain works, like **feasibility reports** should be identified as
   such in the title because this information helps define the purpose
   and scope.
3. Do not indicate dates in the title of a periodic or progress report, but
   put them in a subtitle:

> The Effects of Acid Rain on Red Spruce in the White Mountains
> Quarterly Report
> January–March 19—

T

4. Avoid using **abbreviations, acronyms and initialisms,** chemical formulas, and the like *unless* the work is addressed exclusively to specialists in the field.

5. Do not put titles in sentence form.

      CHANGE   How Residential Passive Solar Heating Could Affect Seven Electric Utilities

          TO   Potential Effects of Residential Passive Solar Heating on Seven Utilities

6. For multivolume publications, repeat the title on each volume. The distinction among the volumes is made by the volume number (Volume 1, 2, 3, and so on), and sometimes each volume will have a different subtitle. For example, Volume 1 could be an **executive summary;** Volume 2 could be the main report; and any subsequent volumes could be lengthy **appendixes.**

(For guidelines on how to capitalize titles and when to use italics and quotation marks, see **capital letters, italics,** and **quotation marks.**)

to/too/two ↻

To, too, and *two* are confused only because they sound alike. *To* is used as a **preposition** or to mark an infinitive.

      EXAMPLES   Send the report *to* the district manager. (preposition)

                      I wish *to* go. (mark of the infinitive)

Too is an **adverb** meaning "excessively" or "also."

      EXAMPLES   The price was *too* high. ("excessively")

                      I, *too,* thought it was high. ("also")

Two is a **number.**

      EXAMPLE   Only *two* buildings have been built this fiscal year.

T

tone ↻

In writing, *tone* is the writer's attitude toward the subject and his or her **readers.** The tone may be casual or serious, enthusiastic or skeptical, friendly or hostile. In technical writing, the tone in **correspondence, memorandums,** and **formal reports** may range widely—depending on the **purpose,** situation, and context. For example, in a memo read only by an associate who is also a friend, your tone might be casual and friendly.

EXAMPLE   I think your proposal to Smith and Sons is great. If we get the contract, I owe you a lunch! I've marked a couple of places where we could cover ourselves on the schedule. See what you think.

In a memo to a superior, however, your tone might be quite different.

EXAMPLE   I think your proposal to Smith and Sons is excellent. I have marked a couple of places for your consideration where we could ensure that we are not committing ourselves to a schedule we might not be able to keep. If I can help in any other way, please let me know.

In a memo that serves as a **report** to numerous readers, the tone would again be different.

EXAMPLE   The Smith and Sons proposal appears complete and thorough, based on our department's evaluation. Several small revisions, however, would ensure that Acme is not committing itself to an unrealistic schedule. These are marked on the copy of the report being circulated.

Your tone will be set by many factors in your writing. As illustrated above, a formal writing **style** will normally have a different tone than will an informal writing style. (See also **technical writing style.**) **Word choice,** the **introduction** or **opening,** and even your **title** all contribute to the overall tone. For instance, a title such as "Some Observations on the Diminishing Oil Reserves in Wyoming" clearly sets a tone quite different from that of "What Happens When We've Pumped Wyoming Dry?" The first title would most likely be appropriate for a report; the second title would be more appropriate for a **newsletter article.** The important thing is to make sure that your tone is the one best suited to your objective, by always keeping your reader in mind. In **correspondence,** tone is particularly important because the letter represents a direct communication between two people. Furthermore, good business letters establish a rapport between your organization and the public, and a positive and considerate tone is essential.

topic sentences (see **paragraphs**)

topics

On the job, the topic of a writing project is usually determined by need. In a college writing course, on the other hand, you may have to select your own topic. If you do, keep the following points in mind:

1. Select a topic that interests you.
2. Select a topic that you can **research** adequately with the facilities available to you.
3. Limit your topic so that its **scope** is small enough to handle within the time you are given. A topic like "Air Pollution," for example, would be too broad. On the other hand, keep the topic broad enough so that you will have enough to write about.
4. Select a topic for which you can make adequate **preparation** to ensure a good final **report.**

toward/towards

Both *toward* and *towards* are acceptable variant spellings of the preposition meaning "in the direction of." *Toward* is more common in the United States, and *towards* is more common in Great Britain.

> EXAMPLES   We walked *toward* the Golden Gate Bridge.
> They were headed *towards* London Bridge.

trade journal articles

A *trade journal article* is an article written on a specific subject for a professional periodical. These periodicals, commonly known as trade journals, are often the official publications of professional societies. *Technical Communication,* for example, is an official voice of the Society for Technical Communication. Other professional publications include *Electrical Engineering Review, Chemical Engineering,* and *Nucleonics Week.* Professional staff people, such as chemists and engineers, regularly contribute articles to trade journals.

From time to time in your career, you may wish to write a trade journal article that would be of interest to others in your field. Such an article makes your work more widely known, provides favorable publicity for your employer, gives you a sense of satisfaction, and may even improve your chances for professional advancement.

Planning the Article

When thinking about writing such an article, ask yourself the following questions:

- Is your work or your knowledge of the subject original? If not, what is there about your approach that justifies publication?

- Will the significance of the article justify the time and effort needed to write it?
- What parts of your work, project, or study are most appropriate to include in the article?

To help you answer these questions, learn about the periodical or periodicals to which you will send the article, and consult your colleagues for advice. Once you have decided on several journals, consider the following factors about each:

- the professional interests of its readership
- the size of its readership
- the professional reputation of the journal
- the appropriateness of your article to the journal's goals, as stated on the "masthead" page
- the frequency with which its articles are cited in other journals

After you have settled on the right journal, read back issues to find out such information as the amount and kind of details that the articles include, their length, and the typical writing style.

If your subject involves a particular project, you should begin work on your article, ideally, when the project is in progress. Doing so will permit you to devote short periods during the project to writing various sections of the article, thereby writing the draft in manageable increments. This practice makes the writing integral to the project and may even reveal any weaknesses in the design or details of the project, such as the need for more data.

As you plan your article, decide whether to invite one or more coauthors to join you. Coauthors can add strength and substance to the paper but can add complications as well. To minimize potential problems, one of you should take the responsibility of being the primary author. That person makes assignments, sets up a schedule, and keeps the various parts in perspective. The primary author must also take responsibility for ensuring that the finished paper reads as smoothly as if it had been written by one person.

Gathering the Data

As you gather information, take notes from all the sources available to you, such as the following:

- published material on the subject
- your own experience
- notebooks recorded during your research

- survey results
- **progress reports**
- performance records and test data
- patent disclosures

You should begin your **research** with a careful **literature review** to establish what has been published about your topic. A review of the relevant information in your field can be insurance against writing an article that in essence has already been published. (Some articles, in fact, begin with a literature review.) As you compile this information, record your references in full; that is, include all the information you would need to document the source (see **documenting sources**).

Organizing the Draft

Some trade journals (particularly in the sciences) use a prescribed **organization** for the major sections, such as the following:

- **Introduction**
- Materials and methods
- Results
- Discussion
- Literature cited

If the major organization is not prescribed, choose and arrange the various sections of the draft in a way that shows your results to best advantage.

The best guarantee of a logically organized article is a good **outline.** In addition to shaping your information in a logical order, outlining helps you organize your thinking. If you have coauthors, it is essential that the writing team work from a common, well-developed outline; otherwise, coordinating the various writers' work will be impossible, and the parts produced by each will not fit together logically as a whole.

Once you have written your outline, consider it to be flexible; you may well change and improve your organization as you write the rough draft.

Preparing Sections of the Article

Introduction. The purpose of an **introduction** is to give your readers enough general information about your topic for them to be able to understand the detailed information in the body of the article. The introduction should include the following:

- the purpose of the article
- a definition of the problem examined
- the **scope** of the article
- the rationale for your approach to the problem or project and the reasons you rejected alternative approaches
- previous work in the field, including other approaches described in previously published articles

Emphasize what is new and different about your approach, especially if you are not dealing with a new concept. Show the overall significance of your project or approach by explaining how it fills a need, solves a current problem, or offers a useful application. For detailed guidance on writing introductions, see **introductions.**

Conclusion. The **conclusion** section pulls together your results and findings and interprets them based on the purpose of the study and the methods used to conduct it. Thus, the conclusion is the focal point of the article, the goal toward which the study was aimed. Your conclusions must grow out of the evidence for the findings in the body of the article. They must also be consistent with the scope of information presented in the introduction.

Abstract. After the article is written, you will need to write an **abstract.** Be sure to follow any **instructions** provided by the journal on writing abstracts, and review previously published abstracts in that journal. Prepare your abstract carefully, since it will be the basis on which many other researchers will decide whether to read the work in full. Abstracts are often published independently in abstract journals and are used as a source of terms (called keywords) used to **index,** by subject, the original article for computerized information-retrieval systems.

Illustrations and Tables. Use **illustrations** and **tables** wherever they are appropriate, but design each for a specific purpose:

- to describe a function
- to show an external appearance
- to show internal construction (with cutaway or exploded-view drawings, for example)
- to display statistical data
- to indicate trends

T

Consider working your rough illustrations into your outline so that you decide on your illustrations as you write your manuscript. Used effectively and appropriately, illustrative material can clarify information in ways that text alone cannot.

Headings. The use of **headings** throughout your manuscript is important because they break the manuscript into manageable portions. They allow journal readers to understand the development of your topic and pinpoint sections of particular interest to them. The main headings in your outline will often become the headings in your final manuscript.

References. This section of the article **lists** the sources used or quoted in the article. The specific **format** for listing these sources varies from field to field. Usually the journal to which the article is submitted will specify the form the editors require for citing sources. (For a detailed discussion of using and citing sources, see **documenting sources** and **quotations.**)

Preparing the Manuscript

Some trade journals make available, or at least recommend, a "style sheet" with detailed guidelines on the style and format of the article. Such style sheets often include specific instructions about how the manuscript should be typed, the style for **abbreviations, symbols,** and units (like the International System of Units), how to handle **mathematical equations,** how many copies to submit, and the like. The following guidelines are typical:

- Double-space the manuscript, using at least one-inch margins all around, and number each page.
- Make sure that the captions for figures and tables are specific, accurate, and self-explanatory. Use labels (known as *callouts*) with those illustrations that need them.
- Provide clean, even line work and accurately worded axes and labels for your drawings if the journal will not be redrawing them.
- Place mathematical equations, either typed or legibly printed in black ink, on separate lines in the text, and number them consecutively.
- If you include **photographs,** use glossy-finish black-and-white prints. Both the photograph and the contrast should be of good quality. Identify each photograph on the back; don't write on the face of it, and don't write on the back of it with a heavy hand. If necessary, indicate which edge of the photograph is the top.

Obtaining Publication Clearance

After the manuscript has been prepared, submit a copy to your employer for review before sending it to the journal. This review will make sure that no proprietary information has been inadvertently revealed or classified or other restricted information has not been disclosed. Likewise, secure ahead of time permission to print information for which someone else holds the copyright.

transition

Transition is the means of achieving a smooth flow of ideas from **sentence** to sentence, **paragraph** to paragraph, and subject to subject. Transition is a two-way indicator of what has been said and what will be said; that is, it provides a means of linking ideas to clarify the relationship between them. You can achieve transition with a word, a **phrase,** a sentence, or even a paragraph. Without the guideposts of transition, **readers** can lose their way. Transition can be quite obvious.

> EXAMPLE   *Having considered* the economic feasibility of this alloy as a transformer core, *we turn now* to the problem of inadequate supply.

Or it can be more subtle.

> EXAMPLE   Even if this alloy is economically feasible as a transformer core, there still remains the problem of inadequate supply.

Either way, you now have your reader's attention fastened on the problem of inadequate supply, exactly what you set out to do.

Certain words and phrases are inherently transitional. Consider the following terms and their functions:

> Result: *therefore, as a result, consequently, thus, hence*
> Example: *for example, for instance, specifically, as an illustration*
> Comparison: *similarly, likewise*
> Contrast: *but, yet, still, however, nevertheless, on the other hand*
> Addition: *moreover, furthermore, also, too, besides, in addition*
> Time: *now, later, meanwhile, since then, after that, before that time*
> Sequence: *first, second, third, then, next, finally*

Within a paragraph, such transitional expressions clarify and smooth the movement from idea to idea. Conversely, the lack of transitional devices can make the going bumpy for the reader. Consider first the following passage, which lacks adequate transition:

> EXAMPLE   People had always hoped to fly. Until 1903 it was only a dream. It was thought by some that human beings were not meant to fly. The Wright brothers launched the world's first heavier-than-air flying machine. The airplane has become a part of our everyday life.

T

Now read the same passage with words and phrases of transition added (in **italics**), and notice how much more smoothly the thoughts flow:

> EXAMPLE   People had always hoped to fly, *but* until 1903 it was only a dream. *Before,* it was thought by some that human beings were not meant to fly. *In 1903* the Wright brothers launched the world's first heavier-than-air flying machine. *Now* the airplane has become a part of our everyday life.

And finally, read the same passage with stronger transition provided:

> EXAMPLE   People had always hoped to fly, *but* until 1903 it was only a dream. *Before that time,* it was thought by some that human beings were not meant to fly. *However,* in 1903 the Wright brothers launched the world's first heavier-than-air flying machine. *Since then* the airplane has become a part of our everyday life.

If your **organization** and **outline** are good, your transitional needs will be less difficult to satisfy (although they must nonetheless be satisfied). Just as the outline is a road map for the writer, transition is a road map for the reader.

Transition between Sentences

In addition to using transitional words and phrases such as those shown above, the writer may achieve effective transition between sentences by repeating key words or ideas from preceding sentences, by using **pronouns** that refer to antecedents in previous sentences, and by using **parallel structure**—that is, by repeating the pattern of a **phrase** or **clause.** Consider the following short paragraph, in which all these means are employed.

> EXAMPLE   Representative of many American university towns is Millville. *This midwestern town,* formerly a *sleepy farming community,* is today the home of a large and bustling *academic community.* Attracting students from all over the Midwest, *this university* has grown very rapidly in the last ten years. *This same decade* has seen a physical expansion of the campus. The state, recognizing *this expansion,* has provided additional funds for the acquisition of land adjacent to the university. *The university* has become Millville's major industry, generating most of *the town's* income—and, of course, many of *its* problems, too.

Another device for achieving transition is enumeration:

> EXAMPLE   The recommendation rests upon *three conditions. First,* the department staff must be expanded to a sufficient size to handle the increased work load. *Second,* sufficient time must be pro-

vided for the training of the new members of the staff. *Third,* a sufficient number of qualified applicants must be available.

Transition between Paragraphs

All the means discussed above for achieving transition between sentences—and especially the repetition of key words or ideas—may also be effective for transition between paragraphs. For paragraphs, however, longer transitional elements are often required. One technique is to use an opening sentence that summarizes the preceding paragraph and then to move ahead to the business of the new paragraph.

EXAMPLE   One property of material considered for manufacturing processes is hardness. Hardness is the internal resistance of the material to the forcing apart or closing together of its molecules. Another property is ductility, the characteristic of material that permits it to be drawn into a wire. The smaller the diameter of the wire into which the material can be drawn, the greater the ductility. Material also may possess malleability, the property that makes it capable of being rolled or hammered into thin sheets of various shapes. Engineers, in selecting materials to employ in manufacturing, must consider these properties before deciding on the most deisrable for use in production.

The requirements of hardness, ductility, and malleability account for the high cost of such materials . . .

Ask a question at the end of one paragraph and answer it at the beginning of the next.

EXAMPLE   Automation has become an ugly word in the American vocabulary because it has at times displaced some jobs. But the all-important fact that is often overlooked is that it invariably creates many more jobs than it eliminates. (The vast number of people employed in the great American automobile industry as compared with the number of people that had been employed in the harness-and-carriage-making business is a classic example.) Almost always, the jobs that have been eliminated by automation have been menial, unskilled jobs, and those who have been displaced have been forced to increase their skills, which resulted in better and higher-paying jobs for them. *In view of these facts, is automation really bad?*

Certainly automation has made our country the most wealthy and technologically advanced nation the world has ever known . . .

A purely transitional paragraph may be inserted to aid readability.

EXAMPLE   . . . that marred the progress of the company.

There were two other setbacks to the company's fortunes that year that also marked the turning of the tide: the loss of many skilled workers through the Early Retirement Program and the intensification of the devastating rate of inflation.

The Early Retirement Program . . .

Checking for Transition during Revision

Check for **unity** and **coherence** to determine whether the transition is effective. If a paragraph has unity and coherence, the sentences and ideas will be tied together and contribute directly to the subject of the paragraph. Look for places where transition is missing, and add it. Look for places where it is weak, and strengthen it.

transmittal letters (see cover letters)

trip reports

Many companies require or encourage reports of the business trips their employees take. A trip report both provides a permanent record of a business trip and its accomplishments and enables many employees to benefit from the information that one employee has gained.

A trip report should normally be in the **format** of a **memorandum,** addressed to your immediate superior. On the subject line give the destination and dates of the trip. The body of the **report** will explain why you made the trip, whom you visited, and what you accomplished. The report should devote a brief section to each major event and may include a **heading** for each section (you needn't give equal space to each event, but instead, elaborate on the more important events). Follow the body of the report with any appropriate **conclusions** and recommendations. The trip report shown in Figure 1 is typical.

T trite language

Trite language is made up of words, **phrases,** or ideas that have been used so often that they are stale.

CHANGE   *It may interest you to know* that all the folks in the branch office are *hale and hearty.* I should finish my report *quick as a wink,* and we should *clean up* on it.

TO   Everyone here at the branch office is well. I should finish my project within a week, and I'm sure it will prove profitable for us.

INTEROFFICE MEMORANDUM

TO: Robert K. Ford, Manager
         Customer Service
FROM: James D. Kerson, Maintenance Specialist *J. D. K.*
DATE: January 13, 19—

SUBJECT: Trip to Smith Electric Co., Huntingon, West Virginia
         January 19—

I visited the Smith Electric Company in Huntington, West Virginia, to determine the cause of a recurring failure in a Model 247 Printer and to fix it.

Problem

The printer stopped printing periodically for no apparent reason. Repeated efforts to bring it back on line eventually succeeded, but the problem recurred at irregular intervals. Neither customer personnel operating the printer nor the local maintenance specialist solved the problem.

Action

On January 3, I met with Ms. Ruth Bernardi, the Office Manager, who explained the problem. My troubleshooting did not reveal the cause of the problem. Only when I tested the logic cable did I find that it contained a broken wire. I replaced the logic cable and then ran all the normal printer test patterns to make sure no other problems existed. All patterns were positive, so I turned the printer over to the customer.

Conclusion

There are over 12,000 of these printers in the field, and to my knowledge, this is the first occurrence of a bad cable. Therefore, I do not believe the logic cable problem found at Smith Electric Company warrants further investigation.

FIGURE 1 Trip Report

Trite language shows that the writer is thoughtless in his or her **word choice** or is not thinking carefully about the **topic** but is relying instead on what others have thought and said. (See also **affectation.**)

trouble reports

The *trouble report* is used to report an accident, an equipment failure, an unplanned work stoppage, and so on. The **report** enables the man-

---

MEMORANDUM

To: James K. Arburg, Safety Officer
From: Lawrence T. Baker, Foreman of Section A-40 LTB
Date: November 30, 19—

Subject: Personal-Injury Accident in Section A-40
              October 10, 19—

On October 10, 19—, at 10:15 p.m., Jim Hollander, operating punch press #16, accidentally brushed the knee switch of his punch press with his right knee as he swung a metal sheet over the punching surface. The switch activated the punching unit, which severed Hollander's left thumb between the first and second joints as his hand passed through the punch station. While an ambulance was being summoned, Margaret Wilson, R.N., administered first aid at the plant dispensary. There were no witnesses to the accident.

The ambulance arrived from Mercy Hospital at 10:45 p.m., and Hollander was admitted to the emergency room at the hospital at 11:00 p.m. He was treated and kept overnight for observation, then released the next morning.

Hollander returned to work one week later, on October 17. He has been given temporary duties in the tool room until his injury heals.

Conclusions About the Cause of the Accident

The Maxwell punch press on which Hollander was working has two switches, a hand switch and a knee switch, and *both* must be pressed to activate the punch mechanism. The hand switch must be pressed first, and then the knee switch, to trip the punch mechanism. The purpose of the knee switch is to leave the operator's hands free to hold the panel being punched. The hand switch, in contrast, is a safety feature. Because the knee switch cannot activate the press until the hand switch has been pressed, the operator cannot trip the punching mechanism by touching the knee switch accidentally.

Inspection of the punch press that Hollander was operating at the time of the accident made it clear that Hollander had taped the hand switch of his machine in the ON position, effectively eliminating its safety function. He could then pick up a panel, swing it onto the machine's punching surface, press the knee switch, stack the newly punched panel, and grab the next unpunched panel, all in one continuous motion—eliminating the need to let go of the panel, after placing it on the punching surface, in order to press the hand switch.

To prevent a recurrence of this accident, I have conducted a brief safety session with all punch press operators, at which I described Hollander's experience and cautioned them against tampering with the safety features of their machines.

mo

FIGURE 1 Trouble Report

agement of an organization to determine the cause of the problem and to make any changes necessary to prevent its recurrence. The trouble report normally follows a simple **format** (and is often a **memorandum**), since it is an internal document and is not large enough in either size or **scope** to require the format of a **formal report.**

In the subject line of the memorandum, state the precise problem you are reporting. Then begin your description of the problem in the body of your report. What happened? Where did it occur? When did it occur? Was anybody hurt? Was there any property damage? Was there a work stoppage? Since insurance claims, worker's compensation awards, and, in some instances, lawsuits may hinge on the information contained in a trouble report, be sure to include precise times, dates, locations, treatment of injuries, names of any witnesses, and any other crucial information. Give a detailed analysis of what caused the problem. Be thorough and accurate in your analysis, and support any judgments or conclusions with facts. Be careful about your **tone;** avoid any condemnation or blame. If you speculate about the cause of the problem, make it clear to your **reader** that you are speculating. In your **conclusion,** state what has been done, what is being done, or what will be done to correct the conditions that led to the problem. This may include training in safety practices, better or improved equipment, protective clothing (for example, shoes or goggles), and so on.

The report shown in Figure 1, about an accident involving personal injury, was written by the foreman of a group of punch press operators for the plant's safety officer, at the safety officer's request. Since there were no witnesses, the foreman obtained the information for the report by talking to the plant nurse, hospital personnel, and the victim—and by inspecting the equipment used by the victim at the time of the accident.

try and

The **phrase** *try and* is colloquial for *try to.* Unless you are writing a casual personal letter, it is better to use *try to.*

> CHANGE   Please *try and* finish the report on time.
> TO   Please *try to* finish the report on time.

T

unity

Unity is singleness of purpose and treatment, the cohesive element that holds a piece of writing together; it means that everything in an article or paper is essentially about one thing or idea.

To achieve unity, the writer must select one **topic** and then treat it with singleness of **purpose**—without digressing into unrelated paths. The prime contributors to unity are a good **outline** and effective **transition.** After you have completed your outline, check it to see that each part relates to your subject. Be certain that your transitional terms make clear the relationship of each part to what precedes it.

Transition dovetails **sentences** and **paragraphs** like the joints of a well-made drawer. Notice, for example, how neatly the sentences in the following paragraph are made to fit together by the italicized words and **phrases** of transition:

EXAMPLE   Any company that operates internationally today faces a host of difficulties. Inflation is worldwide. Most countries are struggling with other economic problems *as well. In addition,* many monetary uncertainties and growing economic nationalism are working against multinational companies. *Yet* ample business is available in most developed countries if you have the right products, services, and marketing organization. To maintain the growth Data Corporation has achieved overseas, we recently restructured our international operations into four major trading areas. *This reorganization* will improve the services and support that the corporation can provide to its subsidiaries around the world. *At the same time,* the reorganization establishes firm management control, ensuring consistent policies around the world. *So* you might say the problems of doing business abroad will be more difficult this year, but we are better organized to meet those problems.

u

The logical sequence provided by a good outline is essential to achieving unity. An outline enables the writer to lay out the most direct route from **introduction** to **conclusion** without digressing into side issues that are not related, or that are only loosely related, to the subject. Without establishing and following such a direct route, a writer cannot achieve unity.

up

Adding the word *up* to **verbs** often creates a redundant **phrase.**

CHANGE   Next open *up* the exhaust valve.
TO   Next open the exhaust valve.

CHANGE   He wrote *up* the report.
TO   He wrote the report.

(See also **conciseness.**)

usage

Usage describes the choices we make among the various words and constructions available in our language. The line between standard English and nonstandard English, or between formal and informal English, is determined by these choices. Your guideline in any situation requiring such choices should be appropriateness: Is the word or expression you use appropriate to your **reader** and subject? When it is, you are practicing good usage.

This book has been designed to help you sort out the appropriate from the inappropriate: Just look up the item in question in the index. A good **dictionary** is also an invaluable aid in your selection of the right word.

utilize

Utilize should not be used as a **long variant** of *use,* which is the general word for "employ for some purpose." When you are tempted to use this term, try to substitute *use.* It will almost always prove a clearer and less pretentious word.

CHANGE   You can *utilize* the fourth elevator to reach the fiftieth floor.
TO   You can *use* the fourth elevator to reach the fiftieth floor.

vague words

A *vague word* is one that is imprecise in the context in which it is used. Some words encompass such a broad range of meanings that there is no focus for their definition. Words such as *real, nice, important, good, bad, contact, thing,* and *fine* are often called "omnibus words" because they can mean everything to everybody. In speech we sometimes use words that are less than precise, but our vocal inflections and the context of our conversation make their meanings clear. Since writing cannot rely on vocal inflections, avoid using vague words. Be concrete and specific. (See also **abstract words/concrete words.**)

> CHANGE It was a *meaningful* meeting, and we got *a lot* done.
> TO The meeting resolved three questions: pay scales, fringe benefits, and work loads.

verb phrases (see **phrases**)

verbals

Verbals, which are derived from **verbs,** function as **nouns, adjectives,** and **adverbs.** There are three types of verbals: gerunds, infinitives, and participles.

A *gerund* is a verbal ending in *-ing* that is used as a noun.

> EXAMPLE *Smelting* is a technique used to extract metal from ore.

A gerund may be used as a subject, a direct **object,** an object of a **preposition,** a subjective **complement,** or an **appositive.**

> EXAMPLES *Estimating* is an important managerial skill. (subject)
> I find *estimating* difficult. (direct object)

We were unprepared for their *coming.* (object of a **preposition**)

Seeing is *believing.* (subjective complement)

My primary departmental function, *programming,* occupies about two-thirds of my time on the job. (appositive)

Only the possessive form of a noun or **pronoun** should precede a gerund.

EXAMPLES   *John's* working has not affected his grades.
His working has not affected his grades.

An *infinitive* is the bare, or uninflected, form of a verb *(go, run, fall, talk, dress, shout)* without the restrictions imposed by **person** and **number.** Along with the gerund and participle, it is one of the nonfinite verb forms. The infinitive is generally preceded by the word *to,* which, although not an inherent part of the infinitive, is considered to be the sign of an infinitive.

EXAMPLES   It is time *to go* to work.
We met in the conference room *to talk* about the new project.

An infinitive is a verbal and may function as a noun, and adjective, or an adverb.

EXAMPLES   *To expand* is not the only objective. (noun)
These are the instructions *to follow.* (adjective)
The company struggled *to survive.* (adverb)

The infinitive may reflect two **tenses:** the present and (with a helping verb) the present perfect.

EXAMPLES   *to go* (present tense)
to have gone (present perfect tense)

The most common mistake made with infinitives is to use the present perfect tense when the simple present tense is sufficient.

CHANGE   I should not have tried *to have gone* so early.
TO   I should not have tried *to go* so early.

Infinitives formed with the root form of transitive verbs can express both active and (with a helping verb) passive **voice.**

EXAMPLES   *to hit* (present tense, active voice)
to have hit (present perfect tense, active voice)
to be hit (present tense, passive voice)
to have been hit (present perfect tense, passive voice)

V

A split infinitive is one in which an adverb is placed between the sign of the infinitive, *to,* and the infinitive itself. Because they make up a grammatical unit, the infinitive and its sign are better left intact than separated by an intervening adverb.

CHANGE   *To* initially *build* the table in the file, you could input transaction records containing the data necessary to construct the record and table.

TO   *To build* the table in the file initially, you could input transaction records containing the data necessary to construct the record and table.

However, it may occasionally be better to split an infinitive than to allow a **sentence** to become awkward, ambiguous, or incoherent.

CHANGE   She agreed immediately *to deliver* the toxic materials. (Could be interpreted to mean that she agreed immediately.)

TO   She agreed *to* immediately *deliver* the toxic materials. (No longer ambiguous.)

A *participle* is a verb form that functions as an **adjective.**

EXAMPLES   The *waiting* driver raced his engine.
Here are the *revised* estimates.
The *completed* report lay on his desk.
Rising costs reduced our profit margin.

A participle cannot be used as the verb of a sentence. Inexperienced writers sometimes make this mistake; the result is a **sentence fragment.**

CHANGE   The committee chairman was responsible. His vote *being* the decisive one.

TO   The committee chairman was responsible, his vote *being* the decisive one.

OR   The committee chairman was responsible. His vote *was* the decisive one.

The present participle ends in *-ing.*

EXAMPLE   *Declining* sales forced us to close one branch office.

The past participle may end in *-ed, -t, -en, -n,* or *-d.*

EXAMPLES   Repair the *bent* lever.
What are the *estimated* costs?
Here is the *broken* calculator.
What are the metal's *known* properties?
The story, *told* many times before, was still interesting.

The perfect participle is formed with the present participle of the helping verb *have* plus the past participle of the main verb.

> **EXAMPLE**   *Having gotten* (perfect participle) a large raise, the *smiling* (present participle), *contented* (past participle) employee worked harder than ever.

verbs

A *verb* is a word, or a group of words, that describes an action (The antelope *bolted* at the sight of the hunters), states the way something or someone is affected by an action (He *was saddened* by the death of his friend), or affirms a state of existence (He *is* a wealthy man now).

Types of Verbs

Verbs may be described as being either transitive verbs or intransitive verbs.

Transitive Verbs. A transitive verb is a verb that requires a direct **object** to complete its meaning.

> **EXAMPLES**   They *laid* the foundation on October 24. *(foundation* is the direct object of the transitive verb *laid)*
>
> George Anderson *wrote* the treasurer a letter. *(letter* is the direct object of the transitive verb *wrote)*

Intransitive Verbs. An intransitive verb is a verb that does not require an object to complete its meaning. It is able to make a full assertion about the subject without assistance (although it may have **modifiers**).

> **EXAMPLES**   The water *boiled.*
> The water *boiled* rapidly.
> The engine *ran.*
> The engine *ran* smoothly and quietly.

Although intransitive verbs do not have an **object,** certain intransitive verbs may take a **complement.** These verbs are called linking verbs because they link the complement to the subject. When the complement is a **noun** (or **pronoun**), it refers to the same person or thing as the noun (or pronoun) that is the subject.

> **EXAMPLES**   The winch *is* rusted. *(Rusted* is an **adjective** modifying *winch.)*
>
> A calculator *remains* a useful tool. *(A useful tool* is a subjective complement renaming *calculator.)*

When the complement is an adjective, it modifies the subject.

> **EXAMPLES**   The study *was* thorough.
> The report *seems* complete.

V

Such intransitive verbs as *be, become, seem,* and *appear* are almost always linking verbs. A number of others, such as *look, sound, taste, smell,* and *feel,* may function as either linking verbs or simple intransitive verbs. If you are unsure about whether one of them is a linking verb, try substituting *seem;* if the sentence still makes sense, the verb is probably a linking verb.

EXAMPLES   Their antennae *feel* delicately. (simple intransitive verb; you could not substitute *seem*)

Their antennae *feel* delicate. (linking verb; you could substitute *seem*)

Some verbs, such as *lie* and *sit,* are inherently intransitive.

EXAMPLE   The patient should *lie* on his back during treatment. *(On his back* is an adverb phrase modifying the verb *lie.)*

Forms of Verbs

By form, verbs may be described as being either finite or nonfinite.

Finite Verbs.   A finite verb is the main verb of a **clause** or sentence. It makes an assertion about its subject and can serve as the only verb in its clause or sentence.

EXAMPLE   The telephone *rang,* and the secretary *answered* it.

A helping verb (sometimes called an *auxiliary verb*) is used in a verb **phrase** to help indicate **mood, tense,** and **voice.**

EXAMPLES   The work *had* begun.
I *am* going.
I *was* going.
I *will* go.
I *should have* gone.
I *must* go.

The most commonly used helping verbs are the various forms of *have (has, had), be (is, are, was,* and so on), *do, (did, does),* and *can (may, might, must, shall, will, would, should* and *could).* **Phrases** that function as helping verbs are often made up of combinations with the sign of the infinitive, *to:* for example, *am going to* and *is about to* (compare *will*), *has to* (compare *must*), and *ought to* (compare *should*).

The helping verb always precedes the main verb, although other words may intervene.

EXAMPLE   Machines *will* never completely *replace* people.

V

Nonfinite Verbs. Nonfinite verbs are **verbals,** which, although they are derived from verbs, actually function as nouns, adjectives, or **adverbs.**
  The *gerund* is the *-ing* form of a verb used as a noun.

  EXAMPLE   *Seeing* is *believing.*

An *infinitive,* which is the root form of a verb (usually preceded by *to*), can be used as a noun, an adverb, or an adjective.

  EXAMPLES   He hates *to complain.* (noun, direct object of *hates*)
                     The valve closes *to stop* the flow. (adverb, modifies *closes*)
                     This is the proposal *to select.* (adjective, modifies *proposal*)

A *participle* is a verb form that functions as an adjective.

  EXAMPLES   The *waiting* driver raced his engine.
                     Here are the *revised* estimates.
                     The *completed* report lay on his desk.
                     Rising costs reduced our profit margin.

(See also **verbals**).

Properties of Verbs

Person is the grammatical term for the form of a personal **pronoun** that indicates whether the pronoun refers to the speaker, the person spoken to, or the person (or thing) spoken about. Verbs change their forms to agree in **person** with their subjects.

  EXAMPLES   I *see* (first person) a yellow tint, but he *sees* (third person) a yellow-green hue.
                     I *am* (first person) convinced, and you *are* (third person) not convinced.

Voice refers to the two forms of a verb that indicate whether the subject of the verb acts or receives the action. If the subject of the verb acts, the verb is in the active voice; if it receives the action, the verb is in the passive voice. (see also **voice.**)

  EXAMPLES   The aerosol bomb *propels* the liquid as a mist. (active)
                     The liquid *is propelled* as a mist by the aerosol bomb. (passive)

V

Number refers to the two forms of a verb that indicate whether the subject of a verb is singular or plural.

  EXAMPLES   The machine *was* in good operating condition. (singular)
                     The machines *were* in good operating condition. (plural)

Most verbs show the singular of the third person, present **tense,** indicative **mood** by adding an *s* or *es.*

EXAMPLES   he *stands,* she *works,* it *goes*

The verb *to be* normally changes its form to indicate the plural.

EXAMPLES   I *am* ready to begin work. (singular)
We *are* ready to begin work. (plural)

If in doubt about the plural form of a word, look it up in a **dictionary.** Most dictionaries give the plural if it is formed in any way other than by adding *s* or *es.*

Tense refers to verb forms that indicate time distinctions. There are six tenses: present, past, future, present perfect, past perfect, and future perfect. (See also **tense.**)

Conjugation of Verbs

The conjugation of a verb arranges all forms of the verb so that the differences caused by the changing of the tense, number, person, and voice are readily apparent. The following is a conjugation of the verb *drive.*

Tense	Number	Person	Active Voice	Passive Voice
Present	Singular	1st	I drive	I am driven
		2nd	You drive	You are driven
		3rd	He drives	He is driven
	Plural	1st	We drive	We are driven
		2nd	You drive	You are driven
		3rd	They drive	They are driven
Progressive Present	Singular	1st	I am driving	I am being driven
		2nd	You are driving	You are being driven
		3rd	He is driving	He is being driven
	Plural	1st	We are driving	We are being driven
		2nd	You are driving	You are being driven
		3rd	They are driving	They are being driven
Past	Singular	1st	I drove	I was driven
		2nd	You drove	You were driven
		3rd	He drove	He was driven
	Plural	1st	We drove	We were driven
		2nd	You drove	You were driven
		3rd	They drove	They were driven

Tense	Number	Person	Active Voice	Passive Voice
		1st	I was driving	I was being driven
	Singular	2nd	You were driving	You were being driven
		3rd	He was driving	He was being driven
Progressive Past		1st	We were driving	We were being driven
	Plural	2nd	You were driving	You were being driven
		3rd	They were driving	They were being driven
		1st	I will drive	I will be driven
	Singular	2nd	You will drive	You will be driven
		3rd	He will drive	He will be driven
Future		1st	We will drive	We will be driven
	Plural	2nd	You will drive	You will be driven
		3rd	They will drive	They will be driven
		1st	I will be driving	I will have been driven
	Singular	2nd	You will be driving	You will have been driven
		3rd	He will be driving	He will have been driven
Progressive Future		1st	We will be driving	We will have been driven
	Plural	2nd	You will be driving	You will have been driven
		3rd	They will be driving	They will have been driven
		1st	I have driven	I have been driven
	Singular	2nd	You have driven	You have been driven
		3rd	He has driven	He has been driven
Present Perfect		1st	We have driven	We have been driven
	Plural	2nd	You have driven	You have been driven
		3rd	They have driven	They have been driven
		1st	I had driven	I had been driven
	Singular	2nd	You had driven	You had been driven
		3rd	He had driven	He had been driven
Past Perfect		1st	We had driven	We had been driven
	Plural	2nd	You had driven	You had been driven
		3rd	They had driven	They had been driven
		1st	I will have driven	I will have been driven
	Singular	2nd	You will have driven	You will have been driven
		3rd	He will have driven	He will have been driven
Future Perfect		1st	We will have driven	We will have been driven
	Plural	2nd	You will have driven	You will have been driven
		3rd	They will have driven	They will have been driven

very

The temptation to overuse **intensifiers** like *very* is great. Evaluate your use of them carefully. When you do use them, clarify their meaning.

> EXAMPLE   Bicycle manufacturers had a *very* good year: sales across the country were up 43 percent over the previous year.

In many sentences, however, the word can simply be deleted.

> CHANGE   The board was *very* angry about the newspaper report.
> TO   The board was angry about the newspaper report.

via

Via is Latin for "by way of."

> EXAMPLE   The equipment is being shipped to Los Angeles *via* Chicago.

The term should be used only in routing instructions.

> CHANGE   His project was funded *via* the recent legislation.
> TO   His project was funded *through* the recent legislation.
> OR   His project was funded *as the result* of the recent legislation.

vogue words

Vogue words are words that suddenly become popular and, because of an intense period of overuse, lose their freshness and preciseness. They may become popular through their association with science, technology, or even sports. We include them in our vocabulary because they seem to give force and vitality to our language. Ordinarily, this language sounds pretentious in our day-to-day writing.

> EXAMPLES   *super, interface* (as a verb), *bottom line, input, mode, variable, state of the art, impact* (as a verb), *cutting edge, parameter, communication, feedback,* and many words ending in *-wise*

Obviously, some of these terms are appropriate in the right context. It is when they are used outside that context that imprecision becomes a problem.

> EXAMPLE   An *interface* was provided between the two companies. (appropriate)
> CHANGE   We must *interface* with the Purchasing Department. (inappropriate)
> TO   We must *cooperate* with the Purchasing Department.

voice

In grammar, *voice* indicates the relation of the subject to the action of the verb. When the **verb** is in the active voice, the subject acts; when it is in the passive voice, the subject is acted upon.

> **EXAMPLES**   David Cohen *wrote* the advertising copy. (active)
> The advertising copy *was written* by David Cohen. (passive)

The sentences say the same thing, but each has a different **emphasis:** In the first sentence emphasis is on the subject, *David Cohen*, whereas in the second sentence the focus is on the object, *the advertising copy*. Notice how much stronger and more forceful the active sentence is.

Always use the active voice unless there is good reason to use the passive. Because they are wordy and indirect, passive sentences are hard for the **reader** to understand.

> **CHANGE**   Things *are seen* by the normal human eye in three dimensions: length, width, and depth.
> **TO**   The human eye *sees* things in three dimensions: length, width, and depth.

Passive-voice sentences are wordy because they always use a helping verb in addition to the main verb, and an extra **preposition** if they identify the doer of the action specified by the main verb. The passive voice version is also indirect because it puts the doer of the action behind the verb instead of in front of it.

> **ACTIVE**   Employees *resent* changes in policy.
> **PASSIVE**   Changes in policy *are resented* by employees.

The active-voice version takes one verb *(resent)* and one preposition *(in)*; the passive-voice version takes two verbs *(are resented)* and two prepositions *(in and by)*. The passive-voice version is also indirect because it puts the doer of the action behind the verb instead of in front of it.

One difficulty with passive sentences is that they can bury the subject, or performer of the action, in **expletives** and prepositional **phrases.**

> **CHANGE**   It *was reported* by Engineering that the new relay is defective.
> **TO**   Engineering *reported* that the new relay is defective.

V

Sometimes writers using the passive voice fail to name the performer—information that might be missed.

> **CHANGE**   The problem *was discovered* yesterday.
> **TO**   The Engineering Department *discovered* the problem yesterday.

Very often the passive voice can be just plain confusing, especially when used in **instructions.**

> CHANGE   Plates B and C should be marked for revision. (Are they already marked?)
>
> TO   You *should mark* plates B and C for revision.
>
> OR   *Mark* plates B and C for revision.

Another problem that sometimes occurs with the passive voice is **dangling modifiers.**

> CHANGE   Hurrying to complete the work, the wires *were connected* improperly. (Who was hurrying, the wires?)
>
> TO   Hurrying to complete the work, the technician improperly *connected* the wires. (Here, *hurrying to complete the work* properly modifies *technician.*)

There are, however, certain instances when the passive voice is effective or even necessary. Indeed, for reasons of tact and diplomacy, you might need to use the passive voice to avoid identifying the doer of the action.

> CHANGE   Your sales force didn't meet the quota last month. (active)
>
> TO   The quota wasn't met last month. (passive)

When the performer of the action is either unknown or unimportant, use the passive voice.

> EXAMPLES   The copper mine *was discovered* in 1929.
>
> Fifty-six barrels *were processed* in two hours.

When the performer of the action is less important than the receiver of that action, the passive voice is sometimes more appropriate.

> EXAMPLE   Ann Bryant *was presented* with an award by the president.

When you are explaining an operation in which the reader is not actively involved (but not when you are giving the reader instructions) or when you are explaining a process or a procedure, the passive voice may be more appropriate.

> EXAMPLE   Area strip mining *is used* in regions of flat-to-gently rolling terrain, like that found in the Midwest and West. Depending on applicable reclamation laws, the topsoil *may be removed* from the area to be mined, *stored,* and later *reapplied* as surface material during reclamation of the mined land. After the removal of the topsoil, a trench *is cut* through the overburden to expose the upper surface of the coal to be mined. The length of the cut generally corresponds to the length of the property or of the

deposit. The overburden from the first cut *is placed* on the unmined land adjacent to the cut. After the first cut *has been completed,* the coal *is removed,* and a second cut *is made* parallel to the first.

Anyone could be the doer of the action in this example; it really doesn't matter who does it. Therefore, it would be pointless for the writer to put this passage in the active voice. Do not, however, simply assume that any such explanation should be in the passive voice. Ask yourself, "Would it be of any advantage to the reader to know the doer of the action?" If you can answer no, then use the passive voice. But if your honest answer is yes, then use the active voice. The following example uses the active voice to explain an operation:

EXAMPLE   In the operation of an internal combustion engine, an explosion in the combustion chamber *forces* the pistons down in the cylinders. The movement of the pistons in the cylinders *turns* the crankshaft. The crankshaft, in turn, *operates* the differential. The differential *turns* the rear axles, which *turn* the wheels and *move* the automobile.

Whether you use the passive or the active voice, however, be careful not to *shift* voices in a sentence.

CHANGE   Ms. McDonald *corrected* the malfunction as soon as it *was identified* by the technician.
TO   Ms. McDonald *corrected* the malfunction as soon as the technician *identified* it.

V

wait for/wait on

Wait on should be restricted in writing to the activities of waitresses and waiters—otherwise, use *wait for.* (See also **idioms.**)

EXAMPLES Be sure to *wait for* Ms. Sturgess to finish reading the report before asking her to make a decision.

Joe's feet ached from *waiting on* tables at the diner.

when and if

When and if (or *if and when*) is a colloquial expression that should not be used in writing.

CHANGE *When and if* your new position is approved, I will see that you receive adequate staff help.

TO *If* your new position is approved, I will see that you receive adequate staff help.

OR *When* your new position is approved, I will see that you receive adequate staff help.

where . . . at

In **phrases** using the *where . . . at* construction, *at* is unnecessary and should be omitted.

CHANGE *Where* is his office *at?*

TO *Where* is his office?

where/that

Do not substitute *where* for *that* to anticipate an idea or fact to follow.

CHANGE   I read in the journal *where* molecules will be used in our process.

TO   I read in the journal *that* molecules will be used in our process.

whether or not

When *whether or not* is used to indicate a choice between alternatives, omit *or not;* it is redundant, since *whether* itself communicates the notion of a choice.

CHANGE   The project director asked *whether or not* the request for proposals had been issued.

TO   The project director asked *whether* the request for proposals had been issued.

(See also **as to whether.**)

while

While, meaning "during an interval of time," is sometimes substituted for connectives like *and, but, although,* and *whereas.* Used as a connective in this way, *while* often causes **ambiguity.**

CHANGE   John Evans is sales manager, *while* Joan Thomas is in charge of research.

TO   John Evans is sales manager, *and* Joan Thomas is in charge of research.

Do not use *while* to mean *although* or *whereas.*

CHANGE   *While* Ryan Patterson wants the job of chief engineer, he has not yet asked for it.

TO   *Although* Ryan Patterson wants the job of chief engineer, he has not yet asked for it.

Restrict *while* to its meaning of "during the time that."

EXAMPLE   I'll have to catch up on my reading *while* I am on vacation.

who/whom (◯)

Writers often have difficulty with the choice of *who* or *whom*. *Who* is the subjective case form, *whom* is the objective **case** form, and *whose* is the possessive case form. When in doubt about which form to use, try substituting a personal **pronoun** to see which one fits. If *he* or *they* fits, use *who*.

> **EXAMPLES**   *Who* is the congressman from the tenth district?
> *He* is the congressman from the tenth district.

If *him* or *them* fits, use *whom*.

> **EXAMPLES**   It depended on *them*.
> It depended on *whom?*

It is becoming common to use *who* for the objective case when it begins a sentence, although some still object to such an ungrammatical construction. The best advice is to know your **reader.**

The parenthetical clause *you thought* tends to attract the form *whom*, a form that appears logical as the object of the verb form *thought*. In fact, *who* is the correct form, since it is the subject of the verb form *were*. Note that the interjected clause *you thought* can be omitted without disturbing the construction of the clause.

The choice between *who* and *whom* is especially complex in sentences with interjected **clauses.**

> **CHANGE**   These are the men *whom* you thought were the architects.
> **TO**   These are the men *who* you thought were the architects.

whose/of which

Whose should normally be used with persons; *of which* should normally be used with inanimate objects.

> **EXAMPLES**   The man *whose* car had been towed away was angry.
> The mantle clock, the parts *of which* work perfectly, is over one hundred years old.

W

If these uses cause a sentence to sound awkward, however, *whose* may be used with inanimate objects.

> **EXAMPLE**   There are added fields, for example, *whose* totals should never be zero.

who's/whose

Who's is a contraction of *who is.*

EXAMPLE   *Who's* scheduled to attend the productivity seminar next month?

Whose is the possessive for *who* or *of which.*

EXAMPLE   *Whose* department will be affected by the budget cuts?

Who's and *whose* are not interchangeable.

-wise

Although the **suffix** *-wise* often seems to provide a tempting shortcut, it leads more often to inept than to economical expression. It is better to rephrase the sentence.

CHANGE   Our department rates high *efficiency wise.*
TO   Our department has a high efficiency rating.

The *-wise* suffix is appropriate, however, to **instructions** that indicate certain space or directional requirements.

EXAMPLES   Fold the paper *lengthwise.*
Turn the adjustment screw half a turn *clockwise.*

word choice

As Mark Twain once said, "The difference between the right word and almost the right word is the difference between 'lightning' and 'lightning bug.'" The most important goal in choosing the right word in technical writing is the preciseness implied by Twain's comment. **Vague words** and **abstract words** defeat preciseness because they do not convey the writer's meaning directly and clearly. Vague words are imprecise because they can mean many different things.

CHANGE   It was a *meaningful* meeting.
TO   The meeting helped both sides understand each other's position.

In the first sentence, *meaningful* ironically conveys no meaning at all. See how the revised sentence says specifically what made the meeting meaningful. Although abstract words may at times be appropriate to your **topic,** their unnecessary use creates dry and lifeless writing.

W

| ABSTRACT | work, fast, food |
| CONCRETE | sawing, 110 m.p.h., steak |

Being aware of the **connotation** and denotation of words will help you anticipate the **reader's** reaction to the words you choose. Connotation is the suggested or implied meaning of a word beyond its dictionary definition. Denotation is the literal, or primary, dictionary meaning of a word.

Understanding **antonyms** and **synonyms** will increase your ability to choose the proper word. Antonyms are words with nearly the opposite meaning *(fresh/stale)*, and synonyms are words with nearly the same meaning *(notorious/infamous)*.

Malapropisms and **trite language** are likely to irritate your readers, by either confusing or boring them. Malapropisms are words that sound like the intended word but have different, sometimes even humorous, meanings.

| CHANGE | The repairperson cleaned the typewriter's *plankton*. |
| TO | The repairperson cleaned the typewriter's *platen*. |

Trite language consists of stale and worn phrases, frequently **clichés.**

| CHANGE | We will finish the project *quick as a flash*. |
| TO | We will finish the project *quickly*. |

Avoid using technical **jargon** unless you are certain that all of your readers understand the terms. Avoid choosing words with the objective of impressing your reader, which is called **affectation;** also avoid using **long variants,** which are elongated forms of words used only to impress *(utilize* for *use, telephonic communication* for *telephone call, analyzation* for *analysis)*. Using a **euphemism** (an inoffensive substitute for a word that is distasteful or offensive) may help you avoid embarrassment, but the overuse of euphemisms becomes affectation. Using more words than necessary *(in the neighborhood of* for *about, for the reason that* for *because, in the event that* for *if)* is certain to interfere with clarity.

A key to choosing the correct and precise word is to keep current in your reading and to be aware of new words in your profession and in the language. Be aware also, in your quest for the right word, that there is no substitute for a good **dictionary.**

word processing

Word processing enables you to enter rough drafts into a computer, edit your text on a screen, print out a well-designed final draft, and store that version electronically for future use or revision. The technology also offers a variety of features that can improve the quality of your writing.

Format and Design Features

Good page design is crucial to the effectiveness of a document. For a **résumé** it signals an all-important first impression of the author. For a **technical manual,** how the information is structured and what's highlighted are crucial guideposts to reader understanding. Used in conjunction with the principles discussed in **design and layout,** this entry will help you master and execute effective design.

Margins. The page margins are usually preset for one inch on all sides of a printed page and are visible only when a page is printed or in preview mode; the margins do not appear on the work screen. Margin settings can be changed for the whole page or for blocks of text within a page.

Columns. Most word processors allow you to arrange the text on the page in two or more columns, a feature that can be used to create newspaper-style columns of text. This feature can also be used to create membership lists, financial statements, rosters, and similar materials without having to use tab stops to separate each row into the appropriate columns.

Margin Alignment. Page columns can be printed with the text aligned (or "justified") on the right margin as well as the left. Be aware, however, that text justified along the right margin will contain extra white space between some words. (See "Line, Word, and Letter Spacing" below.)

Widow and Orphan Control. This feature automatically protects against the awkward appearance of stand-alone words or lines at the top or bottom of a page. A *widow* refers to a word or several words, generally preceding a new paragraph, that appear as the first line on a page. An *orphan* refers to the first line of a paragraph appearing alone at the bottom of a page. These can occur when the word processor divides the text of a document into pages for printing. When you activate this feature, the software recognizes and prints widows on the previous page and orphans on the next page.

Centering. As the name implies, this feature allows you to center a word, line, or block of text on the page. Centering is especially useful for creating titles, letterhead stationery, and captions for tables and figures.

Line, Word, and Letter Spacing. The word processor single-spaces all text unless you instruct it otherwise. However, text can be separated by as many spaces as you choose in full or fractional increments (2, 2.5, 3.33, and so on). This feature, most useful for creating drafts on which you or others will comment, also permits you to vary line spacing from one block of text to another within a document.

You can also change the spacing between words and letters if you have a printer that supports these options. These features are most useful for fine-tuning the look of a block of right-justified text containing too much white space.

Headers and Footers. Word processors allow you to enter the header or footer at the beginning of a document and automatically position it on every page when the document is printed. Headers and footers do not appear on the work screen, although they can be seen in preview mode (see below).

Footnotes. Footnotes may be placed at the foot of a page or at the end of a chapter or document. The software positions and numbers them automatically according to your instructions. Most word processors will update the numbering automatically from the point where a note has been added or deleted. Once a note is linked by the footnote number in text, it will move to the foot of the same page automatically if the text to which it is "tagged" moves.

Page Numbers. The positioning of page numbers, like headers and footers, is done automatically after you specify where you want the numbering to begin and where on the page you want the numbers to appear. By convention, the first page of a letter, chapter, or report is unnumbered. You can instruct the word processor to begin numbering the second page, and it will number all following pages in sequence automatically. The word processor will also number the front matter pages of a **formal report** sequentially with small roman numerals. Once you establish the numbering style (page 3, -3-, #3, and so on), the word processor will repeat it throughout. Page numbers can be centered at the top or bottom of the page or located in the upper or lower right or left corner.

The software will even distinguish between even and odd pages, positioning the page number at the top or bottom left corner for even pages and at the top or bottom right corner for odd pages. The software can also create a **table of contents** after you identify the elements—usually your chapter titles and other headings—to be included in it. Using the same principle, you can create a subject index.

You can either move through the text marking the appropriate words and phrases that will appear as headings and subheadings, or you can create a separate listing of words and phrases and have the software search through the text for them, listing the page number of each occurrence next to it in a separate subject index at the end of your document. (See also **indexing.**)

Type Style Options. You can vary the style and size of type used for text and headings and to highlight important words and phrases by the text font you select. A *font* is a complete set of letters, numbers, and other type characters of distinctive and uniform design. They are divided into families of several sizes (for text, headings, and footnotes), weights (boldface or roman), and other features (*italic* or <u>underscore</u>), that share the same design and that distinguish them from other fonts.

Preview Mode. Your work screen shows only 24 lines of text from top to bottom. Some word processors, however, permit you to view a reduced version of the full $8^1/_2$-by-11-inch page before you print it. Called *preview mode,* this feature lets you see each page and all the format characteristics you created for it—columns, line and letter spacing, justification, pagination, headers, footers, footnotes, and the overall "look" of everything together. It also displays all the fonts your printer can create. If anything looks amiss, you can correct it before you print it.

Style Sheets. After you have created the format features for a document, you can save them for future use as a *style sheet*—a set of user defined text-format codes stored as a separate file with a unique name. They include such features as paragraph indention and spacing; line and page breaks; and line, word, and letter spacing, in addition to type font style and size. Once created, a style sheet can be applied to as many documents as you wish. They are useful for creating recurrent uniform documents, such as monthly reports, form letters, proposals, and newsletters.

Printing

After you have entered, revised, formatted, and previewed the text, you can print one or more copies of the entire document or specific pages. You can print on $8^1/_2$-by-11-inch or $8^1/_2$-by-14-inch plain or letterhead paper. Many printers also print pages in landscape (sideways) orientation for tables and graphics that will not fit on a page in portrait (conventional) orientation. Most printers also have paper-feeding mechanisms for printing nonstandard forms and even address labels.

Microcomputers use *dot matrix, letter quality,* and *laser* printers, each of which produces typefaces of varying quality, as shown in Figure 1. *Dot matrix* printers print characters composed of tiny dots that give them a "fuzzy," indistinct look. They can produce better quality print for memos and informal reports in double-strike or overstrike mode. *Letter quality* printers are slower than dot matrix printers and use a plastic daisy wheel or thimble and ribbon to print a clear, dark typeface. *Laser* printers work on the same principle as photocopy machines by using a jet-spraying device to deposit carbon images directly on paper. Laser printers can produce a great variety of typefaces at a quality comparable to professionally printed material.

```
This is an example of dot matrix printing in draft mode.

This is an example of dot matrix near-letter quality printing.

This is an example of letter quality printing.

This is an example of laser printing with bold, italic, and enlarged typefaces.
```

FIGURE 1 Varying Typeface Quality for Different Printers

Most word processors enable you to merge information from different files and then print them as one document. Using the merge feature, you can combine several sections of a letter, report, or proposal done at different times, each stored as a separate file, and print them as a single document. This feature permits firms to send "personalized" form letters to thousands of potential customers. Such form letters merge standard language with individual names and addresses on a target mailing list.

Word Processing and the Writing Process

Word processing can also help you record ideas quickly and enhance your writing and revising skills. Remember, however, that good writing is still the result of careful planning, constant practice, and thoughtful revision. In some cases, word-processing technology can initially intrude on the writing process and impose certain limitations that many beginners overlook. The ease of making minor, sentence-level changes and the limitation of a 24-line viewing screen, for example, may focus your attention too narrowly on surface problems of the text so that you lose sight of larger problems of scope and organization. The fluid and rapid movement of the text on the screen, together

W

with last-minute editing changes, may allow undetected errors to creep into the text. Also, as you master word processing and become familiar with its revision capabilities, you may begin to "overwrite" your documents: Inserting phrases and rewriting sentences becomes so easy that you may find yourself generating more but saying less. The following guidelines will help you avoid these initial pitfalls and develop writing strategies that take full advantage of the benefits offered by word-processing technology.

1. Do not write first drafts on the computer without any planning or **outlining.** Plan your document carefully by identifying your objective, readers, and scope and by completing your research *before* you begin writing a draft on the computer. You can also copy bibliographic citations to a disk from your library's online search system to incorporate into your document's **bibliography** or list of works cited. (See **library research.**)

2. Use the outline feature to develop a shell that you can fill in, rearrange, and update as you refine your thinking. This feature allows you to experiment with the scope and organization of information in the outline. Using the block and move function makes rearranging sections of the outline much easier than cutting and taping a paper version.

3. When you're ready to begin writing, you can overcome writer's block by practicing free writing on the computer within the structure of your outline. Free writing means typing your thoughts as quickly as possible for five to ten minutes without stopping to correct mistakes or to complete sentences—correct these when you revise.

4. Use the search command to find and delete wordy phrases such as *that is, there are, the fact that,* the overuse of the verb *be,* and unnecessary helping verbs such as *will.*

5. Use a spell checker and other specialized programs to identify and correct typographical errors, misspellings, and grammar and diction problems. Maintain a file of your most frequently misspelled or misused words and use the search command to check them in your documents. Although a spell checker is an invaluable tool, it is not infallible. It cannot tell whether you meant *their* or *there* in a given context, since both words are spelled correctly, and so will pass over each. Nor can a spell checker help you with numbers (except when they're incorrectly used as letters). You can also fine-tune your language by using an online **thesaurus** that searches for and displays synonyms for specific terms you select. However, a thesaurus cannot make the fine discriminations in meaning that the human mind can.

Like a spell checker, the thesaurus matches terms literally and cannot take the context into account.

6. Avoid excessive editing and rewriting on the screen. Print out a double-spaced copy of your draft periodically for major revisions and reorganizations.

7. Always **proofread** your final copy on paper because the fluidity of the viewing screen makes it difficult to catch all the errors in your manuscript. Print out an extra copy of your document for your colleagues to comment on before making final revisions. You can send a file of your draft by **email** to your colleagues or provide them with a disk containing your file.

8. When writing a single document for a variety of readers, use the search command to locate technical terms and other data that may need further explanation for secondary readers or for inclusion in a **glossary.**

9. Use the computer for effective page **design and layout** by emphasizing major **headings** and subheadings with bold print, by using the copy command to create and duplicate parallel headings throughout your text, and by inserting blank lines and tab key spaces in your text to create extra white space around examples and illustrations.

10. Frequently "save" or store your text on disk during long writing sessions, if your computer does not have an automatic storage backup feature. Routinely create an extra or "backup" copy of your documents on floppy disks for safekeeping.

11. Keep the standard version of certain documents, such as your **résumé** and **application letters,** on file so you can revise them to meet the specific needs of each new job opportunity.

writing a draft

You are well prepared to write a rough draft when you have established your **purpose, readers'** needs, and **scope** and when you have done adequate **research** and **outlining.** Writing a rough draft is simply transcribing and expanding the notes from your outline into **paragraphs,** without worrying about **grammar,** refinements of language, or such mechanical aspects of writing as **spelling.** Refinement will come with **revision.**

 Write a rough draft quickly, concentrating on converting your outline into sentences and paragraphs (see also **sentence construction**). Write as though you were explaining your subject to someone across the desk from you. Don't worry about a good opening. Just start. There is no need in a rough draft to be concerned about an **introduction** or **transitions** unless they come easily—concentrate on ideas. Don't try to polish or revise. Writing and revising are different activities. Keep writing quickly to achieve **unity** and proportion.

Even with good preparation, however, writing a draft remains a chore and an obstacle for many writers. Experienced writers use the following tactics to get started and keep moving. Discover which ones are the most helpful to you.

- Set up your writing area with the equipment and materials (paper, dictionary, source books, and so forth) you will need to keep going once you get started.
- Use whatever writing tools—separately or in combination—are most comfortable for you: pencil, felt-tip pen, word processor, or whatever.
- Remind yourself that you are beginning a version of your writing project that no one else will read.
- Remember the writing projects you've finished in the past—you have completed something before, and you will this time.
- Start with the section that seems easiest to you—your reader will not know or care that you first wrote a section in the middle.
- Give yourself a time limit (10 or 15 minutes, for example) in which you will keep your pen or fingers on the keyboard moving, regardless of how good or bad your writing seems to you. The point is to keep moving.
- Don't let anything stop you when you are rolling along easily—if you stop and come back, you may not regain the momentum.
- Stop writing before you're completely exhausted; when you begin again, you may be able to regain your momentum.
- When you finish a section, give yourself a small reward—a short walk, a cup of coffee, a chat with a friend, an easy task, and so on.
- Reread what you've written when you resume writing. Often seeing what you've written will trigger the frame of mind that was productive.

The most effective way to start and keep going, however, is to use a good outline as a springboard and map for your writing. Your outline notes can become the topic sentences for paragraphs in your draft, as shown below. For example, notice that items III.A. and III.B. in the following topic outline become the topic sentences of the succeeding two paragraphs. And notice, too, that the subordinate items in the outline become sentences in the two paragraphs.

W

Outline

> III. Advantages of Chicago as Location for New Plant
>   A. Outstanding Transport Facilities
>   1. Rail
>   2. Air
>   3. Truck
>   4. Sea (except in winter)
>   B. Ample Labor Supply
>   1. Engineering and scientific personnel
>   a. Many similar companies in the area
>   b. Several major universities
>   2. Technical and manufacturing personnel
>   a. Existing programs in community colleges
>   b. Possible special programs designed for us

Resulting Paragraphs

> *Probably the greatest advantage of Chicago as a location for our new plant is its excellent transport facilities.* The city is served by three major railroads. Both domestic and international air cargo service is available at O'Hare International Airport. Chicago is a major hub of the trucking industry, and most of the nation's large freight carriers have terminals there. Finally, except in the winter months when the Great Lakes are frozen, Chicago is a seaport, accessible through the St. Lawrence Seaway.
>
> *A second advantage of Chicago is that it offers a large labor force.* An ample supply of engineering and scientific personnel is assured not only by the presence of many companies engaged in activities similar to ours but also by the presence of several major universities in the metropolitan area. Similarly, technicians and manufacturing personnel are in abundant supply. The seven colleges in the Chicago City College system, as well as half a dozen other two-year colleges in the outlying areas, produce graduates with associate degrees in a wide variety of technical specialties appropriate to our needs. Moreover, three of the outlying colleges have expressed an interest in establishing courses attuned specifically to our requirements.

Remember that as you write, your function as a writer is to communicate certain information to your readers. Do not try to impress them with a fancy writing **style.** Write in a plain and direct style that is comfortable and natural for both you and your reader—the first rule of good writing is to *help the reader.*

As you write your rough draft, keep in mind your reader's level of knowledge of the subject. Doing so will not only help you write directly to your reader, it will also tell you which terms you must define. (See also **defining terms.**)

When you are trying to write quickly and come to something difficult to explain, try to relate the new concept to something with which the reader is already familiar. Although **figures of speech** are not used extensively in technical writing, they can be very useful in explaining a complex process. In the rough draft, a figure of speech might be just the tool you need to keep moving when you encounter a complex concept that must be explained or described.

Above all, don't wait for inspiration to write your rough draft—treat writing a draft in technical writing as you would any on-the-job task.

X-ray

X-ray, usually capitalized, is always hyphenated as an **adjective** and **verb** and usually hyphenated as a **noun.** If you hyphenate the noun, be sure to do so consistently.

EXAMPLES   portable *X-ray* unit (adjective)
The technician *X-rayed* the ankle. (verb)
X-rays and gamma rays (noun)

INDEX

Index terms followed by page numbers refer to the titles of *Handbook* entries. Terms without page numbers refer either to synonyms for entries or to topics discussed within an entry. Index entries that are also *Handbook* entries are boldfaced.